Water Science and Application 8

Riparian Vegetation and Fluvial Geomorphology

Water Science and Application 8

Riparian Vegetation and Fluvial Geomorphology

Sean J. Bennett
Andrew Simon
Editors

American Geophysical Union
Washington, DC

Riparian Vegetation and Fluvial Geomorphology
Water Science and Application 8

Library of Congress Cataloging-in-Publication Data

Riparian vegetation and fluvial geomorphology / Sean J. Bennett, Andrew Simon, editors.

       p. cm. -- (Water science and application series ; 8)

    Includes bibliographical references.

    ISBN 0-87590-357-6

    1. River channels. 2. Riparian plants. 3. Geomorphology. 4. Sediment transport. I. Bennett, Sean J. 1962- II. Simon, Andrew. III. Series: Water science and application ; 8.

    GB562.R56 2004

    551.44'2--dc22   2004043683

ISBN 0-87590-357-6
ISSN 1526-758X

Cover: Aerial photograph of a bend-bar complex on the Animas River, New Mexico, showing sequential establishment and proliferation of riparian vegetation as the meander migrates eastward (toward the bottom of the photograph) (courtesy of Lynette Stevens-Guevara, New Mexico Environment Department Surface Water Quality Bureau).

CONTENTS

Part 5. Floodplains and Watershed Processes

Part 6. Numerical Studies

PREFACE

Rivers and streams are usually in close association with vegetation. This vegetation, loosely termed riparian vegetation for its proximity to the water course, can occupy nearly every geomorphic position within the fluvial environment. Vegetation can cover mid-channel, alternate, and point bars; it can grow on the river bed and at the banktoe, along the face, and on the top of stream banks; and it can populate both terraces and floodplains bordering rivers and streams. Moreover, accumulations of dead vegetation, such as branches and entire logs, can accumulate near riffles and pools, meander bends, and road and bridge crossings.

By its very nature, vegetation is transient. Climate and hydrology can alter growth patterns, colonization rates, diversity, and vegetation density. Vegetation in or near streams is subject to varying flow stages that can inundate vegetation during high flow events or leave it exposed for long periods of time. In-stream vegetal growth or accumulations of debris are subject to unsteady flows that can modify distributions as well as create, destroy, or move such organic material. Human intervention can add to or subtract from vegetation in ways planned or otherwise.

More importantly, riparian vegetation can play a critical role in the physical, biological, and hydraulic function of streams and rivers. Vegetation can affect the transport of water, sediments, and nutrients both within the channel and to or from the riparian zone. These interactions can greatly impact water quality and biologic functionality within river corridors. Vegetation can modulate the pace and characteristics of river channel change. In some cases, vegetation can initiate fluvial adjustment.

Earth scientists, river engineers, and hydrologists now are examining more closely the intimate relationship between fluvial processes, channel shape, and river planform and the role riparian vegetation can play. To further address these issues, the authors of this volume present new research findings and up-to-date critical reviews of the fundamental linkages coupling riparian vegetation and in-stream debris to river flow, form, and process. The volume presents a range of topics from the perspectives of the experimentalist, the field practitioner, the theorist, and the modeler, offering the reader a concise synthesis of recent scientific advances in the area of fluvial geomorphology as well as discussions of unresolved problems and opportunities. In this capacity, the volume is targeted to scientists, researchers, graduate students, and practitioners interested in fluvial geomorphology and hydraulics, stream corridor restoration and design, and watershed management.

Although the volume addresses several key issues in fluvial geomorphology, it is not designed to be exhaustive. The reader is introduced to experimental results and theoretical considerations that examine how vegetation influence flow and sediment transport. This section is followed by field studies that consider both the stream channel and the entire watershed, and increase in scale and complexity. The final chapters propose different numerical frameworks for predicting river flow processes in the presence of vegetated banks and floodplains.

The volume derives from presentations and discussions at a special session devoted to the topic, held at the AGU Fall Meeting (December 2001). Presenters responded enthusiastically when asked to contribute to the volume, and further invitations were extended to other researchers who brought much needed perspective and balance to the discussion.

As editors, we are very grateful to all authors who contributed to the volume and making this publication possible. We gratefully acknowledge all referees, whose names are listed below, for providing constructive and timely reviews of the contributed papers. We also thank the AGU staff, especially Allan Graubard, our acquisitions editor, and Colleen Matan, our production coordinator, for their encouragement, advice, and support. Partial financial support for this book was provided to the editors by the U.S. Department of Agriculture, Agricultural Research Service.

Sean J. Bennett
Department of Geography
University at Buffalo

Andrew Simon
National Sedimentation Laboratory
USDA-ARS

Volume reviewers:

C. Alonso	M. Altinakar	C. Braudrick	J. Castro
M. Daniels	S. Darby	M. Doyle	F. Fitzpatrick
A. Gurnell	G. Hanson	C. Hupp	K. Juracek
A. Khan	E. Langendoen	T. Lisle	F. Magilligan
V. Neary	W. Osterkamp	N. Pollen	M. Rinaldi
K. Schmidt	S. Scott	D. Sear	F.D. Shields, Jr.
S. Smith	M. Singer	E. Wohl	W. Wu

Riparian Vegetation and Fluvial Geomorphology: Problems and Opportunities

Andrew Simon

Channel and Watershed Processes Research Unit, USDA-ARS National Sedimentation Laboratory, Oxford, Mississippi

Sean J. Bennett

Department of Geography, University at Buffalo, Buffalo, New York

Vincent S. Neary

Department of Civil and Environmental Engineering, Tennessee Technological University, Cookeville, Tennessee

Riparian vegetation exerts strong controls on numerous processes in fluvial geomorphology and, in turn, is dependent on many of those same processes. It affects the magnitude and distribution of important hydrologic, hydraulic, and geotechnical variables in river corridors and can therefore ameliorate or exacerbate processes that determine channel morphology. An understanding of the direct and indirect roles that riparian vegetation plays in fluvial geomorphology is critical in designing and conducting research aimed at more accurately quantifying channel hydraulics, sediment transport, and channel morphology.

1. INTRODUCTION

Riparian vegetation is a fundamental component of landscape systems. Its direct and indirect effects range across a broad spectrum of geomorphic processes and scales relating to the hydrologic cycle, water budgets, and soil moisture as well as resistance to overland and concentrated flows in channels and on flood plains [*Thornes*, 1990; *Malanson*, 1993; *Hupp et al.*, 1995]. Riparian vegetation can be regarded both as independent and dependent variables in studies of geomorphic processes and forms. Understanding the role of vegetation in fluvial geomorphology requires a truly inter-disciplinary approach that may involve combinations of disciplines including hydrology, hydraulics, sediment transport, ecology, botany, and geotechnical engineering.

Riparian vegetation can exert strong, direct influences on erosion rates by providing greater hydraulic- and geotechnical-shear strength. For these reasons, vegetation has become a major component in designing erosion control and stream-rehabilitation measures. For instance, riparian buffer strips made from native vegetation along stream channels serve to reduce flow velocities and trap sediment. A byproduct of this application is the reduction of pore-water pressures in streambanks through interception of precipitation and removal of water from the bank mass by evapotranspiration. Large woody debris is used in small- to moderately-sized channels, often in meander bends to protect bank toes, induce deposition, and halt lateral migration. Still, much of what was known about the effects of riparian vegetation has been semi-quantitative at best. Successes and failures in schemes relying on riparian

Riparian Vegetation and Fluvial Geomorphology
Water Science and Application 8
10.1029/008WSA01

vegetation are often reported as case studies without physically-based rationale for their performance. The role of vegetation in controlling sediment transport and channel adjustment has certainly been acknowledged empirically, yet a physically-based numerical understanding of the hydraulic, hydrologic, and mechanical controls remains incomplete.

2. FLOW RESISTANCE, FLOW VELOCITY, AND TURBULENCE

Flow in rivers and streams may encounter vegetation at various geomorphic positions within a corridor. Vegetation can occur on floodplains and near the tops of streambanks, on streambank faces below bankfull stage, and within the channel on point bars, alternate bars, and mid-channel bars. This riparian vegetation may be rigid or flexible, it may be tall (emergent) or short (submerged) relative to the flow depth, and it may be alive or dead (either free standing or accumulated in the channel as debris). Seasons, variations in surface and subsurface hydrology, and land use will affect the growth, distribution, density, and propagation of riparian vegetation. The effects vegetation will have on stream flow processes are further complicated by the temporal and spatial variations of flow stage, uniformity, and steadiness.

The characteristics of the flow are altered in areas where streams encounter vegetation. In general, vegetation causes the flow to decelerate due to an increase in flow resistance and a disruption of the flow path. Because of the deceleration of flow, vegetation can effectively trap and sort sediment [*Lowrance et al.*, 1988; *Tsujimoto*, 1999]. Vegetative barriers, grasses, and filter strips are commonly used in upland areas for trapping fine-grained sediments [*Tollner et al.*, 1976; *Dabney et al.*, 1995]. Willows emplaced as dormant posts are commonly used for bank stabilization and sediment trapping [*Watson et al.*, 1997]. Flow disruption by vegetation has been observed in experimental channels with in-stream vegetation [*Bennett et al.*, 2002] and in natural channels with vegetated streambanks [*Fukuoka and Watanabe*, 1997].

A critical issue in river engineering is the prediction of the magnitude of flow resistance associated with vegetation. Vegetation provides an additional source of momentum loss in river channel and floodplain flows. To address this issue, an additional term can be added to the partitioning of drag and resistance in momentum balance equations, and this term may include characteristics of the vegetation such as height, rigidity, and spacing or density as well as the presence of branches or leaves [*Kouwan et al.*, 1969; *Petryk and Bosmajian*, 1975; *Thompson and Roberson*, 1976; *Kouwan*, 1988; *Masterman and Thorne*, 1992]. These analytical approaches provide modified friction factors or roughness heights that can be incorporated into hydrologic flow models [*Darby*, 1999].

Analytical procedures also have focused on the drag coefficient of an individual vegetal element within vegetated stream channels and floodplains. For an infinitely long cylinder unencumbered by the channel boundaries in a turbulent flow, the drag coefficient is about one. This assertion has been supported by experimental work [*Thompson et al.*, 2003; *García et al.*, this volume], and several studies have incorporated these drag determinations into analytical procedures. However, deviations from ideal conditions do occur in practice, and these deviations can markedly alter the drag coefficient of single elements. Higher drag coefficients reported by *Wallerstein et al.* [2002] were related to the formation of standing waves near the water surface, and *Alonso* [this volume] discusses in detail the effects of element geometry, element orientation, and the processes of flow separation, vortex shedding, and flow unsteadiness on both instantaneous and mean drag and lift coefficients. Lower drag coefficients for entire vegetation populations and individual elements have been observed in laboratory studies when the density is relatively high [*Li and Shen*, 1973; see *Nepf*, 1999]. Field studies examining drag coefficients are more limited. *Arcement and Schneider* [1989], following *Petryk and Bosmajian* [1975], derived apparent drag coefficients on the order of 5 to 10 for flows through vegetation on floodplains.

The distribution of velocity within submerged riparian vegetation shows two characteristic layers. The velocity gradient is low to negligible within the vegetation zone from the canopy top to the bed for both flexible and rigid elements [*López and García*, 2001; *Righetti and Armanini*, 2002; *Nepf and Vivoni*, 2000; *Stephen and Gutknecht*, 2002]. Above the vegetation, the velocity distribution assumes a logarithmic profile from the canopy top to the water surface. Reynolds stress decreases linearly from near zero at the water surface to a maximum at the top of vegetation, typical of flat-bed turbulent boundaries. Reynolds stress decreases from a maximum near the canopy top to near zero at the bed within the submerged vegetation, and this distribution is more non-linear. The effect of submerged vegetation in open channels is the creation of two boundary layers, one within the vegetation (from the bed to the top of the canopy) and one above the vegetation (from the canopy top to the water surface). In general, turbulence intensities are increased near the canopy tops for submerged vegetation and along the interface between emergent riparian vegetation and the main channel [*Tsujimoto*, 1999; *López and García*, 2001; *Nepf and Vivoni*, 2000].

3. BANK EROSION PROCESSES

Streambank retreat occurs by a combination of hydraulic-induced bank-toe erosion and mass failure of the upper part of the bank. In addition to its effects in modifying hydraulic scour

Table 1. Potential stabilizing and destabilizing effects of riparian vegetation on bank stability.

Effects	Hydrologic	Mechanical
Stabilizing	Canopy interception Transpiration	Root reinforcement
Destabilizing	Increased infiltration rate and capacity	Surcharge

through changes in flow resistance and velocity, and its benefits to environmental quality, vegetation is widely believed to increase the stability of streambanks [*Thorne*, 1990; *Simon* 1999]. Stabilizing effects include reinforcement of the soil by the root system and the reduction of soil moisture content because of canopy interception and evapotranspiration. However, studies of vegetation's impact on the stability of hillslopes have highlighted the potential for some destabilizing effects [*Greenway*, 1987; *Collison and Anderson*, 1996]. These include higher near-surface moisture contents during and after rainfall events due to increased infiltration capacity, and surcharge due to the weight of trees. Although many authors have evaluated the mechanical benefits of vegetation on slope stability [see *Gray*, 1978] few studies have specifically addressed the coupling of streambank processes and riparian vegetation [*Abernethy and Rutherfurd*, 2001; *Simon and Collison*, 2002]. Still fewer studies have quantified the hydrologic effects of riparian vegetation, or considered the balance between potential stabilizing and destabilizing effects under different precipitation and flow scenarios [*Simon and Collison*, 2002]. A summary of the stabilizing and destabilizing effects of riparian vegetation on bank stability is shown in Table 1.

3.1. Mechanical Effects

Soil is generally strong in compression, but weak in tension. The fibrous roots of trees and herbaceous species are strong in tension but weak in compression. Root-permeated soil, therefore, makes up a composite material that has enhanced strength [*Thorne*, 1990]. Numerous authors have quantified this enhancement using a mixture of field and laboratory experiments. *Endo and Tsuruta* [1969] used *in situ* shear boxes to measure the strength difference between soil and soil with roots. *Gray and Leiser* [1982] and *Wu* [1984] used laboratory-grown plants and quantified root strength in large shear boxes. *Abernethy and Rutherfurd* [2001] measured the tensile strength of Australian riparian tree roots *in situ* while *Simon and Collison* [2002] measured a variety of American riparian species using the same technique. Most authors note a non-linear, inverse relationship between root diameter and strength, with smaller roots contributing more strength per unit root area. *Wu et al.* [1979] developed a widely-used equation that estimates the increase in soil strength as a function

of root tensile strength, areal density, and root distortion during shear. This equation was used in conjunction with a bank-stability model to simulate the increase in shear strength and bank stability using tensile strength and root density data from various species of riparian vegetation [*Simon and Collison*, 2002]. *Pollen et al.* [this volume] found that the *Wu et al.* [1979] equation overestimates the increase in bank shear strength provided by roots and proposes a new model of root reinforcement that does not require the assumptions that all roots break simultaneously and at the same displacement as when peak soil strength is reached.

3.2. Hydrologic Effects

Vegetation increases bank stability by intercepting rainfall that would otherwise have infiltrated into the bank, and by extracting soil moisture for transpiration. Both processes enhance shear strength by reducing positive pore-water pressure and encouraging the development of matric suction, thereby enhancing streambank stability [*Simon et al.*, 2000]. However, the hydrologic effects of riparian vegetation are even less well quantified than the mechanical effects. Although data are available on canopy-interception rates for many riparian tree species, there is little useful data on the degree to which vegetation dries out the material comprising streambanks. Canopy interception for deciduous tree species is typically in the range of 10 to 20%, [*Coppin and Richards*, 1990], but these figures represent annual averages. A point often overlooked is that most bank failures occur during the winter or early spring, when deciduous vegetation is dormant and canopies have been shed. In addition, the high rainfall events likely to be associated with bank failures tend to have the lowest canopy interception rates, since canopy interception is inversely proportional to rainfall intensity and duration. Likewise, transpiration does not generally have much impact on soil moisture until mid-spring. The beneficial hydrologic effects of riparian vegetation on bank stability were found in most cases to exceed those due to mechanical reinforcement provided by roots [*Simon and Collison*, 2002]. An important avenue of future research is to determine the timing and relative magnitudes of water withdrawal from streambanks for a range of riparian species in diverse fluvial environments.

4. LARGE WOODY DEBRIS AND RIVER RESTORATION

The introduction or recruitment of large woody debris (LWD; or fallen trees) in rivers and streams can occur through a variety of processes including natural (tree death), biological (beavers or other wildlife), mass wasting (debris flows or streambank failure), or anthropogenic (timber harvesting or land use changes; *Keller and Swanson*, 1979; *Downs and Simon*, 2001; *Gurnell et al.*, 2002; *May and Gresswell*, 2003). Trees like the large redwoods in Northern California may remain where they fell for long periods of time [*Keller and Tally*, 1979]. *Gippel et al.* [1992] suggested that minimum residences times can range from 40 to 200 years depending upon tree rate of decay.

The presence of LWD in river corridors can have a marked effect of channel hydraulics, form, and process. LWD can divert channel flow, cause local scour, plunge pool development, and channel widening, create sites of sediment storage, and increase the number and spacing of bars [*Keller and Swanson*, 1979; *Montgomery et al.*, 1995; *Abbe and Montgomery*, 1996; *Buffington and Montgomery*, 1999]. *Wallerstein and Thorne* [in press] describe different types of debris jams formed in the sand-bedded streams of Mississippi, showing that jam type depends on the relative stability of the river channel. *Gurnell et al.* [2002] and *Montgomery and Piégay* [2003] describe short and long-term impacts of wood in rivers as well as the effect of scale on wood delivery, retention, and geomorphic response.

Because of its hydraulic and geomorphic effects, LWD provides a number of benefits to aquatic biota. These benefits include the creation of fish and invertebrate refugia, deep pools that increase habitat diversity, stable surfaces for micro- and macro-organism growth and development, and canopy cover for reduced water temperatures [see review in *D'Aoust and Millar*, 1999; *Shields and Cooper*, 2000]. Environmental engineers have attempted to combine the grade-control characteristics of LWD with the ecological benefits in stream restoration programs by designing engineered log jams [*Abbe et al.*, 1997; *D'Aoust and Millar*, 1999; *Shields and Cooper*, 2000]. *D'Aoust and Millar* [1999] and *Shields and Wood* [in review] present various engineered log jams and make specific recommendations regarding design, site location, construction, and maintenance. Quantifiable design criteria for calculations of available forces and resistance are, however, still incomplete.

5. RIVER RESTORATION WITH WOODY VEGETATION

River restoration programs seek to return biological functionality to degraded stream corridors primarily through enhancement of habitat, habitat resources, and stream channel stability [*Brookes and Shields*, 1996; *Federal Interagency Stream Restoration Working Group (FISRWG)*, 1998]. Recreational opportunities and aesthetic beauty also are improved as a result. Restoration projects, along with streambank stabilization and protection programs, have used vegetation extensively to accomplish these goals [e.g., *USDA-NRCS*, 1996; *FISRWG*, 1998]. There are many different techniques and designs for using vegetation in such projects ranging from managed plantings of grasses and woody vegetation on eroded streambanks to a combination of vegetation and rock, wood, or manufactured structures such as revetments and fencing. Guidelines are available to choose appropriate vegetation species based on geography, climate, and stream-corridor characteristics [*Volny*, 1984; *USDA-NRCS*, 1996].

A common restoration technique in the U.S. is to plant dormant willow (*Salix* spp.) cuttings, 0.1 to 0.3 m in diameter and 2 to 4 m long, along the face and toe of streambanks and sandbars [*Watson et al.*, 1997]. The growth and propagation of these trees stabilize and strengthen banks, increase vegetative cover, and increase sandbar stability while encouraging colonization of natural vegetation species. Other species, however, may provide greater mechanical strength [*Simon and Collsion*, 2002; *Pollen et al.*, this volume]. Design issues seem to depend on water availability (drought and flooding cause high rates of mortality; see *Shields et al.*, 1995; *Shields and Cooper*, 2000) and on the tensile strength and density of roots [*Simon and Collison*, 2002].

6. RIPARIAN VEGETATION, SEDIMENT YIELD, AND CHANNEL EVOLUTION

Because riparian vegetation exerts profound influences on hydrologic processes in the drainage basin and on the fluvial-geomorphic processes described earlier, it strongly affects sediment yields, sediment contributions from channels and styles of morphologic adjustment. Vegetation is the most important intermediary through which climate and land-use modify geomorphic processes and landforms [*Kirkby*, 1995]. Annual sediment yields vary with effective precipitation, peaking at about 0.3 m [*Langbein and Schumm*, 1958]. Reductions in sediment yields for regions with greater amounts of effective precipitation are directly related to the introduction of grasses. Even lower sediment yields occur in wetter forested regions. Analysis of suspended-sediment concentrations at the 1.5-year recurrence interval for more than 2,900 sites across the United States also show peak values in the semi-arid regions, reflecting the lack of vegetative cover [*Simon et al.*, in press]. These spatially varying results also have broad implications to temporal variations through the direct effects of climate change and water use on vegetation distribution and sediment yields.

Spatial patterns of riparian vegetation are a result of and have a control on ecologic, geomorphic, and hydrologic processes active along river corridors [*Malanson*, 1993]. Expansion of riparian habitat coincides with decreases in discharge and channel narrowing as water development surged in the Midwestern United States [*Williams*, 1978; *Johnson*, 1994]. In contrast, removal of riparian vegetation leads to higher rates of runoff and erosion, and the extension of channel networks. Vegetation community structure is closely tied to a range of fluvial landforms relating to the magnitude and frequency of inundation [*Hupp and Osterkamp*, 1996], and the degree of channel incision and adjustment [*Simon and Hupp*, 1992; *Hupp*, 1992]. The establishment of riparian vegetation, particularly woody vegetation is an important diagnostic criterion in interpreting dominant channel processes and relative channel stability in fluvial systems. Conceptual channel-evolution models rely heavily on the state of woody, riparian vegetation to infer processes and stage of channel evolution because of its important controls on flow resistance and soil moisture [*Schumm and Hadley*, 1957; *Schumm et al.*, 1984; *Simon and Hupp*, 1986; *Simon*, 1989].

7. NUMERICAL MODELING

Numerical modeling of vegetation effects on open channel flow can follow one of three approaches. Each approach allows a specific range of flow features to be simulated. Computational hydraulics models can be constructed to solve one –dimensional (1D) averaged flow momentum and continuity equations. These models can simulate the effects of vegetative resistance on bulk flow velocity and depth (de Saint-Venant equations). Computational fluid dynamics (CFD) models can be constructed to solve the 1D to 3D steady Reynolds-averaged-Navier-Stokes (RANS) equations. These models can resolve local flow and turbulence features of the temporally averaged turbulent flow field. Finally, unsteady RANS (URANS) and Large eddy simulation CFD models can be constructed to solve the unsteady 3D Navier-Stokes equations. These models can provide a complete description of the instantaneous unsteady 3D turbulent flow field, capturing organized large-scale unsteadiness and asymmetries (coherent structures) resulting from flow instabilities.

The characterization of vegetative flow resistance in these models has and will continue to command the attention of both researchers and practitioners alike. For flow through vegetation, where the ratio of plant height K to flow depth d is greater than 0.5, resistance is generally due more to form drag of the vegetation than from bed shear. Emergent vegetation can also induce wave resistance from free surface distortion. Plant properties that affect form drag include the ratio K/d, the relative submergence ($K \geq d$), plant density, distribution, and flexibility. Further complicating matters, unsteady nonuniform flow conditions often prevail, wake interference effects can reduce drag, and a variety of different riparian plant species are typically found in combination, which causes the spatial distribution of plant properties to vary greatly.

While it is important to consider the various complexities of flow resistance encountered in fluvial channels, most of our current knowledge on vegetative flow resistance is derived from laboratory flume experiments of steady fully developed flow through simulated vegetation of uniform density within rigid boundary rectangular flumes. These investigations have related vegetative resistance parameters, such as drag coefficients, Manning's n values, and friction factors f, to plant properties, including height, density, and flexibility [e.g. *Kouwen and Unny*, 1973; *Kouwen and Fathi-Moghadam*, 2000; *Wu et al.*, 2000; *Stone and Shen*, 2002].

Presently, computational hydraulics and steady RANS models are the most practical approaches for high Reynolds number fluvial hydraulics applications despite the rapid advancements in computational power and numerical algorithm development. Computational hydraulics models, although limited to the computation of bulk flow properties, are usually sufficient for flood studies. For these models, the bulk flow resistance parameter (e.g., Manning's n or the Darcy-Weisbach friction factor, f) can be modified to account for the measurable physical properties of vegetation based on empirical formulas [*Darby*, 1999].

Although computationally more intensive, steady RANS models allow resolution of the time-averaged turbulent flow field by adding source terms to the RANS and turbulence transport equations to account for vegetative drag effects. Steady RANS models have simulated 1D laboratory flume flows through simulated rigid vegetation corresponding to the laboratory measurements reported by *Shimizu and Tsujimoto* [1994] and *López and García* [1997, 1998; see also *López and García*, 2001; *Neary*, 2000, 2003; *Choi and Kang*, 2001]. *Tsujimoto and Kitamura* [1998] have incorporated a stem deformation model to extend 1D RANS simulations to flexible vegetation. *Naot et al.* [1996] and *Fischer-Antze et al.* [2001] have developed 3D RANS models for vegetated flows in compound channels with vegetation zones in riparian areas and flood plains. These models have enabled prediction of the effects of vegetation on sediment transport in fluvial channels [e.g., *Okabe et al.*, 1997; *López and García*, 1998].

Mean flow features resolved by the steady RANS models include: (1) the suppression of the streamwise velocity profile in the vegetated zone, (2) the inflection of the velocity profile at the top of the vegetation zone, and (3) the vertical distribution of the streamwise Reynolds stress (turbulent shear), with its maximum value at the top of the vegetation zone. However, for some of the experimental test cases, these models have been less successful at predicting the streamwise turbulence intensity. Also, the bulge in the velocity profile that is sometimes present near the bed cannot be resolved. This feature has been observed for some test cases reported by *Shimizu and Tsujimoto* [1994] and *Fairbanks and Diplas* [1998] despite a uniform vertical plant density distribution.

The present limitations of the RANS models are due mainly to spatial and temporal averaging, and possibly failure to model the effects of turbulence anisotropy. Some of these deficiencies may be offset somewhat through the treatment of the drag and weighting coefficients in the governing equations that account for vegetative drag effects. However, adopting non-universal drag coefficients or non-theoretical based weighting coefficients to make up for model deficiencies is not particularly desirable [see *López and García*, 1997; *Neary*, 2003].

The 1D RANS models eliminate streamwise or spanwise gradients in the flow field and vegetation layer by spatial averaging. The 3D models, while not spatial averaging, distribute the drag uniformly throughout the vegetation layer by introducing body force terms in the RANS equations. To date, neither 1D nor 3D models have actually simulated flow around individual stems. Due to this simplification, streamwise vortices (secondary motion), a suspected mechanism for momentum transfer that produces the near bed velocity bulge [*Neary*, 2000, 2003], cannot be simulated with any of the present RANS models.

As a result of time averaging, RANS models also cannot capture the organized large-scale unsteadiness and asymmetries (coherent structures) resulting from turbulent flow instabilities due to unsteady shear and pressure gradients induced by vegetation. These coherent structures include: (1) the transverse and other secondary vortices described by *Finnegan* [2000], which occur at the top of the vegetation layer as a result of a Kelvin-Helmholtz instability due to the inflection of the streamwise velocity profile, and (2) 3D vortices produced by the complex interaction of the approach flow with the stem (e.g., horseshoe and necklace vortices) and the oblique vortex shedding in the wake of the stem due to spanwise pressure gradients. These unsteady vortices would also contribute, or possibly play a dominant role, in redistributing momentum and producing the near bed velocity bulge.

The use of Reynolds stress transport (RST) modeling to account for turbulence anisotropy and its effects has received only limited numerical investigation [*Choi and Kang*, 2001] and its benefits are not yet apparent. The laboratory experiments by *Nezu and Onitzuka* [2001] demonstrate that riparian vegetation has significant effects on secondary currents due

to turbulence anisotropy, which increases with Froude number. However, coherent structures may account for a significantly larger percentage of the total Reynolds stresses and anisotropy [*Ge et al.*, 2003]. Under such circumstances, RST modeling would only have limited value.

Future numerical modeling efforts will focus on advanced CFD modeling techniques—namely statistical turbulence models that directly resolve large scale, organized, unsteady structures in the flow and advanced numerical techniques for simulating flows around multiple flexible bodies. These would include unsteady 3D Reynolds-averaged Navier-Stokes models [URANS; *Paik et al.*, 2003; *Ge et al.*, 2003] and large eddy simulation models [*Cui and Neary*, 2002]. Such techniques will elucidate the large-scale coherent structures described above, their important role in vegetative resistance, and their interaction and feedback with Reynolds stresses and lift forces that initiate sediment transport and bed form development.

8. RESEARCH PARADIGMS FOR COUPLING RIPARIAN VEGETATION AND FLUVIAL GEOMORPHOLOGY

While this volume was not intended to be an exhaustive treatise, it does provide new insights and future research opportunities into the linkages between vegetation and fluvial processes, landscape development and evolution. The research presented also provides a successful paradigm for research as well as demonstrating the strength of cross-disciplinary programs of study.

8.1. Society and Management

The recent explosion of research-related activities examining riparian vegetation in stream corridors has been buoyed largely by stream restoration and rehabilitation programs. These programs often employ "environmentally-friendly" techniques that rely on the use of riparian vegetation. With this clear application, experimental and field programs such as *Bennett, Bunn and Montgomery, Daniels and Rhoads, Garcia et al., Gray and Barker, Griffin and Smith, Ikeda et al., Pollen et al., Ross et al., Rutherfurd and Grove, Tal et al.*, and *Trimble* [all this volume] have examined more closely the important physical, hydrodynamic, and geomorphic processes associated with vegetation. These studies have highlighted (1) the strong physical interaction between streamflow conveyance and vegetation, including the transfer of biologically important nutrients, (2) the modulating effects vegetation can have on fluid forces driving sediment transport and river channel change, and (3) the role vegetation can play in channel form and process.

The results of such research are typically transferred to society in the form of management tools and models. The models proposed by *Kean and Smith*, *Smith*, *Van de Wiel and Darby*, and *Wu and Wang* [all this volume] demonstrate the successful development of appropriate technology. These models were able to simulate flow and sediment transport processes on flood plains, within straight channels, and within meandering channels in the presence of vegetation varying in size, shape, rigidity, and density. This paradigm for research—societal need for management tools and the reaction of researchers—is clearly evidenced in the contents of this volume and the targeted audience.

8.2. Cross-disciplinary Technology Transfer

Significant advances in science and research often occur as a result of cross-discipline transfer of technology. Several examples of this successful transfer process are provided herein. *Alonso* [this volume] used phenomenological studies from coastal engineering applications to assess the fluid forces acting on LWD, forces which define the stability of engineered log jams. *Pollen et al.* [this volume] apply fiber-bundle methodologies to quantify streambank erosion and the mechanical reinforcement by roots. *Gurnell et al.* [this volume] discuss biotic mechanisms of vegetation, namely reproductive strategies adopted by the plant species and propagule dispersal, and their important linkages to fluvial geomorphic processes and landscape stability. In each case, the key scientific advance was the recognition of existing knowledge from an allied discipline, its transfer to the riparian corridor, and its geomorphic application.

9. CONCLUSIONS

Understanding interactions between riparian vegetation and fluvial geomorphology is critical to determining processes and forms in alluvial channels. The interdisciplinary nature of the problems associated with quantifying these interactions requires knowledge from a broad range of disciplines including hydrology, hydraulics, sediment transport, ecology, botany, and geotechnical engineering. Because riparian vegetation serves as both an independent and dependent variable, studies are conducted at varying spatial and temporal scales depending on the processes being evaluated. Erosion-control and stream-restoration activities using riparian vegetation in place of "hard" structures are becoming commonplace, often without proper quantification of hydraulic forces and resistance and the effects on hydraulic and sediment-transport processes. Lessons learned from both successful and failed schemes need to be incorporated into new research programs with both fundamental and applied foci.

REFERENCES

Abbe, T. B., and D. R. Montgomery, Interaction of large woody debris, channel hydraulics, and habitat formation in large rivers, *Regulated Rivers: Research and Management*, 12, 201–221, 1996.

Abbe, T. B., D. R. Montgomery, and C. Petroff, C., Design of stable in-channel wood debris structures for bank protection and habitat restoration: An example from the Cowlitz River, WA, in *Management of Landscapes Disturbed by Channel Incision: Stabilization, Rehabilitation, Restoration*, edited by S. S. Y. Wang, E. J. Langendoen, and F. D. Shields, Jr., pp. 809–815, University of Mississippi, University, MS, 1997.

Abernethy, B., and I. D. Rutherfurd, The distribution and strength of riparian tree roots in relation to riverbank reinforcement, *Hydrological Processes*, 15, 63–79, 2001.

Arcement, G. J., and V. R. Schneider, Guide for selecting Manning's roughness coefficients for natural channels and flood plains, *United States Geological Survey Water-Supply Paper 2339*, 38 pp., U.S. Government Printing Office, Denver, CO, 1989.

Bennett, S. J., T. Pirim, and B. D. Barkdoll, Using simulated emergent vegetation to alter stream flow direction within a straight experimental channel, *Geomorphology*, 44, 115–126, 2002.

Brookes, A., and F. D. Shields, Jr., (Eds.), *River Channel Restoration*, John Wiley and Sons, Chichester, 433 pp., 1996.

Buffington, J. M., and D. R. Montgomery, Effects of hydraulic roughness on surface textures of gravel-bed rivers, *Water Resources Research*, 35, 3507–3522, 1999.

Choi, S.-U., and H. Kang, Reynolds stress modeling of vegetated open channel flows, in *Proceedings, XXIX Inter. Assoc. Hydraul. Res. Congress Conference*, Theme D., vol 1, pp. 264–269, Beijing, China, 2001.

Collison, A. J. C., and M. G. Anderson, Using a combined slope hydrology/stability model to identify suitable conditions for landslide prevention by vegetation in the humid tropics, *Earth Surface Processes and Landforms*, 21, 737–747, 1996.

Coppin, N. J., and I. G. Richards, *Use of Vegetation in Civil Engineering*, 292 pp., Butterworth, London, 1990.

Cui, J., and V. S. Neary, Large eddy simulation (LES) of fully developed flow through vegetation, in *Proceedings, 5th International Conference on Hydroinformatics*, Inter. Assoc. Hydraul. Res., Cardiff, UK, 2002.

Dabney, S. M., L. D. Meyer, W. C. Harmon, C. V. Alonso, and G. R. Foster, Depositional patterns of sediment trapped by grass hedges, *Transactions of the ASAE*, 38, 1719–1729, 1995.

D'Aoust, S. G., and R. G. Millar, Large woody debris fish habitat structure performance and ballasting requirements, *Watershed Restoration Management Report No. 8.*, 119 pp., Watershed Restoration Program, Ministry of Environment, Lands and Parks and Ministry of Forests, British Columbia, CA, 1999.

Darby, S. E., Effect of riparian vegetation on flow resistance and flood potential, *Journal of Hydraulic Engineering*, 125, 443–454, 1999.

Downs, P. W., and A. Simon, Fluvial geomorphological analysis of the recruitment of large woody debris in the Yalobusha River network, Central Mississippi, USA, *Geomorphology*, 37, 65–91, 2001.

Endo, T., and T. Tsuruta, On the effect of tree roots upon the shearing strength of soil, in *Annual Report of the Hokkaido Branch, Forest Place Experimental Station*, Sapporo, Japan, pp. 167–183, 1969.

Fairbanks, J. D., and P. Diplas, Turbulence characteristics of flows through partially and fully submerged vegetation, in *Proceedings, Engineering Approaches to Ecosystem Restoration, Wetlands Engineering and River Restoration Conference, ASCE*, Denver, CO, 1998.

Federal Interagency Stream Restoration Working Group (FISRWG), *Stream Corridor Restoration: Principles, Processes and Practices*, National Technical Information Service, U.S. Department of Commerce, Springfield, VA, 1998.

Finnigan, J., Turbulence in plant canopies, *Annual Review of Fluid Mechanics*, 32, 519–571, 2000.

Fischer-Antze, T., T. Stoesser, P. B. Bates, and N. R. Olsen, 3D numerical modelling of open-channel flow with submerged vegetation, *Journal of Hydraulic Research*, 39, 303–310, 2001.

Fukuoka, S., and A. Watanabe, Horizontal structure of flood flow with dense vegetation clusters along main channel banks, in *Proceedings, Environmental and Coastal Hydraulics: Protecting the Aquatic Habitat, The 27th Congress of the Int. Assoc. Hydraul. Res.*, vol. 2, pp. 1408–1413, 1997.

Ge, L., J. Paik, S. C. Jones, and F. Sotiropoulos, Unsteady RANS of complex 3D flows using overset grids, in *Proceedings, 3rd International Symposium on Turbulent and Shear Flow Phenomena*, Sendai, Japan, 2003.

Gippel, C. J., I. C. O'Neill, and B. L. Finlayson, The hydraulic basis of snag management, *Centre for Environmental Applied Hydrology*, Department of Civil and Agricultural Engineering, and the Department of Geography, 116 pp., University of Melbourne, Australia, 1992.

Gray, D. H., and A. J. Leiser, *Biotechnical Slope Protection and Erosion Control*, Van Nostrand Reinhold, New York, 400 pp., 1982.

Greenway, D. R., Vegetation and slope stability, in *Slope Stability*, edited by M. G. Anderson and K. S. Richards, pp. 187–230, John Wiley and Sons, Chichester, 1987.

Gurnell, A. M., H. Piégay, F. J. Swanson, and S. V. Gregory, Large wood and fluvial processes, *Freshwater Biology*, 47, 601–619, 2002.

Hupp, C. R., Riparian vegetation recovery patterns following stream channelization: A geomorphic perspective, *Ecology*, 73, 1209–1226, 1992.

Hupp, C. R., and W. R. Osterkamp, Riparian vegetation and fluvial geomorphic processes, *Geomorphology*, 14, 277–295, 1996.

Hupp, C. R., W. R. Osterkamp, and A. D. Howard, (Eds.), *Biogeomorphology, Terrestrial and Freshwater Systems, Geomorphology*, 13, 347 pp., 1995.

Johnson, W. C., Woodland expansion in the Platte River, Nebraska: Patterns and causes, *Ecological Monographs*, 64, 45–84, 1994.

Keller, E. A., and F. J. Swanson, Effects of large organic debris on channel form and fluvial process, *Earth Surface Processes and Landforms*, 4, 361–380, 1979.

Keller, E. A., and T. Tally, Effects of large organic debris on channel form and fluvial processes in the coastal redwood environment, in *Adjustments of the Fluvial System*, edited by D. D. Rhodes, and

G. P. Williams, pp. 169–197, George Allen and Unwin, London, 1979.

Kirkby, M. J., Modeling the links between vegetation and geomorphology, *Geomorphology* 13, 319–335, 1995.

Kouwan, N., Field estimation of the biomechanical properties of grass, *Journal of Hydraulic Research*, 26, 559-568, 1988.

Kouwen, N., and M. Fathi-Moghadam, Friction factors for coniferous trees along rivers, *Journal of Hydraulic Engineering*, 126, 732–740, 2000.

Kouwen, N., and T. E. Unny, Flexible roughness in open channel, *Journal of the Hydraulics Division, ASCE*, 99, 713–728, 1973.

Kouwen, N., T. E. Unny, and A. M. Hill, Flow retardance in vegetated channels, *Journal of the Irrigation and Drainage Division, ASCE*, 95, 329–342, 1969.

Langbein, W. B., and S. A. Schumm, Yield of sediment in relation to mean annual precipitation. *American Geophysical Union, Trans.*, 39, 1076–1084, 1958.

Li, R. M., and H. W. Shen, Effect of tall vegetations on flow and sediment, *Journal of the Hydraulics Division, ASCE*, 99, 793-814, 1973.

López., F., and M. García, Open channel flow through simulated vegetation: Turbulence modeling and sediment transport, *Wetlands Research Technical Report WRP-CP-10*, U.S. Army Corps of Engineers, 106 pp, Waterways Experiment Station, Vicksburg, 1997.

López., F., and M. García, Open-channel flow through simulated vegetation: Suspended sediment transport modeling, *Water Resources Research*, 34, 2341–2352, 1998.

López., F., and M. García, Mean flow and turbulence structure of open-channel flow through non-emergent vegetation, *Journal of Hydraulic Engineering*, 127, 392–402, 2001.

Lowrance, R. R., S. McIntyre, and C. Lance, Erosion and deposition in a field/forest system estimated using Cesium-137 activity, *Journal of Soil and Water Conservation*, 43, 195–199, 1988.

Malanson, G. P., *Riparian Landscapes*, Cambridge University Press, Cambridge, 296 pp., 1993.

Masterman, R., and C. R. Thorne, Predicting influence of bank vegetation of channel capacity, *Journal of Hydraulic Engineering*, 118, 1052–1058, 1992.

May, C. L., and R. E. Gresswell, Processes and rates of sediment and wood accumulation in headwater streams of the Oregon Coast Range, USA, *Earth Surface Processes and Landforms*, 28, 409–424, 2003.

Montgomery, D. R., J. M. Buffington, R. D. Smith, K. M. Schmidt, and G. Pess, Pool spacing in forested channels, *Water Resources Research*, 31, 1097–1105, 1995.

Montgomery, D. R., and H. Piégay, Wood in rivers: Interactions with channel morphology and processes, *Geomorphology*, 51, 1–5, 2003.

Naot, D., I. Nezu, and H. Nakagawa, Hydrodynamic behavior of partly vegetated open-channels, *Journal of Hydraulic Engineering*, 122, 625-633, 1996.

Neary, V. S., Numerical model for open channel flow with vegetative resistance, in *Proceedings, 4th International Conference on Hydroinformatics, Inter. Assoc. Hydraul. Res.*, Cedar Rapids, IA, 2000.

Neary, V. S., Numerical solution of fully-developed flow with vegetative resistance, *Journal of Engineering Mechanics*, 129, 558–563, 2003.

Nepf, H. M., Drag, turbulence, and diffusion in flow through emergent vegetation, *Water Resources Research*, 35, 479–489, 1999.

Nepf, H. M., and E. R. Vivoni, Flow structure in depth-limited, vegetated flow, *Journal of Geophysical Research*, 105(C12), 28,547–28,557, 2000.

Nezu I., and K. Onitsuka, Turbulent structures in partly vegetated open-channel flows with LDA and PIV measurements, *Journal of Hydraulic Research*, 39, 629–642, 2001.

Okabe, T., T. Yuuki, and M. Kojima, Bed-load rate on movable beds covered by vegetation, in *Proceedings, Environmental and Coastal Hydraulics: Protecting the Aquatic Habitat, The 27th Congress of the Int. Assoc. Hydraul. Res.*, vol. 2, pp. 1397–1401, 1997.

Paik, J., S. C. Jones, and F. Sotiropoulos, DES and URANS of turbulent boundary layer on a concave wall, in *Proceedings, 3rd International Symposium on Turbulent and Shear Flow Phenomena*, Sendai, Japan, 2003.

Petryk, S., and G. Bosmjian, III., Analysis of flow through vegetation, *Journal of the Hydraulics Division, ASCE*, 101, 871–884, 1975.

Righetti, M., and A. Armanini, Flow resistance in open channel flows with sparsely distributed bushes, *Journal of Hydrology*, 269, 55–64, 2002.

Schumm, S. A., and R. F. Hadley, Arroyos and the semiarid cycle of erosion, *American Journal of Science*, 225, 161–174, 1957.

Schumm, S. A., M. D. Harvey, and C. C. Watson, *Incised Channels: Morphology, Dynamics, and Control*, Water Resources Publications, Littleton, CO, 200 pp., 1984.

Shields, F. D., Jr., and C. M. Cooper, Woody vegetation and debris for in-channel sediment control, *International Journal of Sediment Research*, 15, 83–92, 2000.

Shields, F. D., Jr., A. J. Bowie, and C. M. Cooper, Control of streambank erosion due to bed degradation with vegetation and structure, *Water Resources Bulletin*, 31, 475–489, 1995.

Shields, F. D., Jr., and A. D. Wood, The use of large woody material for habitat and bank protection, Section 12, Chapter 9, *USDA-NRCS Stream Design Guide*, U.S. Department of Agriculture, Natural Resources Conservation Service, Washington, D. C., in review.

Shimizu, Y., and T. Tsujimoto, Numerical analysis of turbulent open-channel flow over vegetation layer using a k-? turbulence model, *Journal of Hydroscience and Hydraulic Engineering, JSCE*, 11, 57–67, 1994.

Simon, A., Shear-strength determination and stream-bank instability in loess-derived alluvium, West Tennessee, USA, in *Applied Quaternary Research*, edited by E. J. DeMulder and B. P. Hageman, pp. 129–146, A. A. Balkema Publications, Rotterdam, 1989.

Simon, A., The nature and significance of incised river channels, in *Incised River Channels: Processes, Forms, Engineering and Management*, edited by S. E. Darby and A. Simon, pp. 3–18, John Wiley and Sons, London, 1999.

Simon A., and A. J. Collison, Quantifying the mechanical and hydrologic effects of riparian vegetation on streambank stability, *Earth Surface Processes and Landforms*, 27, 527–546, 2002.

Simon, A., A. Curini, S.E. Darby, and E.J. Langendoen, Bank and near-bank processes in an incised channel. *Geomorphology* 35:193–217, 2000.

Simon, A., A. Heins, and W. Dickerson, Suspended-sediment transport rates at the 1.5 year recurrence interval: Transport conditions at the bankfull and effective discharge? *Geomorphology*, in press.

Simon, A., and C. R. Hupp, Geomorphic and vegetative recovery processes along modified stream channels of West Tennessee, *U.S. Geological Survey Open-File Report 91-502*, 142 pp., U.S. Government Printing Office, Denver, CO, 1992.

Stephan, U., and D. Gutknecht, Hydraulic resistance of submerged flexible vegetation, *Journal of Hydrology*, 269, 27–43, 2002.

Stone, B. M., and H. T. Shen, Hydraulic resistance of low in channels with cylindrical roughness, *Journal of Hydraulic Engineering*, 128, 500–506, 2002.

Thompson, G. T., and J. A. Roberson, A theory for flow resistance for vegetated channels, *Transactions of the ASAE*, 19, 288–293, 1976.

Thompson, A. M., B. N. Wilson, and T. Hustrulid, Instrumentation to measure drag on idealized vegetal elements in overland flow, *Transactions of the ASAE*, 46, 295–302, 2003.

Thorne, C. R., Effects of vegetation on riverbank erosion and stability, in *Vegetation and Erosion*, edited by J. B. Thornes, pp. 125–143, John Wiley and Sons, Chichester, 1990.

Thornes, J. B., (Ed.), *Vegetation and Erosion: Processes and Environments*, John Wiley and Sons, 518 pp., 1990.

Tollner, E. W., B. J. Barfield, C. T. Haan, and T. Y. Kao, Suspended sediment filtration capacity of simulated vegetation, *Transactions of the ASAE*, 19, 678–682, 1976.

Tsujimoto, T., Fluvial processes in streams with vegetation, *Journal of Hydraulic Research*, 37, 789–803, 1999.

Tsujimoto, T., and T. Kitamura, A model for flow over flexible vegetation-covered bed, in *Proceedings, International Water Resources Engineering Conference, ASCE*, New York, pp. 1380–1385, 1998.

U.S. Department of Agriculture, Natural Resources Conservation Service (USDA-NRCS), Part 650, Chapter 16, Streambank and Shoreline Protection, *National Engineering Handbook Series*, U.S. Government Printing Office, Washington, D.C., 1996.

Volny, S., Riparian stands, in *Forest Amelioration*, edited by O. Riedl and D. Zachar, pp. 423–453, Elsevier, Amsterdam, 1984.

Wallerstein, N., C. V. Alonso, S. J. Bennett, and C. R. Thorne, Surface wave forces acting on submerged logs, *Journal of Hydraulic Engineering*, 128, 349–353, 2002.

Wallerstein, N., and C. R. Thorne, Influence of large woody debris on morphological evolution of incised, sand-bed channels, *Geomorphology*, in press.

Watson, C. C., S. R. Abt, and D. Derrick, Willow posts bank stabilization, *Journal of the American Water Resources Association*, 33, 293–300, 1997.

Williams, G.P., The case of the shrinking channels—The North Platte and Platte Rivers in Nebraska: *U.S. Geological Survey Circular 781*, 48 pp., U.S. Government Printing Office, Denver, CO, 1978.

Wu, T. H., Effect of vegetation on slope stability, *Trans. Res. Record 965*, Trans. Res. Board, Washington, D.C., 37–46, 1984.

Wu, T. H., W. P. McKinnell, and D. N. Swanston, Strength of tree roots and landslides on Prince of Wales Island, Alaska, *Canadian Geotechnical Journal*, 16, 19–33, 1979.

Wu, F. C., H. W. Shen, and Y. J. Chou, Variation of roughness coefficients for unsubmerged and submerged vegetation, *Journal of Hydraulic Engineering*, 125, 934–942, 2000.

S. J. Bennett, Department of Geography, University at Buffalo, Buffalo, NY 14261-0055.

V. S. Neary, Department of Civil and Environmental Engineering, Tennessee Technological University, Cookeville, TN 38505-0001.

A. Simon, USDA-ARS National Sedimentation Laboratory, P.O. Box 1157, Oxford, MS 38655.

Flow, Turbulence, and Resistance in a Flume with Simulated Vegetation

Marcelo H. García[1], Fabián López[1], Chad Dunn[1], Carlos V. Alonso[2]

[1]*Ven Te Chow Hydrosystems Laboratory, Department of Civil and Environmental Engineering*
University of Illinois at Urbana-Champaign, Urbana, Illinois

[2]*U.S. Department of Agriculture-Agricultural Research Service, National Sedimentation Laboratory,*
Oxford, Mississippi

Open-channel flow models through vegetative canopies require a quantitative measure of the ability of plants to absorb momentum by form drag, which is commonly characterized in terms of a drag coefficient. An experimental study was performed to investigate the flow structure and drag coefficients in an open channel with both rigid and flexible simulated vegetation under uniform flow conditions. Acoustic Doppler velocimetry was employed to measure three-dimensional velocity components in and above a cylinder canopy. Local values of the drag coefficient were determined from a horizontal momentum balance, which allowed for the first measurements of the vertical variation of the vegetation-induced drag coefficient in open-channel flow. Results show that the drag coefficient is not constant in the vertical, as many models have assumed, but instead, varies throughout the height of the water column.

1. INTRODUCTION

Historically, vegetation growth in open-channel waterways has been seen as a nuisance primarily because of the resulting reduction in discharge capacity and has consequently been removed. However, attitudes toward river and wetland management have been changing to recognize the considerable environmental benefits that vegetation brings to an aquatic ecosystem. Vegetation cover is known to increase bank stability, reduce erosion and water turbidity, provide habitat for aquatic and terrestrial wildlife, attenuate floods, provide aesthetic properties, and filter pollutants carried by runoff. Engineering

Riparian Vegetation and Fluvial Geomorphology
Water Science and Application 8
Copyright 2004 by the American Geophysical Union
10.1029/008WSA02

practices have been changing to include the use of vegetative linings as effective alternatives in river restoration projects.

This new attitude towards waterway engineering brings about an increased need for understanding open-channel flow through and above vegetation. Vegetative linings influence not only the flow resistance of streams and the habitat quality, but also affect transport processes by reducing the entrainment capability of sediment into suspension and by altering the mixing properties of the stream. Therefore, more conclusive knowledge of the hydraulic properties of channels with vegetation is essential for their effective engineering design and for accurately assessing the influence of vegetation on the total quality and effectiveness of a stream.

Pioneering work into the topic of flow through vegetated channels was performed by *Ree and Palmer* [1949]. The aim of these researchers was to establish a method to determine the discharge capacity of a channel by adjusting Manning's coefficient to account for the increased roughness and subsequent decreased flow conveyance of the channel. More recently, research emphasis has shifted from work primarily aimed at determining empirical methods of design towards investigations oriented at providing physical explanations of the flow phenomenon. These modes include those proposed by *Li and Shen* [1973], *Reid and Whitaker* [1976], *Burke and Stolzenbach* [1983], *Christensen* [1985], and *Saowapon and Kouwen* [1989]. Each of these models requires some quantitative estimation of the ability of plants to absorb momentum by form and viscous drag imposed by the plant leaves and stems.

Field and laboratory investigations have related vegetative resistance parameters, such as drag coefficients and Manning's n values, to plant properties, including height, density, and flexibility [e.g. *Kouwen and Unny*, 1973, *Wu et al.* 2000, *Stone and Shen* 2002]. This information has led to the development of semi-empirical formulas for calculating bulk flow parameters and spatially averaged velocity profiles [e.g. *Klopstra et al.*, 1997; *Freeman et al.*, 1998]. *Shimizu and Tsujimoto* [1994] measured vertical profiles of velocity, Reynolds stress, and turbulence intensity for submerged rigid vegetation of various heights and densities. *Fairbanks and Diplas* [1998] pointed out that despite the attempts of recent laboratory studies to describe the flow through vegetation by a single *spatially-averaged* velocity or turbulence intensity profile (i.e. spatially averaged in the stream and spanwise directions), measured profiles for these parameters may not be representative of the conditions for all locations in the flow.

Numerical modeling studies by *Shimizu and Tsujimoto* [1994] and *Lopez and Garcia* [1997, 1998, 2001] are representative of recent attempts to simulate steady uniform flows through vegetation of uniform density. These models close the Reynolds-Averaged Navier-Stokes Equations (RANS) equations using the standard k-ε turbulence model with wall-functions. They introduce a sink term representing vegetative drag to the RANS equations, $F = \frac{1}{2}C_D\lambda|U|U$; where U is local time-averaged velocity, C_D is local drag coefficient, and λ is local resistance area per unit volume of flow. The k- (turbulent kinetic energy) and ε- (dissipation) transport equations were also modified by introducing the drag-related turbulence production terms $C_{fk}FU$ and $C_{fe}FU$. More recently, Large Eddy Simulations (LES) of fully-developed flow through vegetation have provided further insights into the effects of vegetation on the turbulence field, including turbulence anisotropies, as well as ejection and sweep events [*Cui and Neary*, 2002].

At high Reynolds numbers, the form drag provides the dominant resistance in the channel [*Yen*, 2002]. A major obstacle in the study of vegetated open-channel flow has been the parameterization of the form drag through a drag coefficient C_D. Typically, the entire plant canopy has been characterized by a bulk value of the drag coefficient, however, the k-ε model first pro-

posed by *Burke and Stolzenbach* [1983] is significant in that it allows the drag coefficient to be specified locally within the plant canopy, although no measurements of the vertical variation of the drag coefficient existed at the time of its conception.

To address these needs, an investigation was undertaken to measure mean flow and turbulence characteristics within and above a simulated plant canopy in a steady, uniform open-channel flow at the University of Illinois. A 3-D acoustic Doppler velocimeter (ADV) was used for these measurements. This article reports on the measurements described above and explains in detail a streamwise momentum balance that is used in conjunction with measured velocity and Reynolds stress profiles to determine vertical distributions of the drag coefficient. Values of the bulk drag coefficient are presented as well as a discussion of the effects of flow and channel vegetation characteristics on the flow structure and the value of the bulk drag coefficient.

2. STREAMWISE MOMENTUM BALANCE

Consider the time-averaged horizontal momentum equation for incompressible turbulent flow:

$$0 = \rho g So + \frac{\partial}{\partial x}\left(\mu\frac{\partial \overline{u}}{\partial x} - \rho\overline{u'^2}\right) + \frac{\partial}{\partial y}\left(\mu\frac{\partial \overline{u}}{\partial y} - \rho\overline{u'v'}\right)$$
$$+ \frac{\partial}{\partial z}\left(\mu\frac{\partial \overline{u}}{\partial z} - \rho\overline{u'w'}\right) \tag{1}$$

where ρ is the water density, g is gravitational acceleration; u, v, and w, are the instantaneous velocities in the streamwise (x), spanwise (z), and bed-normal (y) directions, respectively, So is the bed slope, and μ is the absolute dynamic viscosity of the water. Overbars represent the usual Reynolds average operator and prime marks refer to temporal fluctuations around mean values.

The stress associated with x-y plane normal to the bed is dominant in fully turbulent uniform flow and all non-associated terms can be ignored. Therefore, equation (1) can be reduced to:

$$\frac{d\left(-\rho\overline{u'v'}\right)}{dy} = -\rho g So \tag{2}$$

First *Wilson and Shaw* [1977] and later *Raupach and Shaw* [1982] realized the need for introducing both horizontal and temporal averages of the governing equations in order to get a proper representation of flow through vegetation in a one-dimensional frame. Equation (2) above is only temporally averaged and therefore, does not completely represent the problem at hand. *Raupach and Shaw* [1982] give a detailed description of the momentum and energy equations for atmospheric boundary layers over plant canopies. Applying a similar procedure for steady, uniform turbulent flow in a mildly sloped channel, the resulting time and space-averaged horizontal momentum equation can be approximated by [*Lopez and Garcia*, 2001]:

$$\frac{\partial}{\partial y}\left\langle -\rho \rangle \overline{u'v'} \langle \right\rangle - \left\rangle \frac{\partial p''}{\partial x} \left\langle = -\rho g So \right. \tag{3}$$

Angle brackets denote spatial averages in a horizontal plane and double primes refer to spatial fluctuations around the mean value. The second term on the left hand side of equation (3) represents the form drag introduced by the vegetation. It is commonly parameterized using a drag coefficient, C_D, as follows:

$$\left\rangle \frac{\partial p''}{\partial x} \left\langle = \frac{1}{2}\rho C_D a \rangle u \langle^2 \right. \tag{4}$$

Where a is the density of the vegetation expressed as frontal area per unit volume such that for cylindrically shaped, equally spaced vegetation:

$$a = D\big/L^2 \tag{5}$$

In equation (5), D is the plant diameter and L^2 is the horizontal area of influence of each plant. The vegetation density, a, therefore has units of per length. Equation (3) for the total shear stress may now be rewritten as:

$$\frac{d\rangle\overline{u'v'}\langle}{dy} = -\frac{1}{2}C_D a\rangle u\langle^2 + gSo \tag{6}$$

From which the drag coefficient is expressed as:

$$C_D = 2\frac{gSo - \dfrac{\partial\rangle\overline{u'v'}\langle}{\partial y}}{a\rangle u\langle^2} \tag{7}$$

Equation (7) allows for the estimation of vertical profiles of the drag coefficient by measuring the channel slope, the obstruction density, and profiles of the mainstream velocity and dominant Reynolds stress under uniform, steady flow conditions. An experimental study was designed and performed with these specific intentions. Details describing the set-up, measurement procedures, and results are discussed in the remainder of this paper. More details can be found in *Dunn et al.* [1996].

2.1. Experimental Study

2.1.1 Dimensional Analysis. In an attempt to determine the relevant parameters in this complex flow situation, the following variables were considered to influence the flow of water in a vegetated open channel: U, H, So, g, D, a, h, μ, ρ, C_D, α. Here, U is the mean mainstream channel velocity, H is the flow depth, h is the plant height, and α is a non-dimensional parameter characterizing the plant flexibility. The rest of the variables are as defined previously. The resistance of the bed was considered negligible when compared to the resistance imposed by the cylinders. A dimensional analysis with U, ρ, and H as fundamental variables reveals the following functional relationship for C_D.

$$C_D = \phi\left(\frac{\rho UH}{\mu}, \frac{U}{\sqrt{gH}}, S, \frac{D}{H}, Ha, \frac{h_p}{H}, \alpha\right) \tag{8}$$

where $\rho UH/\mu$ is the Reynolds number and $U/(gH)^{1/2}$ is the Froude number. Equation (8) reveals the dimensionless quantities that play a role in the determination of the drag coefficient. Each of these dimensionless parameters was varied in the experimental investigation.

2.1.2 Experimental Set-Up. The experimental investigation was performed in a 19.5 m long by 0.91 m wide by 0.61 m deep tilting flume in the Ven Te Chow Hydrosystems Laboratory at the University of Illinois. Both rigid and flexible cylinders were used to simulate vegetation. Rigid plants were simulated with 6 inch-long, ¼ inch diameter wooden dowels. Flexible plants were simulated with ¼ inch diameter by 7-¾ inch long plastic commercial drinking straws. In order to place the straws and dowels in the channel, a false bottom was constructed with sheets of ¼ inch thick pegboard consisting of ¼ inch diameter holes spaced at 1 inch on centers.

The height of the cylinders extending into the channel varied. For the rigid cylinder experiments, the dowels did not deflect and their average height within the flow of water, h, was measured to be 0.118 ± 0.0167 meters. For the flexible cylinder experiments, the average non-deformed straw height, h, was 0.169 ± 0.0161 meters. Of course the flexible nature of the straws allowed them to deform to a new average height when placed in the water flow.

For all of the experiments, the cylinders were arranged in a staggered pattern and spaced at 2, 3, 4, or 6 inches on centers. Additional details on the experimental facilities and set-up are described in *Dunn et al.* [1996].

2.1.3 Flow Velocity Measurements. The three Cartesian components of velocity were measured with a Son-Tek acoustic

Doppler velocimeter (ADV) at a sampling rate of 25 Hz. The ADV is a point-type current meter based on the acoustic Doppler shift velocity measurement principle. Acoustic pulses are sent out from the ADV and are reflected off small particles suspended in the flow, which are assumed to move at the same velocity as the flow. The reflected signals are then captured by the receivers and processed by computer software. The ADV samples an elliptically shaped volume of less than 1 cm^3, measuring 9 mm along the vertical axis and 4 mm along the horizontal axis. For further information regarding the operational and technical aspects of the ADV refer to the paper by *Kraus et al.* [1994].

The ADV was used in this study primarily because it allowed for the measurement of three velocity components, and consequently Reynolds stresses. However, *Lohrmann et al.* [1995] reported that ADV measurements of Reynolds stress at low velocities of less than 0.01 m/s result in values that have a slight positive bias. This positive offset is caused by variations in the sensitivities of the three ADV receivers, which leads to differences in the magnitude of the noise terms. At higher velocities, the noise terms become negligible and the positive offset of the Reynolds stress does not occur. Preliminary measurements conducted for this study indicated that the Doppler noise was equally balanced in all three channels. Therefore, for the flow velocity ranges considered in these experiments, which were well above 10 cm/s, the Reynolds stress estimates were believed to be unbiased.

Unfortunately, using the ADV in these experiments to measure such a shallow water column resulted in a substantial loss in the measurable depth of the water column. This unavoidable loss of up to 0.01 m resulted from the required distance needed to submerge the ADV and measure the sampling volume. However, the ability to easily and effectively measure velocities and Reynolds stresses more than justified the use of the device in the experiments.

2.1.4 Experimental Procedure. Eighteen experiments were performed to investigate the profiles of velocity, Reynolds stresses, turbulence intensity, and drag in a simulated vegetated channel. Cross-section averaged profiles were determined by measuring at random locations within the cylinders and averaging the measurements together. The controllable variables in the laboratory were flow discharge, channel bed slope, cylinder spacing, and cylinder flexibility. By controlling these four variables, all of the relevant variables discussed previously could be changed. Table 1 shows the combinations of the four variables used for each experiment.

Each of the 18 experiments consisted of four measured profiles with 10 points per profile resulting in a total of 40 measuring points per experiment. The position of each profile with

Table 1. Experimental conditions.

Exp. #	Q (L/s)	So (%)	L (cm)	F
1	179	0.36	7.62	rigid
2	88	0.36	7.62	rigid
3	46	0.36	7.62	rigid
4	178	0.76	7.62	rigid
5	98	0.76	7.62	rigid
6	178	0.36	15.24	rigid
7	95	0.36	15.24	rigid
8	180	0.36	5.08	rigid
9	58	0.36	5.08	rigid
10	180	1.61	5.08	rigid
11	177	0.36	5.08	rigid
12	181	1.08	5.08	rigid
13	179	0.36	7.62	flexible
14	180	1.01	7.62	flexible
15	93	0.36	7.62	flexible
16	179	0.36	15.24	flexible
17	78	0.36	5.08	flexible
18	179	1.01	5.08	flexible

Exp. # – Experiment number; *Q* – Discharge; *So* – Bed Slope; *L* – Cylinder Spacing; F – Cylinder Flexibility

respect to adjacent cylinders and the height of each of the adjacent cylinders were measured and recorded.

Profiles were not measured directly behind a cylinder, because these measurements resulted in profiles that were significantly altered when compared to those measured elsewhere in the channel. Therefore, including these measurements in the averaging process would unfairly influence the averaged profiles. However, to obtain the most representative horizontal average of the flow, profiles were taken within cylinder wakes. *Seginer et al.* [1976] found that profiles measured within cylinder wakes were not significantly different from those measured outside the wakes. The profiles measured in this investigation agree with this observation.

For the flexible cylinder experiments, a video camera was used to record the deflection of the cylinders. Then, imaging software was used to accurately measure the deflection angle of the cylinders. An estimate of the average deflected cylinder height was also obtained by randomly measuring a sample of cylinders and averaging this group.

Each point of the experiments was sampled and averaged for 3 minutes to obtain accurate measurements of Reynolds stress. Preliminary measurements showed that the time-averaged values of mainstream velocity, $\bar{u}$, Reynolds stress, $-\rho\overline{u'v'}$, and mainstream turbulence intensity, $\sqrt{\overline{u'^2}}$, were highly dependent on the total averaging time of the ADV record. As the averaging time was extended, the values of the above parameters became relatively constant.

Figure 1 shows the variation of the velocity, turbulence intensity, and Reynolds stress with the dimensionless total averaging time, t^*. These statistics have been normalized with their values after 10 minutes of averaging, $\bar{u}_{10}$, $\overline{u'v'}_{10}$, and $\sqrt{\overline{u'^2}}_{10}$. The averaging time has been made dimensionless so that $t^* = t\,\bar{u}_{10}/h_p$. The canopy height, h_p,

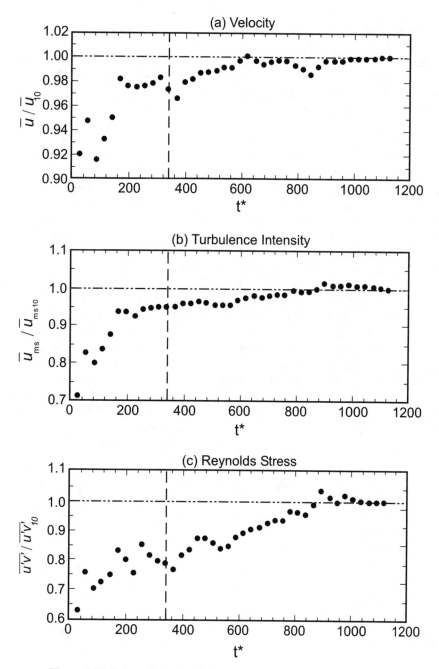

Figure 1. Variation of mean statistics with dimensionless averaging time.

was chosen to normalize the time scale, because the size of eddies within the canopy is determined by the characteristic length of the cylinders. The mean velocity, $\bar{u}_{10}$, was chosen because the convection velocity of the eddies is dependent on the measured velocity at that point in the channel. These measurements were taken within the canopy under extreme flow conditions where the canopy velocity was low because of a high cylinder density; therefore, the convection velocity of eddies within the flow was relatively small and the averaging period was large.

A compromise had to be reached between the need to extend the record length, so as to increase the measurement accuracy, and the need to shorten the record length, so that 40 measurement points could be measured in a reasonable amount of time. The dimensionless time of 340 was chosen as the appropriate averaging interval. For the flow conditions of Figure 1, this time corresponds to 3 minutes.

Averaging for this abbreviated amount of time does introduce some error into the computations of the statistics, especially for the computation of the Reynolds stress. Figure 1 demonstrates that when averaging the Reynolds stress even over an extremely long interval of time, the relative uncertainty in the measurements will be no better than 5%. For a 3 minute averaging interval, there was about a 15 percent error in the measurement of Reynolds stress. This error estimation is the worst case because when sampling at lower cylinder densities, the convection velocity of eddies would be greater and the chosen averaging interval would result in smaller errors. In fact, at the lowest tested cylinder density, the error in the measurement of the Reynolds stress was below the minimum uncertainty.

It is important to point out here that the small errors in the Reynolds stress measurements were not believed to cause equal errors in the computation of the drag coefficient. In the computation of the drag coefficient, only the gradient of the Reynolds stress profiles, and not the actual magnitude, is of importance. Therefore, even though up to a 15 percent error may have existed in the measurements of the Reynolds stress, this error is not believed to have significantly affected the computations of the drag coefficient.

2.2. Experimental Results

2.2.1 Velocity Profiles.
The instantaneous raw velocity data was corrected for small ADV tilt angles and rotation errors before analyzing the data to avoid significant errors in the estimation of Reynolds stress [Lohrman et al., 1995]. Tilt and rotation corrections were very small, because every effort was made to eliminate tilt when the probe was set up in the laboratory; however, small corrections were made [Dunn et al., 1996].

The corrected instantaneous velocities were used to compute time-averaged values of the velocities: $\bar{u}$, $\bar{v}$, and $\bar{w}$. Then, the values of mainstream velocity, $\bar{u}$, for all of the verticals were averaged at each distance from the bed resulting in one-dimensional profiles of spatially-averaged mainstream velocity.

The space and time-averaged velocity profiles through rigid cylinders had a characteristic shape that was dependent on the cylinder density. Figure 2 shows the shapes of the velocity profiles with $\rangle u \langle$ made dimensionless with respect to the velocity at the top of the cylinder canopy, u_{ref}, and the distance from the bed, y, made dimensionless with the average cylinder height, h_p. Four different cylinder densities were tested resulting in four different dimensionless profiles. As Figure 2 illustrates, the dimensionless velocity profiles collapsed extremely well.

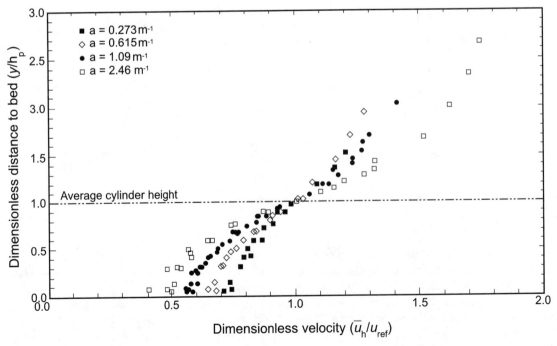

Figure 2. Dimensionless velocity profiles for flow through rigid cylinders.

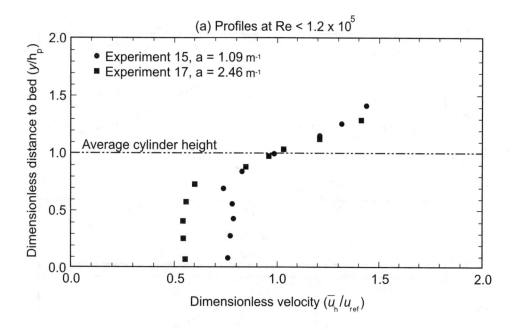

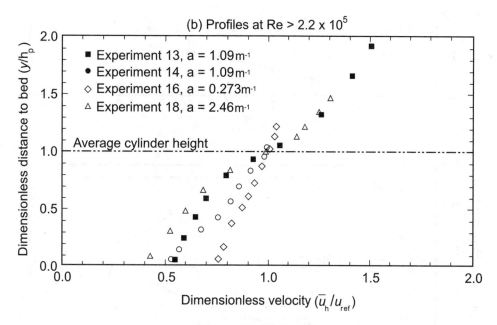

Figure 3. Dimensionless velocity profiles for flow through flexible cylinders.

The results of the experiments with flexible cylinders are markedly different as illustrated in Figure 3. Three different cylinder densities were tested. The shapes of these dimensionless profiles indicated that the cylinder density still played a major role, but because the profiles did not collapse well at all, there was substantial evidence that the cylinder density was not the only parameter that significantly affected the shapes of the velocity profiles. The cylinder density was obvi-

ously important, but so were the cylinder flexibility and the canopy velocity, which determined the deflection of the cylinders. The two profiles with $Re < 1.2 \times 10^5$ were of significantly different shape than the three with $Re > 2.2 \times 10^5$ as illustrated in Figure 3.

The profiles measured in the present study agreed qualitatively with the velocity profiles measured by *Tsujimoto et al.* [1991] in and above simulated vegetation. A quantitative com-

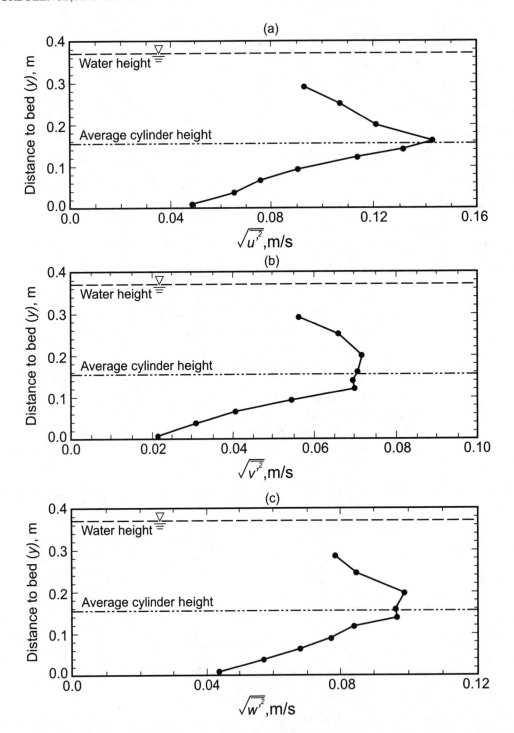

Figure 4. Turbulence intensity profiles. H = 0.368 m, hp = 0.152 m and a = 1.09 m^{-1}. Profile 1.

parison was not possible, because of the varying channel and flow conditions, but the shapes of these profiles were in good agreement. In particular, both investigations found that as the canopy became sparser, the velocity profiles became more like the typical profiles for regular open-channels. Profiles in denser cylinder arrangements deviated from the standard open-channel shape and became concave down within the canopy, with an inflection point near the top of the canopy.

Seginer et al. [1976] reported a similar finding in a wind tunnel and *Hartog and Shaw* [1975] measured similar profiles in an atmospheric field study.

2.2.2 Turbulence Intensity Profiles. The time and space-averaged values of the turbulence intensities $\rangle\sqrt{\overline{u'^2}}\langle$, $\rangle\sqrt{\overline{v'^2}}\langle$, and $\rangle\sqrt{\overline{w'^2}}\langle$ were also computed. The averaged profiles of turbulence intensity had a typical shape that is illustrated on Figure 4. The profiles showed no obvious differences between flow through rigid or flexible cylinders; in both cases, the turbulence intensities were suppressed inside the canopy.

An interesting tendency was discovered in the course of the analysis. The mean turbulence intensity within the canopy, which is defined by equation (9), was strongly related to the dimensionless variable *Ha*, which is the product of flow depth and cylinder density. This is illustrated in Figure 5, where the mean turbulence intensity is defined as:

$$\bar{i}_c = \frac{1}{h_p} \int_0^{h_p} \frac{\rangle\sqrt{\overline{u'^2}}\langle}{\rangle\bar{u}\langle} dy \qquad (9)$$

Such correlation was not totally unexpected however, since an increase in cylinder density was expected to increase the level of turbulence in the canopy as discussed by *Seginer et al.* [1976].

However, the dimensionless variable *Ha* was found to be more strongly correlated to than to the cylinder density, *a*, alone.

Measurements of turbulence intensity inside a simulated vegetated canopy have also been performed by *Seginer et al.* [1976] in an air flow and *Tsujimoto et al.* [1991] in a water flow. Their measurements showed profiles that reached a maximum near the top of the canopy and were in good qualitative agreement with those of the present study.

2.2.3 Reynolds Stress Profiles. Values of the Reynolds stresses per unit density and $-\overline{u'v'}$, $-\overline{u'w'}$, $-\overline{v'w'}$ and were also computed and spatially averaged to determine one-dimensional profiles. The space- and time-averaged profiles of Reynolds stress were much like the profiles of turbulence intensity since they showed a maximum value near the top of the canopy, and lower values inside the canopy. The dimensionless averaged profiles for rigid cylinders are shown on Figure 6. The profiles have been made dimensionless by dividing the Reynolds stress $\rangle-\overline{u'v'}\langle$ by the maximum Reynolds stress for such profile $\rangle-\overline{u'v'}\langle_{max}$. Typically, the maximum Reynolds stresses occurred at the top of the cylinder canopy. The y-axis was made dimensionless by dividing the distance to the bed, *y*, by the average cylinder height, *h*, in the region within the canopy, and dividing the quantity (*y-h*) by (*H-h*) above the canopy. These profiles

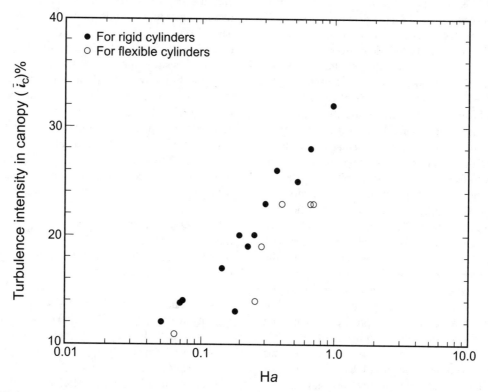

Figure 5. Correlation between the turbulence intensity in the canopy and the dimensionless parameter Ha.

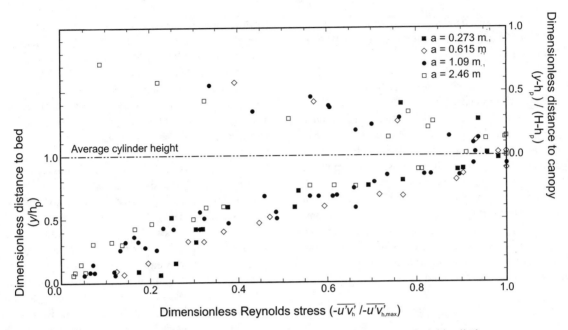

Figure 6. Dimensionless Reynolds stress profile profiles for flow through rigid cylinders.

collapsed for equal plant densities, although considerable scatter existed. It is evident though, that attempts with lower cylinder densities produced profiles with higher dimensionless Reynolds stresses. This was especially evident closer to the bed. This trend occurs because most of the resistive force is supplied by the vegetation and not by the bottom friction in a densely vegetated channel. Therefore, the turbulent stress near the bed is lower in this type of channel. As the vegetation density decreases, the measured Reynolds stress near the bed approaches the theoretical value for open-channels without vegetation.

The profiles of dimensionless Reynolds stress for flow through flexible cylinders shown in Figure 7 has the same general trend as those through rigid cylinders in that the Reynolds stress was damped inside of the cylinder canopy, but the degree to which the Reynolds stress was suppressed was quite different. Again, the cylinder density appeared to be a major factor in the shape and magnitude of the profiles. When the cylinder density was low, and the Reynolds number and deflection angles were high, the drag imposed by the cylinders was reduced and the profiles within the canopy moved towards the theoretical profile for open-channel flow without vegetation.

The profiles for experiments 14 and 16 shown in Figure 7(b) are good examples of this phenomenon. Unlike all of the other measured profiles, these two profiles appear to be concave up. In addition, the velocity profiles of these two experiments were the least affected by the vegetation, thus possibly explaining the different shapes of these two Reynolds stress profiles. The limited experimental data and its relatively high scatter, along with the complexity of this flow con-

dition made it difficult to determine what variables were relevant in determining the shapes of the Reynolds stress profiles for flexible vegetation. Cylinder density and flexibility clearly played a major role in the Reynolds stress profiles; however, factors such as Reynolds number and Froude number may have also played an important part. Further experimentation is required to determine this conclusively.

Reynolds stress measurements performed by *Tsujimoto et al.* [1991] in and above simulated vegetation agreed with those of the present study in that the stress within the canopy was significantly suppressed. The Reynolds stress profile reached a maximum at the top of the canopy and was not noticeably affected above the top of the cylinders in both studies. The findings of *Seginer et al.* [1976] were also in agreement. Each of these studies indicated that the Reynolds stress curve was concave down within the canopy.

2.2.4 Drag Coefficient Profiles. For each experiment, a third-order polynomial was fit through the spatially and temporally averaged Reynolds stress points below the top of the cylinders. Then, the derivative of this polynomial was computed and its value at each of the measured depths was calculated. The value of the derivative, along with the mean velocity averaged in the horizontal plane was used in equation (7) to compute the local drag coefficient at each depth. The result was a horizontally averaged vertical profile of the drag coefficient for each experiment.

These computed drag coefficient profiles were not constant within the canopy, as many researchers have assumed, but

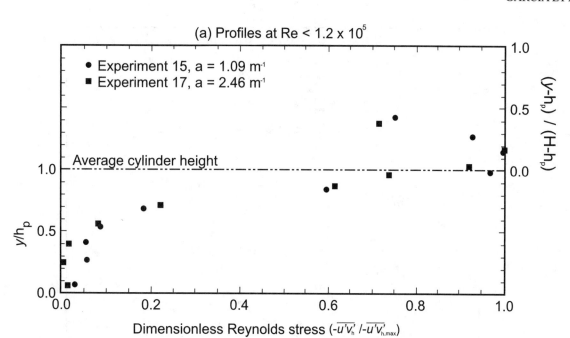

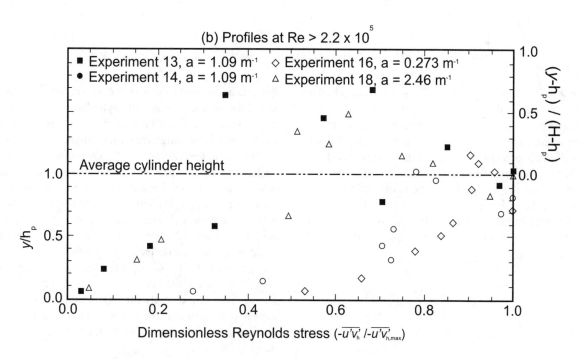

Figure 7. Dimensionless Reynolds stress profile profiles for flow through flexible cylinders.

instead typically reached a maximum within the canopy and diminished towards a minimum at the top of canopy. At the top of the canopy, there was a discontinuity in the value of the drag coefficient. This discontinuity occurred because of the discontinuity in the profiles of the Reynolds stresses. Above the

canopy, the values of C_D computed from equation (7) were nearly equal to zero, because the gradients of the measured Reynolds stress profiles above the canopy were approximately linear and nearly equal to the theoretical gradient of the total shear stress per unit mass (gSo).

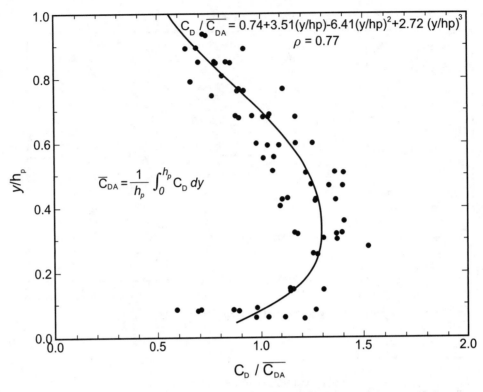

Figure 8. Vertical variation of the drag coefficients inside the canopy for rigid cylinders.

For flow through rigid cylinders, C_D generally reached a maximum around the dimensionless height of 0.38, but this value ranged from 0.25 to 0.50. The maximum value of the drag coefficient reached in each of the profiles ranged from 1.32 to 1.86. The mean of these maximum values was 1.55 ± 0.18. For a given profile, the maximum value was up to 50% greater than the mean. All of the values of C_D computed for flow through rigid cylinders are shown in dimensionless form in Figure 8. A third degree polynomial has been fit through the points with a correlation coefficient of 0.77 and is presented below:

$$\frac{C_D}{\frac{1}{h_p}\int_0^{h_p} C_D dy} = 0.74 + 3.51\left(\frac{y}{h_p}\right) - 6.41\left(\frac{y}{h_p}\right)^2$$

$$+ 2.72\left(\frac{y}{h_p}\right)^3 \qquad (9)$$

The profiles of C_D for flexible vegetation revealed two general profile shapes. Two measured profiles are shown in Figure 9 and exhibit the shapes that were commonly found. The profile of Experiment 13, Figure 9(a), reached a maximum within the canopy, much like those for rigid cylinders, although

the maximum was higher in the profile than that typically found for rigid cylinders. The cylinders in the experimental runs resulting in this characteristic drag coefficient profile swayed, but never deflected by more than 45 degrees

The second general shape of the C_D profiles is illustrated by Experiment 16 in Figure 9(b). In this case, C_D had a maximum near the bottom and decreased as the distance from the bed increased. The profiles with this shape resulted from experiments where the cylinder deflected by at least 50 degrees. With the flexible vegetation, the value of h was the one corresponding with the deflected plant height.

Seginer et al. [1976] reported that the drag coefficient slightly increased with height in the canopy. Their results were from a wind tunnel study through simulated vegetation and were highly scattered. They did not find that drag coefficient profiles reached a maximum within the canopy as the present investigation did.

3. BULK DRAG COEFFICIENTS

Typically, the resistive forces induced by vegetation on the flow are quantified by a single bulk value of the drag coefficient, which is constant anywhere within the plant canopy and represents both a horizontally and vertically averaged value. For submerged vegetation, as modeled in the investi-

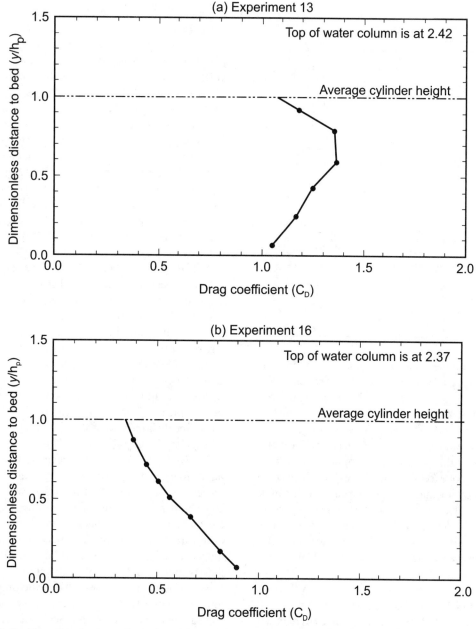

Figure 9. Profiles of the horizontally averaged drag coefficient for two experimentally runs through flexible cylinders.

gation described in this paper, the most appropriate definition for the bulk drag coefficient is:

$$\overline{C_D} = \frac{\int_0^{hp} C_D \big\rangle \overline{u}\big\langle^2 dy}{\int_0^{hp} \big\rangle \overline{u}\big\langle^2 dy} \qquad (10)$$

The bulk drag coefficient, $\overline{C_D}$, defined by equation (10) above was computed for each horizontally averaged profile.

The computed value of $\overline{C_D}$ for each experiment is reported in Tables 2 and 3.

The computed values of $\overline{C_D}$ for flexible cylinders indicate that when the cylinders became highly deflected, $\overline{C_D}$ significantly decreased. Attempts to determine the effects of various channel and flow parameters on the value of the bulk drag coefficient revealed highly scattered results from which it was difficult to make any definite inferences. However, for flexible cylinders, it was apparent that the Froude number

Table 2. Results for rigid cylinders.

Experiment	$\overline{C_D}$	β	Manning's n $(m^{1/6})$
1	1.01	1.10	0.034
2	0.95	1.07	0.041
3	0.86	1.09	0.048
4	1.29	1.08	0.038
5	1.18	1.12	0.045
6	1.46	1.05	0.025
7	1.39	1.05	0.027
8	0.94	1.15	0.042
9	1.13	1.15	0.056
10	1.19	1.14	0.052
11	1.06	1.06	0.031
12	1.14	1.12	0.036

Table 3. Results for flexible cylinders.

Experiment	$\overline{C_D}$	β	Manning's n $(m^{1/6})$
13	1.13	1.13	0.039
14	0.33	1.09	0.034
15	1.45	1.13	0.045
16	0.55	1.02	0.020
17	1.19	1.27	0.061
18	0.59	1.16	0.046

had some influence on the value of the bulk drag coefficient; $\overline{C_D}$ significantly decreased with increasing Froude number. For rigid cylinders, no parameter was found to have significant effect on the value of $\overline{C_D}$.

The mean value of $\overline{C_D}$ for flow through rigid cylinders was found to be 1.13 ± 0.18. All of the values of $\overline{C_D}$ were within 30% of the computed mean. This amount of scatter is not excessive however, especially in comparison to other experimental attempts at measuring the bulk drag coefficient. *Klaassen and Van Der Zwaard's* [1974] measurements of the bulk drag coefficient varied by as much as 100% and *Reid and Whitaker's* [1976] estimates varied by 20% from the mean even though they only computed three values of the bulk drag coefficient.

Li and Shen [1973] computationally estimated values of $\overline{C_D}$ between 1.1 and 1.2 for emergent rigid cylinders at a cylinder Reynolds number of 9 x 10³. These values were based on a local drag coefficient of 1.2 for an idealized two-dimensional flow. Because the flow investigated in the present study was for slightly lower cylinder Reynolds numbers of between 1 x 10³ and 5 x 10³, the local drag coefficient read from the standard cylinder drag curve will be slightly smaller, thus slightly reducing the expected value of $\overline{C_D}$. However, the values of $\overline{C_D}$ reported in this paper are in good agreement with those of *Li and Shen* [1973]. In fact, the mean of all of the values for rigid cylinders was 1.13, the same value reported by these researchers.

Klaassen [1974] criticized *Li and Shen's* values of $\overline{C_D}$ as being too low. He reported values of $\overline{C_D}$ between 0.8 and 3.0 for a range of Re_D values between 1 x 10³ and 9 x 10³.

The mean of his $\overline{C_D}$ measurements was well above 1.2, at approximately 1.5. Although the range of $\overline{C_D}$ measured by *Klaassen* was quite wide, the values seem to be in fair agreement with those found in this investigation. The findings of this work indicate that *Klaassen's* values were often too high. *Klaassen* concluded that his higher $\overline{C_D}$ values might have resulted from higher turbulence intensities inside the canopy, although no turbulence measurements were taken to support this conclusion. The findings of the present study and the study by *Seginer et al.* [1976] contradict *Klaassen's* conclusion. Both the present investigation and *Seginer et al.'s* found that increasing the turbulence intensity within the canopy resulted in decreasing bulk drag coefficients.

Li and Shen's and *Klaassen's* results did agree that the value of $\overline{C_D}$ increased as the obstruction density increased. However, *Seginer et al.* reported the opposite trend: as density increased, the bulk drag coefficient decreased. The results of the present study are unclear, but seem to indicate that the bulk drag coefficient decreased as the cylinder density increased.

Reid and Whitaker [1976] found values of $\overline{C_D}$ between 1.40 and 2.05 with a mean of 1.77 from experimental data. From best fits of their model to various laboratory and field data, *Burke and Stolzenbach* [1983] determined a $\overline{C_D}$ value of 2.5. Their model did have the capability to consider varying vertical profiles of the drag coefficient; however, there were no measurements of drag coefficient profiles for them to utilize at the time their model was developed. The values of $\overline{C_D}$ used in these studies by *Reid and Whitaker* and by *Burke and Stolzenbach* [1983] were significantly higher than any of the values found in the present investigation.

Kouwen and Unny [1973] described two separate hydraulic flow regimes: an erect or waving regime, and a prone regime. They found that the roughness imposed by these two regimes varied from one another. The results of this experimental study seemed to support the findings of *Kouwen and Unny*. This is indicated by many of the similarities found in the present study between the profiles and drag coefficients of rigid and slightly swaying flexible cylinders when contrasted to the results for highly deformed flexible cylinders. For instance the Reynolds stress and drag coefficient profiles for the rigid and slightly deformed cylinders were similar, as shown in Figures 6 and 7, and 8 and 9. Whereas, the Reynolds stress profiles for the highly deformed cylinders were significantly altered. A more telling example however, is in the values of the bulk drag coefficient $\overline{C_D}$, which was considerably smaller when the deflection angle was high. It is difficult to determine exactly when the swaying flow regime becomes prone, but it is clear from the findings of this report and those of *Kouwen and Unny* that these two separate hydraulic regimes do exist and affect the flow in different ways.

4. OTHER RESULTS

The momentum, or Boussinesq coefficient, β, was also computed for each experiment and is defined below as:

$$\beta = \int_0^1 \left(\frac{\rangle\overline{u}\langle}{U} \right)^2 d\left(\frac{y}{H} \right) \qquad (11)$$

This integration was only an estimate however, because the velocity profiles were not complete due to the fact that much of the water column was lost to the submergence depth of the ADV. The top and bottom points of each velocity profile were estimated by the simplest assumption possible: a linear extension of the measured profile. Values of β are reported in Tables 2 and 3.

Values of Manning's n were also computed and reported in Tables 2 and 3. Preliminary measurements in the channel when no cylinders were present indicated a Manning's n value of 0.011 for the smooth flume bed.

The results listed in Tables 2 and 3 indicate that β is slightly greater than 1.0 and is dependent on the density of the cylinders in the channel. When the cylinder density was less sparse, the Boussinesq coefficient approached a value of 1.02. For rigid cylinders at a given density, the Boussinesq coefficient was essentially constant and as the cylinder density increased, so did the values of the Boussinesq coefficient. The lowest density resulted in a Boussinesq coefficient of 1.05, whereas the greatest density yielded a value of 1.15.

For the flexible cylinders, the same trends were evident; however, Boussinesq coefficients varied by greater amounts. At the lowest density, β was 1.02, while at the maximum density β increased by 1.27. Unlike the Boussinesq coefficients for flow through rigid cylinders, those for flow through flexible cylinders were not constant at a given cylinder density. There was an obvious trend for the Boussinesq coefficient to decrease as the magnitude of the flow velocity increased for a given cylinder density. This occurred because the cylinders deflected more under higher velocities and offered less resistance to the flow. The resulting velocity profile was more similar to that of a non-vegetated open-channel flow yielding a Boussinesq coefficient that was closer to 1.02. Although a limited number of experiments through flexible cylinders was performed, the Boussinesq coefficient varied by as much as 9 percent for a given cylinder density.

The results described in this paper were obtained where the relevant dimensionless parameters varied in the following ranges: $0.57 \times 10^5 < Re < 2.58 \times 10^5$; $0.18 < Fr < 0.62$; $0.0036 < So < 0.0161$; $0.0173 < D/H < 0.0387$; $0.073 < Ha < 0.699$; $0.300 < h/H < 0.714$. The values of the dimensionless parameters for each experiment were reported in Dunn et al. [1996].

5. EFFECTIVENESS OF MODELING PROCEDURES AND COMPUTATIONAL METHODS

The results described in this paper allow for some remarks about the general effectiveness of modeling techniques and computational methods utilized in the study.

5.1. Modeling Flexible Vegetation

Using drinking straws to simulate flexible vegetation was an effective method of introducing the flexibility parameter into the system. The results presented above show that the flexibility of the cylinders significantly affected the various profiles and the values of the bulk drag coefficients. This indicates that care should be taken when extending the results of rigid obstruction flow experiments to flow through flexible vegetation.

5.2. Computing Drag Coefficients

The validity of the computational method for computing drag coefficients is dependent on the assumptions required in the derivation of equation (7). The primary assumption in this derivation is that $-\rho\overline{u'v'}$ is the dominant Reynolds stress and that all other Reynolds stresses are not important. This assumption is validated in the work reported in Dunn et al. [1996]. The authors found that transverse gradients of the secondary Reynolds stresses were negligible within the canopy and secondary currents were unimportant. This might not be the case if the channel is not straight or the vegetation is not uniform.

An observation of some of the measured profiles allowed for an easy check of the computational method. For flow through rigid cylinders, some profiles showed almost uniform flow conditions inside the canopy with negligible values of Reynolds stress, thus approaching ideal uniform flow conditions. Some level of turbulence intensity was present ($\sqrt{\overline{u'^2}}/\rangle\overline{u}\langle \cong 3\%$, $\sqrt{\overline{v'^2}}/\rangle\overline{v}\langle \cong 1\%$ to 3%, and $\sqrt{\overline{w'^2}}/\rangle\overline{w}\langle \cong 2.5\%$ to 3.5%). Under the given conditions, the estimate of the drag coefficient from equation (7) should be very near to the value read from the standard cylinder drag curve for a single cylinder in a uniform flow. Table 4 reports the estimates of $\overline{C_D}$ from equation (7) and those from the standard drag coefficient curve. These values agree within 10%. This excellent agreement supports the validity of our method for computing drag coefficients.

Table 4. Drag coefficients for approximately ideal flow conditions.

Exp. #	$\overline{u_c}$ (m/s)	CR	$C_D(7)$	C_D
15	0.248	1,868	1.05	0.97
17	0.155	1,143	1.20	1.0

Exp. # – Experiment number; $\overline{u_c}$ – velocity; CR – Cylinder Reynolds number; $C_D(7)$ – C_D from equation (7); C_D – from standard drag curve

6. CONCLUSIONS

The following conclusions are drawn from the work described in this paper concerning the velocity, Reynolds stress, turbulence intensity, and drag coefficient profiles and bulk drag coefficients for flow through simulated vegetation.

The horizontally averaged velocity profiles through the rigid cylinders had a characteristic shape that was dependent on the cylinder density. For each cylinder density, a constant Boussinesq coefficient was observed. This characteristic shape was absent for flow through flexible cylinders and consequently, the computed values of the Boussinesq coefficient varied for a given cylinder density. These results suggest that the plant density is of primary importance to the shape of the velocity profile in a vegetated channel. In a channel with a flexible lining, other parameters play a crucial role, particularly the flexibility of the plants that make up the lining and the cylinder Reynolds number.

The measured profiles of Reynolds stress and turbulence intensity show that these values reach a maximum near the top of the canopy and are significantly suppressed inside the canopy. A significant tendency was observed for the average turbulence intensity inside the canopy to increase as the dimensionless parameter Ha increased.

In flow through rigid cylinders the drag coefficient was not constant throughout the canopy as many researchers have assumed, but instead reached a maximum at about one-third of the canopy height and diminished towards a minimum at the top of the canopy. For flow through flexible cylinders, two separate shapes of the drag coefficient profiles were observed, possibly indicating the existence of two separate roughness conditions. The measured drag coefficient profiles are directly applicable to the k-ε model for obstructed low-Reynolds number flows introduced by *Burke and Stolzenbach* [1983] and the *k-ε* model for high-Reynolds number flows proposed by *López and García* [2001].

The computed values of the bulk drag coefficient for flexible cylinders indicated that when the cylinders became highly deflected, $\overline{C_D}$ significantly decreased. The mean value of $\overline{C_D}$ for flow through rigid cylinders was 1.13 ± 0.18. The Froude number had some influence on the value of $\overline{C_D}$ for flow through flexible simulated vegetation: $\overline{C_D}$ decreased with increasing Froude number. For flow through rigid cylinder, no parameter was found to have a significant effect on the value of $\overline{C_D}$.

Acknowledgments. This work was partially supported by the Wetlands Research Program of the US Army Corps of Engineers Waterways Experiment Station through Research Grant DACW39-94-K-0010 with Brad Hall as Project Monitor. The work was completed with support from the US Department of Agriculture, National Sedimentation Laboratory, Agricultural Research Service, through Research Cooperation Agreement AG58-6408-1-134, with Sean Bennett as Project Monitor. All the financial and intellectual support is gratefully acknowledged. The views and findings presented here are only those of the authors and should not be interpreted as being the opinion of any of the Federal Agencies that have financially supported the work.

REFERENCES

Burke, R. W., and K. D. Stolzenbach, Free surface flow through salt marsh grass, *MIT-Sea Grant Report MITSG 83-16*, 252 pp., Cambridge, Mass., 1983.

Christensen, B. A., Open channel and sheet flow over flexible roughness, *International Association for Hydraulic Research, 21st Congress*, 463–467, 1985.

Cui, J., and V. S. Neary, Large eddy simulation (LES) of fully developed flow through vegetation, *IAHR's 5th International Conference on Hydroinformatics*, Cardiff, UK, July 1-5, 2002.

Den Hartog, G., and R. H. Shaw, A field study of atmospheric exchange processes within a vegetative canopy, in *Heat and Mass Transfer in the Biosphere*, edited by D. A. deVries and N. H. Afgan, John Wiley and Sons, 1975.

Dunn, C. J., F. Lopez, and M. H. Garcia, Mean flow and turbulence in a laboratory channel with simulated vegetation, *Hydraulic Engineering Series Report No. 51, UILU-ENG 96-2009*, 148 pp., University of Illinois, Urbana, 1996.

Fairbanks, J. D. and P. Diplas, Turbulence characteristics of flows through partially and fully submerged vegetation, Engineering approaches to ecosystem restoration, *Proc. Wetlands Engrg. and River Restoration Conf,* Denver, CO, 1998.

Freeman, G. E., R. E. Copeland, W. Rahmeyer, and D. L. Derrick, Field determination of Manning's n value for shrubs and woody vegetation, Engineering approaches to ecosystem restoration, *Proc, Wetlands Engrg. and River Restoration Conf,* Denver, CO, 1998.

Klaassen, G. J., Discussion of *Effect of Tall Vegetations on Flow and Sediment*, by R. M. Li and H. W. Shen, 1973. *Journal of the Hydraulics Division, ASCE*, 100, 495–497, 1974.

Klaassen, G. J., and J. J. Van Der Zwaard, Roughness coefficients of vegetated flood plains, *Journal of Hydraulic Research*, 12, 43–63, 1974.

Klopstra, D., H. J. Barneveld, J. M. Van Noortwijk, and E. H. Van Velzen, Analytical model for hydraulic roughness of submerged vegetation, in *Proceedings, Managing Water: Coping with Scarcity and Abundance, Proc. 27th Congress of the Intl. Assoc. of Hydraulic Research*, pp. 775-780, 1998.

Kouwen, N., and T. E. Unny, Flexible roughness in open channels, *Journal of the Hydraulics Division, ASCE*, 99, 723–728, 1973.

Kraus, N. C., Lohrmann, A., and Cabrera, R., New acoustic meter for measuring 3D laboratory flows, *Journal of Hydraulic Engineering*, 120, 406–412, 1992.

Li, R. M., and H. W. Shen, Effect of tall vegetations on flow and sediment, *Journal of the Hydraulics Division, ACSE,* 99, 793-813, 1973.

Lohrmann, A., R. Cabrera, G. Gelfenbaum, and J. Haines, Direct measurements of reynolds stress with an acoustic doppler velocimeter, *IEEE,* 205–210, 1995.

López, F., and M. García, Open channel flow through simulated vegetation: Turbulence modeling and sediment transport, *Wetlands Research Technical Report WRP-CP-10,* 106 pp., Waterways Experiment Station, Vicksburg, 1997.

López, F., and M. García, Open channel flow through simulated vegetation: Suspended sediment transport, *Water Resources Research,* 34, 2341–2352, 1998.

López, F., and M. García, Mean flow and turbulence structure of open channel flow through non-emergent vegetation, *J. Hydraulic Engrg.,* 127, 392–402, 2001.

Raupach, M. R., and R. H. Shaw, Averaging procedures for flow within vegetation canopies, *Boundary-Layer Meteorology,* 22, 79–90, 1982.

Ree, W. O., and V. J. Palmer, Flow of water in channels protected by vegetative linings, *U.S. Soil Conservation Bulletin No. 967,* 115, 1949.

Reid, R. O., and R. E. Whitaker, Wind-driven flow of water influenced by a canopy, *Journal of the Waterways, Harbors, and Coastal Engineering Division, ACSE,* 102, 61–77, 1976.

Saowapon, C., and N. Kouwen, A physically based model for determining flow resistance and velocity profiles in vegetated channels, *Symposium on Manning's Equation,* edited by B. C. Yen, pp. 559-568, Virginia, 1989.

Seginer, I., P. J. Mulhearn, E. F. Bradley, and J. J. Finnigan, Turbulent flow in a model plant canopy, *Boundary-Layer Meteorology,* 10, 423–453, 1976.

Shimizu, Y., and T. Tsujimoto, Numerical analysis of turbulent open channel flow over vegetation layer using a *k-ε* turbulence model, *J. of Hydroscience and Hydraulic Engrg.,* 11, 57–67, 1994.

Stone, B. M., and H. T. Shen, Hydraulic resistance of flow in channels with cylindrical roughness, *J. Hydraulic Engrg.,* 128, 500–506, 2002.

Tsujimoto, T., T. Okada, and T. Kitamura, Turbulent flow over flexible vegetation-covered bed in open channels, *KHL-Progressive Report,* Kanazawa University, 2, 31–40, 1991.

Wilson N. R., and R. H. Shaw, A higher order closure model for canopy flow, *Applied Meteorology,* 16, 1198, 1977.

Wu, F. C., H. W. Shen, and Y. J. Chou, Variation of roughness coefficients for unsubmerged and submerged vegetation, *J. Hydraulic Engrg.,* 125, 934–942, 2000.

Yen, B. C., Open channel flow resistance, *J. Hydraulic Engrg.,* 128, 20–39, 2002.

C.V. Alonso, USDA-ARS National Sedimentation Laboratory, P.O. Box 1157, Oxford, MS 38655.

M. García, F. López, and C. Dunn, Ven Te Chow Hydrosystems Laboratory, Department of Civil and Environmental Engineering, University of Illinois at Urbana-Champaign, Urbana, IL 61801.

Effects of Emergent Riparian Vegetation on Spatially Averaged and Turbulent Flow Within an Experimental Channel

Sean J. Bennett

U.S. Department of Agriculture-Agricultural Research Service, National Sedimentation Laboratory, Oxford, Mississippi[1]

Interest has been renewed in understanding the interactions between flow, vegetation, and sediment transport because of the growing popularity of river restoration and streambank stability programs utilizing vegetation. To further this research, an experimental channel was systematically vegetated with emergent, wooden dowels of varying density to document how the vegetation alters both the spatially averaged and turbulent flow. Results show that (1) surface waves, vortical structures, flow separation, and dead zones were associated with the vegetation zones, and these turbulent flow structures greatly enhanced fluid mixing processes, and (2) as vegetation density increased, flow resistance, bed shear stress, flow depth, and thalweg sinuosity increased, while flow velocity decreased. Design considerations for stream restoration programs using managed vegetation plantings to trigger desired morphologic and ecologic responses are presented and discussed.

1. INTRODUCTION

By the turn of the 20th century, human disturbance to the land due to urban and agricultural development resulted in upland erosion and excessive sedimentation within river channels of the highly erodible loess region of the south-central U.S. [see reviews in *Schumm et al.*, 1984; *Simon and Rinaldi*, 2000]. Subsequent action to correct the problem, including channelization and dredging, further exacerbated erosion processes by creating stream corridors in disequilibrium with prevailing hydrologic conditions. At present, many streams in northern Mississippi are characterized by flashy hydrographs, vertical banks prone to failure, excessive sediment loads, channels that are straight and wide, and denuded riparian zones [*Shields et al.*, 1995b, c]. These streams would be considered impaired (degraded) using the water quality standards of the *U.S. EPA* [1998, 2000].

River restoration programs were designed to address these issues of stream degradation (impairment) and return biological functionality primarily through enhancement of habitat, habitat resources, and stream channel stability while improving recreational opportunities and aesthetic beauty [*Brookes and Shields*, 1996; *FISRWG*, 1998]. Restoration projects along with streambank stabilization and protection programs have used vegetation extensively to achieve these goals [e.g., *USDA-NRCS*, 1996; *FISRWG*, 1998].

Because bioengineering and stream restoration techniques have such practical and environmental appeal, there is renewed interest in understanding the interactions between flow, sediment transport, and riparian vegetation in river channels. These studies have shown that (1) flow velocities and near-bed shear stresses are reduced within vegetation zones, (2) vegetation can increase the local and boundary flow resistance, (3) turbulence intensities are increased near the canopy tops for submerged vegetation and along the interface between emergent riparian vegetation and the main channel, (4) vegetation can reduce the transport capacity of the flow and cause sorting and deposition of sediment, and (5) vegetation can create secondary circulation patterns [*Kouwen and Unny*, 1973; *Li and Shen*, 1973; *Thompson and Roberson*, 1976; *Tollner et al.*, 1976; *López and García*, 1996, 1997, 1998,

[1]Currently at: Department of Geography/University of Buffalo, Buffalo, New York

Riparian Vegetation and Fluvial Geomorphology
Water Science and Application 8
This paper is not subject to U.S. copyright. Published in 2004 by the American Geophysical Union
10.1029/008WSA03

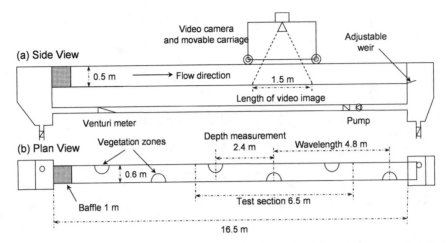

Figure 1. Schematic diagram of flume facility showing (a) side and (b) plan views [*Bennett et al., 2002*].

2001; *Okabe et al.*, 1997; *Nepf*, 1999; *Tsujimoto*, 1999; *Wu et al.*, 1999; *Järvelä*, 2002; *Righetti and Armanini*, 2002; *Stephan and Gutknecht*, 2002; *Stone and Shen*, 2002]. The magnitude of these effects depends on the characteristics of the flow and vegetation including size, shape, flexibility, orientation, concentration (density), and degree of submergence.

Coherent turbulent flow structures also appear to be associated with vegetated stream channels. *Ikeda and Kanazawa* [1996] described vortices generated near the top of submerged, flexible vegetation, and concluded that the wavy motion of the vegetation, commonly referred to as monami, was induced by the observed vortices. *Ghisalberti and Nepf* [2002] expanded *Ikeda and Kanazawa's* [1996] work and showed that the monami is caused by Kelvin-Helmholtz instabilities at the top of the vegetative canopy. *Tsujimoto* [1996] observed organized fluctuations of the water surface in an experimental channel covered by emergent vegetation over one-half of its width. These low-frequency water surface fluctuations, caused by the shear instabilities at the vegetation-free stream interface, produced a net flux of bedload into the vegetated zone. *Nezu and Onitsuka* [2001] observed large-scale, vertically-oriented vortical structures in compound channels with vegetated floodplains. The formation and translation of these eddies enhanced fluid mixing processes and caused meandering in the surface velocity vector [*Fukuoka et al.* 1994; *Fukuoka and Watanabe*, 1997].

The interactions between stream flow and riparian vegetation remain poorly understood, especially how rigid, emergent vegetation alters the mean and turbulent flow field. Such information is critical in the effective design of stream restoration programs. To this end, the objectives of the present study were (1) to transform a straight, degraded experimental stream corridor typical of the south-central U.S. into a meandering channel though the use of vegetation and (2) to document the

effects of vegetation density on turbulent flow structure and bulk hydraulic parameters. Hemispherical vegetation zones were placed on alternate sides of the experimental channel at the approximate spacing of meanders that would be in equilibrium in a sand-bedded stream with the imposed flow rate. This study employed particle image velocimetry to quantify the surface flow field of the experimental channel. Design considerations for stream restoration programs also are presented and discussed.

2. EXPERIMENTAL METHODS

2.1. Flume

Experiments were conducted at the USDA-ARS National Sedimentation Laboratory using a tilting recirculating flume 16.5 m long and 0.6 m wide (Figure 1). Flow discharge was measured with an inclined mercury manometer connected to a calibrated, in-line Venturi meter. Hydraulic conditions chosen for these experiments were based on meander wavelength determinations discussed in *Bennett et al.* [2002]. Flow discharge Q was kept constant at 0.0042 m^3 s^{-1}. Flow depth was measured using a point gauge mounted to a movable carriage that rode along the flume rails. Depths were measured where possible (unencumbered by the vegetation) over a 2.4-m longitudinal section (one-half of a meander wavelength, apex to apex; Figure 1) at downstream intervals of 0.24 m and cross-stream intervals of 0.06 m. Depth data were used to calculate a spatially averaged flow depth d and flow velocity u from u = Q/dw where w is flume width (Table 1). Spatially averaged surface flow velocity within each vegetation zone u_v (except for the highest density) was determined using the particle image velocimetry measurements described below. Centerline (cross-over) measurements of the dynamic water surface

Table 1. Summary of experimental parameters (defined in text).

Vegetation Zones			Flow Parameters								
N^a	Dist., mm^b	VD, m^{-1}	Q, $m^3 s^{-1}$	d, m	u, $m s^{-1}$	u_V, $m s^{-1}$	S	τ, Pa	C_D	Sn	Re_C
0	n.a.	0.0	0.0042	0.023	0.307	0.256	0.0016	0.35	0.004	1.00	n.a.
7	144	0.011	0.0042	0.023	0.298	0.234	0.0015	0.35	0.004	1.04	702
30	72	0.047	0.0042	0.025	0.280	0.183	0.0016	0.39	0.005	1.07	549
113	36	0.176	0.0042	0.026	0.267	0.171	0.0015	0.39	0.005	1.10	513
441	18	0.689	0.0042	0.032	0.220	0.033	0.0015	0.47	0.010	1.12	99
1753	9	2.739	0.0042	0.036	0.196	n.a.	0.0020	0.68	0.018	1.20	n.a.

[a]Number of dowels per vegetation zone
[b]Minimum distance between rows and columns

elevation (relative to a horizontal datum between successive vegetation zones) were used to determine an average water surface slope S. Mean bed shear stress τ and mean boundary drag coefficient C_D were determined using $\tau = \rho gdS$ and $C_D = u_*^2/u^2$ respectively, where ρ is the density of water, g is gravitational acceleration, and $u_* = (\tau/\rho)^{0.5}$. Uniform flow conditions for the straight, unvegetated channel were Q = 0.0042 $m^3 s^{-1}$, d = 0.023 m, u = 0.307 $m s^{-1}$, τ = 0.35 Pa, and C_D = 0.004 (Table 1).

2.2. Vegetation Zones

The spacing for alternate vegetation zones, i.e. meander wavelength λ, was determined using

$$\lambda = 49.53 Fr^{0.427} A^{0.5} \qquad (1)$$

$$\lambda = 37.79 Q^{0.476} \qquad (2)$$

empirically derived by *Ackers and Charlton* [1970], where A is cross-sectional flow area, Fr is Froude number (Fr = $u/(gd)^{0.5}$), and equations are in English units. These formulae gave values of 4.82 and 4.64 m, respectively, and λ = 4.8 m was used here. This relation was chosen amongst many because the discharge used herein was within the range of the values employed in the derivation of the equation, it was derived for sand-bedded channels, and it maximized the number of vegetation zones allowable in the flume [see *Bennett et al.*, 2002]. Six vegetation zones were spaced along the entire flume length (Figure 1). As point bars typically occupy a large proportion of the channel width [see, for example, *Richards*, 1982, p. 207] and to ensure significant velocity diversion, the radius for each hemisphere was set equal to 0.3 m or 0.5w. Emergent vegetation was simulated using wooden dowels 3.2 mm in diameter D and approximately 70 mm in height. The plywood floor of the flume was drilled and each dowel was placed individually, perpendicular to the bed, and in a staggered

arrangement to maximize flow resistance [*Li and Shen*, 1973]. The smallest spacing between successive rows and columns was 9 mm or 3D (Table 1). Following *López and García* [1997] and others, the density of vegetation for one meander wavelength VD (m^{-1}), defined as the ratio between the sum of the frontal areas of the vegetation divided by the volume of fluid (depth of water over an area encompassing one meander wavelength), was determined using

$$VD = NDd/dw\lambda \qquad (3)$$

where N is the number of elements.

To vary the density of the vegetation zones, dowels were harvested in successive experiments by removing alternate rows of dowels and alternate dowels within the remaining rows. This systematic harvesting doubled the distance between successive vegetation rows and columns while retaining the staggered arrangement (Table 1). While vegetation on natural point bars can be randomly dispersed, the design used herein conserved the geometric similarity of the vegetation to document precisely the effect of vegetation density on the flow field, maximized flow resistance by using a staggered arrangement of vegetal elements, and maintained the general shape of the hemisphere (hence the shape of the point bar) during successive harvests. Vegetation density and the number of dowels for each configuration are given in Table 1.

2.3. Particle Image Velocimetry

Particle image velocimetry (PIV) characterized the surface flow field for each vegetation density and for all flow structures. An S-VHS video camera was mounted to a movable carriage that rode on the flume rails (Figure 1). In a downward-looking, orthogonal orientation, the camera image covered ~1.5 m of flume length, hence one meander wavelength (4.8 m) was divided into five sections with approximately 10% overlap. Each flow section for each vegetation density was

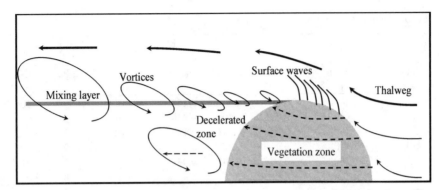

Figure 2. Schematic diagram of vegetation-related flow structures.

recorded to video tape for two to three minutes while buoyant, black plastic seed particles 2 mm in diameter were added in an even distribution to the flow upstream of the region being recorded.

For spatially averaged flow analysis, two successive images (fields), separated in time by 0.0167 s, were grabbed at ~1 s intervals for a total of 60 s for each flow section and for each vegetation density. This resulted in 60 pairs of images, each pair separated by 1 s. For flow structure analysis, up to three groups of 10 successive fields were grabbed over a period of a few seconds. Each video pair was analyzed using a commercially available PIV software package. The software package applies a cross-correlation technique to track flow-related quantities forming recognizable gray-level patterns, such as particles or groups of particles, from one image to another within a user-defined area of interrogation [e.g. *Fujita et al.*, 1998; *Raffel et al.*, 1998]. The resulting flow vector is deduced by determining the location of the maximum correlation coefficient within each area of interrogation.

At the prescribed camera focal length, spatial resolution of each video image defined by the pixel size was 2.5 mm. The size of the interrogation area was set to 32 pixels or 200 mm². The vector grid spacing was set to 16 pixels or 39.4 mm. The vector plot was scaled to actual spatial units using images of a grid placed into the flume at the exact height of the seed particles and at the same focal length as the flow images. Variations in grid line spacing and image distortion due to lens curvature caused an average spatial error of 1.6%.

2.4. Procedure

Uniform flow conditions for the unvegetated channel were determined prior to the experiment. The channel was vegetated to the maximum vegetation density with the wooden dowels using the geometry described above. Without altering the flow rate, bed slope, or downstream weir, the pump was turned on and flow was recirculated through the flume. Seed particles were added to the flow and each of the five flow zones were

video recorded for the prescribed period of time. Measurements of flow depth were also obtained. Once completed, vegetation was systematically harvested to achieve the next vegetation density while flow continued to be recirculated. Video recording, depth measurement, and vegetation harvesting continued until all elements were removed and the initial, unvegetated channel and flow conditions were established. The entire experiment was completed in one day with a team of six individuals.

3. RESULTS

3.1. General Observations

The emergent vegetation markedly affected the pattern of flow within the experimental channel. In general, the highest-density vegetation zones severely reduced flow velocities just upstream and downstream in the near-vegetation regions, they caused a rapid acceleration of flow around each zone, and they diverted the thalweg or the trace of the maximum surface velocity toward the unvegetated bank. As flow moving downstream encountered a vegetation zone on the opposite bank, these flow patterns were destroyed and recreated in the opposite direction, and a meandering thalweg with a wavelength of 4.8 m was produced [for further details, see *Bennett et al.*, 2002]. These effects became less pronounced as vegetation density decreased. At VD ≤ 0.047 m⁻¹, the flow was unencumbered by the vegetation present. Several distinct vegetation-related flow structures were identified, and these phenomena as well as the effects of vegetation density on spatially averaged flow are described below.

3.2. Effect of Vegetation on Turbulent Flow Structure

A number of vegetation-related turbulent flow structures were observed during the experiment. These structures included surface waves, vortices, flow separation, and dead zones, as shown schematically in Figure 2.

3.2.1 Surface waves. Surface waves were observed along-side each vegetation zone for all vegetation densities (Figures 2 and 3). At higher vegetation densities, the surface waves had angles of propagation α of about 50°, they were restricted in space to the upstream half of the vegetation zone, and the spacing between waves was relatively small (ca. millimeters). At lower vegetation densities, the surface waves had angles of propagation of about 60°, they could be found in space all along the vegetation zone, and the spacing between waves was relatively large (ca. centimeters). In all cases, these sur-

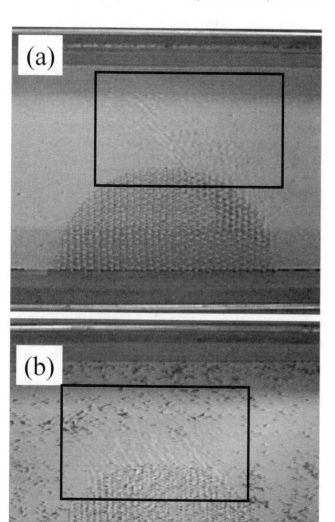

Figure 3. Pictures of the flow field near the vegetation zones showing the surface waves (see insets) for (a) VD = 0.689 m^{-1} and (b) VD = 0.176 m^{-1}. Flow is from right to left and wetted flume width is 0.6 m.

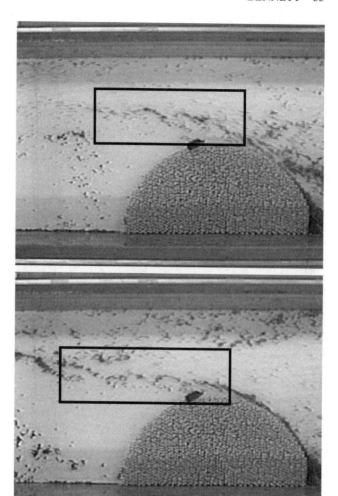

Figure 4. Pictures of the flow field near the vegetation zones show-ing the small-scale vortices (see insets) for VD = 2.739 m^{-1}. Flow is from right to left and wetted flume width is 0.6 m.

face waves extended about 0.3w into the accelerated flow zone opposite the vegetation (Figure 3).

The cause for these waves was the creation of wakes and stacked wakes due to flow impinging the emergent vegetation, where wave spacing was a function of vegetation density. The propagation velocity c of small amplitude, progressive waves over a frictionless bottom is given by

$$c = \sqrt{gd} \qquad (4)$$

[e.g. *van Rijn*, 1994]. For the range of vegetation densities, c is expected to vary from 0.48 at VD = 0.011 m^{-1} to 0.59 for VD = 2.739 m^{-1} (refer to Table 1). Wave propagation angle can be expected to be a function of the wave celerity and local flow velocity, defined as

$$\tan \alpha = c/u_a \qquad (5)$$

where u_a is the maximum thalweg flow velocity near the apex of the vegetation zone, which varies from about 0.3 for VD = 0.011 m⁻¹ to about 0.45 for VD = 2.739 m⁻¹ [*Bennett et al.*, 2002]. Based on (4) and (5), α should range from about 58° for VD = 0.011 m⁻¹ to 53° for VD = 2.739 m⁻¹, which agrees well with the observed values.

3.2.2 Dead zones, flow separation, and vortical structures. For VD = 2.739 m⁻¹, flow through the vegetation was greatly reduced as compared to the mean flow and thalweg velocity. A relatively small dead zone, ca. 0.002 m², was created at the intersection of the wall and the vegetation zone (Figure 2). A relatively large near-dead or decelerated zone, ca. 0.4 m², was created downstream of the vegetation (Figure 2). It was near-dead because this zone had a measurable translation velocity on the order of 10 to 20 mm s⁻¹ at VD = 2.739 m⁻¹. As vegetation density decreased, the velocity of the decelerated zone downstream of the vegetation increased, assumed equal to the spatially averaged velocity within the vegetation zone u_v (Table 1). Accelerated, diverted flow around the vegetation zone separated just upstream of the zone apex, but only for VD = 2.739 m⁻¹. Flow reattached at about 1.97 m, or 0.41λ, downstream.

A mixing layer was created downstream of the vegetation zone and was associated with both small-scale and large-scale, horizontal vortical structures (axis of rotation normal to bed; Figure 2). Small-scale vortices formed near and just downstream of the vegetation apex for VD ≥ 0.176 m⁻¹ (Figures 4 and 5). At VD = 2.739 m⁻¹ where the vortices were most clearly visible, these flow structures were elliptical in shape, were about 0.05 to 0.13 m long and 0.04 m wide, and were shed nearly continuously from the vegetation zone (Figures 4 and 5). Their direction of rotation was toward the vegetation zone from which they were derived. An estimate of the frequency of vortex shedding f_v, which PIV was unable to accomplish due to the small spatial scale of the vortex, can be derived from the vortex length scale V_L and flow velocity using

$$f_v = u/V_L \qquad (6)$$

By employing $V_L \approx 0.5$ to 0.13 m and $u \approx 0.3$ m s⁻¹, $f_v \approx 2$ to 4 Hz for VD = 2.739 m⁻¹. It was observed that as VD decreased from 2.739 to 0.176 m⁻¹, the frequency and size of these small-scale vortices decreased significantly. No small-scale vortices were observed when VD < 0.176 m⁻¹.

At a distance of about 1 m downstream of the vegetation zone for VD = 2.739 m⁻¹, larger-scale vortices were observed (Figure 6). These vortices spanned the entire width of the

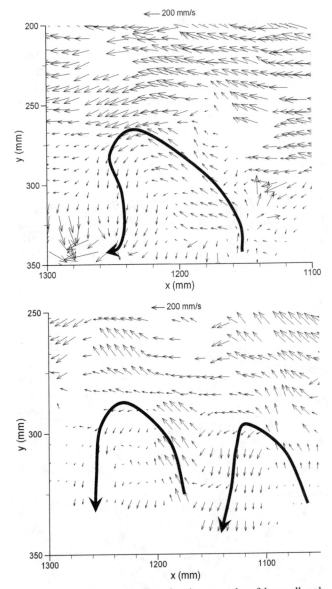

Figure 5. Flow-field vector plots showing examples of the small-scale vortices observed for VD = 2.739 m⁻¹ (positions shown with arrows). Flow is from right to left and reference vector is provided.

flume, they were of similar size in the downstream direction, and their direction of rotation was toward the vegetation zone immediately upstream (Figures 6 and 7). The frequency of these larger-scale vortices based on the video recordings was 0.19±0.18 Hz (n = 34) for VD = 2.739 m⁻¹ and 0.15±0.07 Hz (n = 13) for VD = 0.689 m⁻¹. These larger-scale vortices were similar to those described by *Fukuoka et al.* [1994] and *Nezu and Onitsuka* [2001] in experimental channels with vegetation and *Schmidt et al.* [1993] in an experimental channel with a flow obstruction. Although flow was affected by the presence of the vegetation, evidenced by the wavy streamlines in Fig-

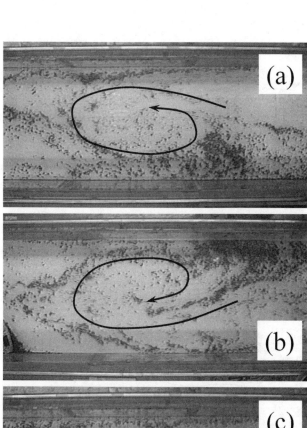

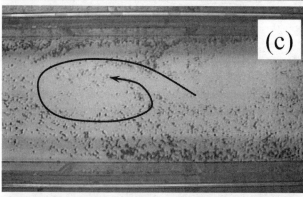

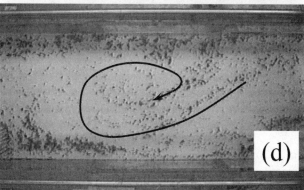

Figure 6. Pictures of the flow field downstream of the vegetation zones showing the large-scale vortices for VD = 0.689 m⁻¹ with (a) counter-clockwise and (b) clockwise rotation, and for VD = 2.739 m⁻¹ with (c) counter-clockwise and (d) clockwise rotation. Flow is from right to left, wetted flume width is 0.6 m, and arrows show general flow pattern.

ures 7c, 7d, and 7e, no large-scale vortices were observed for VD < 0.689 m⁻¹ (Figure 7).

Mixing layers or free-shear layers such as those described here are characterized by turbulent motions caused by Kelvin-Helmholtz wave instabilities, which grow until roller-type vortices are developed [e.g. *Ho and Huerre*, 1984]. These vortical structures dominate the transfer of mass and momentum through the mixing layer, and this is evident in the present experiments. According to *Ho and Huerre* [1984; see also *Rogers and Moser*, 1992; *Ghisalberti and Nepf*, 2002], the frequency of the Kelvin-Helmholtz instability f_{KH} is given by

$$f_{KH} = 0.032 (U_A / \theta) \tag{7}$$

$$\theta = \int_{-\infty}^{\infty} \left[\frac{1}{4} - \left(\frac{u_z - U_A}{U_2 - U_1} \right)^2 \right] dz \tag{8}$$

where $U_A = 1/2(U_1 + U_2)$, U_1 is the mean flow velocity of the slower fluid layer, U_2 is the mean flow velocity of the faster fluid layer, θ is the momentum thickness, u_z is the downstream velocity at the cross-stream position z (note that equations (7) and (8) were derived originally for mixing of two fluids within a vertical plane). These relations were applied to the present dataset from the vegetation apex to the midpoint between two successive vegetation zones. Values of f_{KH} and θ ranged from 0.1 to 0.2 Hz and 0.03 to 0.04 m, respectively, for VD > 0.689 m⁻¹.

Alternatively, *Levi* [1991] proposed a universal Strouhal law for the frequency of vortex shedding f_S by various objects in turbulent water flows, given by

$$0.159 = f_S D_L / u_S \tag{9}$$

where D_L is the flow length scale before separation, taken here as approximately equal to the constricted flow width (also equal to the radius of the vegetation zone), and u_S is the approach flow velocity. Using equation (9), $D_L \approx 0.3$ m, and $u_S \approx 0.3$ m s⁻¹, $f_S \approx 0.16$ Hz. In both examples, the frequency of the Kelvin-Helmholtz instabilities f_{KH} and vortex shedding given by the Strouhal law f_S are in close agreement to the observed frequency of the large-scale vortices. The observed small-scale vortices, however, occur nearly 10 to 20 times more frequently than the large-scale vortices. It is unknown (1) if the small-scale vortices interact, disintegrate, or amalgamate to form the larger-scale vortices, (2) if every 10th or 20th small-scale vortex grows into a large-scale vortex, or (3) if the larger-scale vortices are more akin to low-frequency shear-layer flapping as described *Simpson* [1989] rather than Kelvin-Helmholtz instabilities.

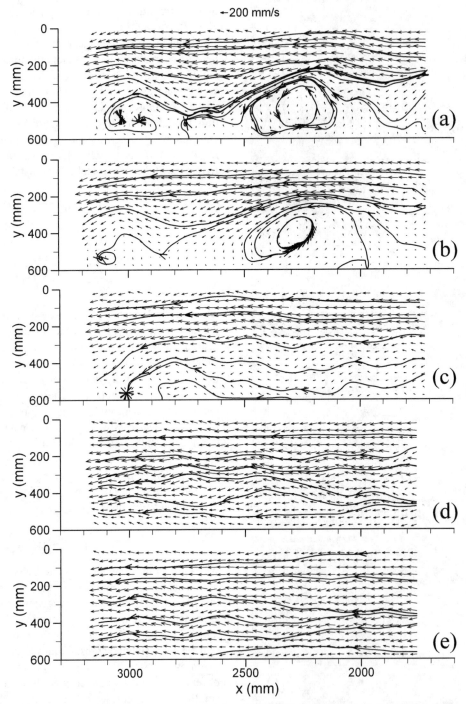

Figure 7. Flow-field vector plots with streamline traces showing examples of the large-scale vortices observed for vegetation densities (a) 2.739 m⁻¹ and (b) 0.689 m⁻¹. Also shown are the flow-field vectors for vegetation densities (c) 0.176 m⁻¹, (d) 0.047 m⁻¹, and (e) 0.011 m⁻¹. Flow is from right to left and reference vector is provided.

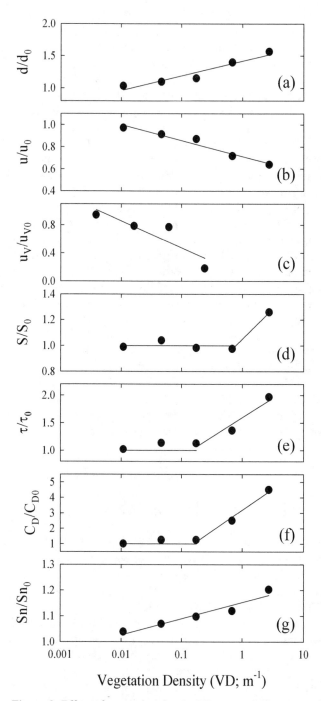

Figure 8. Effect of vegetation density VD on spatially averaged (a) flow depth d, (b) flow velocity u, and (c) surface flow velocity within a vegetation zone u_V, (d) water surface slope S, (e) mean boundary shear stress τ, (f) mean boundary drag coefficient C_D, and (g) thalweg sinuosity Sn [*Bennett et al., 2002*]. Each parameter is normalized by the value without vegetation present (subscript 0), and trend lines are also shown.

3.3. Effect of Vegetation Density on Spatially Averaged Flow

The simulated vegetation also markedly affected spatially averaged flow, and Figure 8 and Table 1 summarize these effects. As vegetation density increased, the bulk flow resistance increased, raising mean flow depth d (Figure 8a) and mean boundary drag coefficient C_D (Figure 8f) and lowering the mean flow velocity u (Figure 8b). The decrease in spatially averaged flow velocity within an individual vegetation zone u_V (Figure 8c) decreased more markedly than the velocity averaged over the entire channel (Figure 8b). The magnitude of thalweg meandering or flow sinuosity Sn increased with vegetation density (Figure 8g). These results are consistent with the previous numerical work of *Li and Shen* [1973].

A threshold vegetation density is apparent from these data, above which flow resistance becomes significantly affected. Both mean boundary shear stress τ (Figure 8e) and mean boundary drag coefficient C_D (Figure 8f) were nearly invariant at VD < 0.176 m^{-1}, as determined in the cross-over section (see Figure 1). Water surface slope S was nearly invariant at VD < 0.689 m^{-1} (Figure 8d). At VD > 0.176 m^{-1}, τ and C_D increased in a linear fashion as the flow become progressively nonuniform and three-dimensional due to thalweg meandering. In an experimental channel, *López and García* [1997, 2001] observed a similar threshold for the effect of vegetation density on flow resistance and flow depth, above which both increased linearly. Their observed threshold occurred at a vegetation density VD of 0.01 to 0.04 m^{-1}, lower than reported here. Compared to the present study, *López and García* [1997, 2001] used greater flow rates (0.05 to 0.18 $m^3\ s^{-1}$) and depths (0.164 to 0.335 m) and larger diameter dowels (5 mm) that were submerged [see also *Dunn et al.*, 1996].

Figure 9 shows the variation in normalized cross-stream flow depth d/d_0 as a function of vegetation density and downstream position x/λ, where vegetation apices occur at $x/\lambda = 0$ and $x/\lambda = 1$. These depth data show that near the upstream vegetation zone, $x/\lambda = 0$ and $x/\lambda = 0.2$, representing the apex and just downstream, flow depth increases monotonically with vegetation density. In the cross-over region where water surface slope was determined, $x/\lambda = 0.4$ and $x/\lambda = 0.6$, and upstream of the next vegetation zone, $x/\lambda = 0.8$ and $x/\lambda = 1.0$, there is a step increase in flow depth as VD increases from 0.176 to 0.689 m^{-1}, corresponding to the threshold described above (Figure 9). It is unclear why d/d_0 at $x/\lambda = 0$ does not mirror d/d_0 at $x/\lambda = 1.0$, although this may be attributable to experimental error.

There are two potential causes for the apparent threshold shown in Figures 8 and 9. First, it may be related to the location of the water surface slope measurements relative to the size of flow area impacted by the vegetation. At low vegetation densities, the size of the flow area impacted by the vegeta-

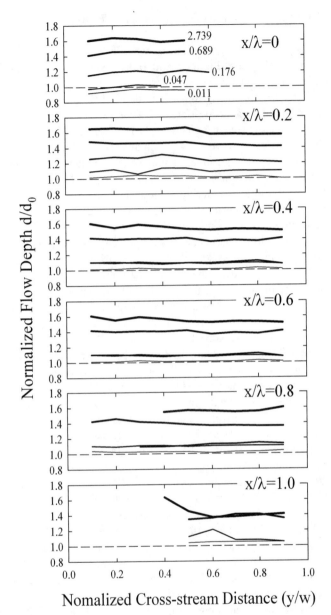

Figure 9. Normalized flow depth (d/d_0) across the channel (y/w) as a function of vegetation zone position (x/λ) and vegetation density (curves). The apices of the vegetation zones occur at y/w = 0.5 at x/λ = 0 and x/λ = 1.

vegetative zone in ways that are still unclear. It has been shown that in turbulent shear flows, vortex shedding from cylinders depends on the cylinder Reynolds number $Re_C = u_v D/v$ and the shear parameter $K = u_* D/u_v$, where v is the fluid kinematic viscosity [*Kiya et al.*, 1980]. The experimental data of *Kiya et al.* [1980] show that vortex shedding occurs at $Re_C > 50$ at $K = 0$, and this threshold increases linearly from $Re_C > 50$ at $K = 0.05$ to $Re_C > 300$ at $K = 0.25$. *Nepf et al.* [1997] observed vortex shedding from a circular array of submerged vegetation stems at $Re_C \approx 150$ to 200, and attributed this delay of shedding to shear associated with upstream wakes. In the present experiment, $Re_C > 150$ when VD < 0.689, exactly at the threshold observed (Table 1). At VD < 0.689, one would expect vortex shedding from any cylinder placed in the flow. *Nepf et al.* [1997] and *Nepf* [1999] showed that such vortex shedding can (1) increase horizontal diffusivity by an order of magnitude, (2) enhance smaller-scale turbulence intensities, and (3) break apart larger-scale turbulent motions or eddies. Moreover, *Shen* [1973] described how wakes generated by flow separation around cylinders can interact, couple, and become superposed. While it is unclear how turbulent wakes impact spatially averaged flow, the linear variations of mean flow depth and velocity with vegetation density shown in Figure 8, which were spatially averaged, and the changes in flow depth shown in Figure 9 appear to support the zone of influence interpretation for this apparent threshold rather than wake-induced flow modulation.

4. DISCUSSION

4.1. Fluid Mixing Processes in Rivers with Vegetation

As vegetation density increased in the experimental channel, the frequency and magnitude of vortical structures and thalweg sinuosity increased. For flow through rigid, emergent vegetation, *Nepf et al.* [1997] showed that when $Re_C \approx$ 150 to 200 (1) the transverse fluid diffusivity was an order of magnitude larger than that observed without vegetation for comparatively similar flows, (2) the transverse fluid diffusivity scaled with vegetation diameter, and (3) the vertical fluid diffusivity was four times less than the transverse diffusivity. At channel width and reach scales, it was shown here that both the transverse and longitudinal mixing processes, which are intimately related, were greatly enhanced by the presence of vegetation and increased with vegetation density.

For straight, rectangular channels, transverse mixing processes can be defined using the dimensionless transverse mixing coefficient k_z [*Rutherford*, 1994] and the dimensional transverse mixing coefficient L_C (the distance required for a

tion was relatively small. As vegetation density increased, the zone of influence also increased. At VD > 0.176 m^{-1} (Figure 8), this zone of influence was large enough to affect the water surface slope measurements, and hence τ and C_D, which were restricted to the flume centerline within x/λ = 0.4 to 0.6 and equally spaced between successive vegetation zones. Second, these observations may be further complicated by vortex shedding from individual elements within the bank-side source to first reach the opposite bank; *Holley et al.*, 1972), defined as

$$k_z/u_*d = 0.1 \text{ to } 0.3 \qquad (10)$$

$$L_C = 0.54\left(uw^2/k_z\right) \qquad (11)$$

For the flow without vegetation, values of k_z and L_C range from about 0.00004 to 0.0001 m^2 s^{-1} and about 60 to 150 m, respectively. However, for the same flow rate with VD = 2.739 m^{-1}, $L_C \approx 0.5$w or 2.4 m, which is a decrease of more than an order of magnitude, $k_z \approx 0.002$ m^2 s^{-1}, and $k_z/u_*d \approx 2$, which is an increase of an order of magnitude. Adding vegetation to rivers will greatly affect the dispersion of fluid, sediment, nutrients and pollutants important to discharge conveyance, river channel change, and aquatic ecology.

4.2. Stream Restoration and Channel Design

The rigid, emergent vegetation elements used herein were chosen to simulate dormant willow posts commonly employed in stream restoration and bank stabilization programs [*Shields et al.*, 1995a; *Watson et al.*, 1997]. These posts are typically 0.1 to 0.3 m in diameter, 2 to 4 m long, and are planted along the face and toe of banks and on sandbars. While reported success rates for dormant willow post plantings on sandbars in northern Mississippi is about 50% [*Shields et al.*, 1995a], the surviving willows increased the percentage of vegetative cover on the sandbars and significantly increased sandbar stability. It appears that managed plantings of willow posts on alternate sides of stream channel, as suggested herein, could attain the desired point bar sedimentation and geomorphic adjustment.

Hasfurther [1985], *Brookes and Sear* [1996], and *Shields* [1996] describe four methods used to design stable meandering channels for the purpose of stream restoration. These include: (i) the carbon copy technique that replaces meanders exactly as they were found before channel disturbance, (ii) empirical relationships that predict meander characteristics based on the characteristics of the designed channel, (iii) the natural approach that allows prevailing processes to alter planform geometry, and (iv) a systems approach that involves a complete analysis of the stream catchment as a whole (geology, soils, vegetation, hydrology, etc.) to determine meander characteristics.

Based on the experimental results described here, a hybrid method is suggested for the design of a meandering stream corridor using vegetation. For a design flow such as bankfull, an equilibrium meander wavelength is derived using empirical methods (method (ii) above). At the intended point bar loca-

tions, dormant willow posts or like vegetation are planted in a hemispherical shape with a radius up to one-half the channel width. The vegetation should be planted in staggered arrangement at a vegetation density greater than 0.7 m^{-1}. This hybrid technique should (1) cost less than conventional construction designs since vegetation plantings are less expensive than channelization, and (2) work with and not against natural flow processes that will alter planform geometry thereby allowing the meanders to develop in concert with the imposed boundary conditions and hydrology (methods (iii) and (iv) above). While the future meandering planform may look markedly different compared to the original corridor design, the final planform would be both stable and appropriate for the drainage system since it could freely adjust to the imposed boundary conditions and hydrology. This conceptual method of river training or triggering desirable changes in planform geometry was proposed by *Nunnally* [1978] and *Brookes* [1987] in designing stream meanders and by *Keller* [1978] in designing pool and riffle sequences.

5. CONCLUSIONS

River restoration programs seek to return functionality to degraded or impaired stream corridors. The use of vegetation in restoration design and stream stabilization projects would greatly enhance the functionality of such corridors while improving recreational opportunities and aesthetic beauty. Moreover, riparian vegetation can induce the formation of large-scale vortical structures within river channels, enhancing fluid mixing processes. An experimental channel was systematically vegetated to assess the effect of vegetation on spatially averaged and turbulent flow within a straight, degraded stream corridor. Wooden dowels simulating vegetation were placed in the channel in staggered arrangement and the spacing between successive zones was identical to the equilibrium meander wavelength for the imposed flow rate.

Various turbulent flow structures were created in association with the vegetation zones, and these included surface waves, dead zones, flow separation, and small scale and large scale vortices. These turbulent flow structures significantly increased fluid mixing processes within the channel. This study also showed that flow velocity decelerated within and near the vegetation zones, flow accelerated and diverted toward the opposite bank, and the channel become sinuous in response to the added vegetation. The magnitude of these effects was controlled by the vegetation density. This method of river training—triggering desirable changes in planform geometry—could have merit in field applications if the morphological effects and ecological impacts of the vegetation program are both envisioned and desired.

Acknowledgments. Taner Pirim, Carlos Alonso, and Collin Anderson provided assistance during various stages of this work. Funding for this work was provided by the USDA-CSREES National Research Initiative Competitive Grants Program. Greg Hanson and an anonymous referee reviewed this paper and made many helpful suggestions.

REFERENCES

Ackers, P., and F. G. Charlton, The slope and resistance of small meandering channels, *Proc., The Institution of Civ. Eng.*, Supplement 15, Paper 7362 S, London, 349–370, 1970.

Bennett, S. J., T. Pirim, and B. D. Barkdoll, Using simulated emergent vegetation to alter stream flow direction within a straight experimental channel, *Geomorphology*, 44, 115–126, 2002.

Brookes, A., Restoring the sinuosity of artificially straightened stream channels, *Environ. Geol. and Water Sci.*, 10, 33–41, 1987.

Brookes, A., and D. A. Sear, Geomorphological principles for restoring channels, in *River Channel Restoration*, edited by A. Brookes and F. D. Shields, Jr., pp. 75–101, John Wiley and Sons, Chichester, 1996.

Brookes, A. and F. D. Shields, Jr., (Eds.), *River Channel Restoration*, 433 pp., John Wiley and Sons, Chichester, 1996.

Dunn, C., F. López, and M. García, Mean flow and turbulence measurements in a laboratory channel with simulated vegetation, *Hydrosystems Laboratory Hydraulic Engineering Series* 51, 148 pp., University of Illinois, Urbana, 1996.

Federal Interagency Stream Restoration Working Group (FISRWG), *Stream Corridor Restoration: Principles, Processes and Practices*, National Technical Information Service, U.S. Department of Commerce, Springfield, 1998.

Fujita, I., M. Muste, and A. Kruger, Large-scale particle image velocimetry for flow analysis in hydraulic engineering applications, *J. Hydraul. Res.*, 36, 397–414, 1998.

Fukuoka, S., and A. Watanabe, Horizontal structure of flood flow with dense vegetation clusters along main channel banks, in *Proceedings, Environmental and Coastal Hydraulics: Protecting the Aquatic Habitat, The 27th Congress of the Int. Assoc. Hydraul. Res.*, vol. 2, pp. 1408–1413, 1997.

Fukuoka, S., A. Watanabe, and T. Tsumari, Structure of horizontal shear flow in open channel with vegetation and its analysis. *Proc., Japan. Soc. Civ. Eng.*, 491, 41–50, 1994.

Ghisalberti, M., and H. M. Nepf, Mixing layers and coherent structures in vegetated aquatic flows, *J. Geophys. Res.*, 107(C2), 1–11, 2002.

Hasfurther, V. R., The use of meander parameters in restoring hydrologic balance to reclaimed stream beds, in *The Restoration of Rivers and Streams, Theories and Experience*, edited by J. A. Gore, pp. 21–40, Butterworth Publishers, Boston, 1985.

Ho, C.-M., and P. Huerre, Perturbed free shear layers, *Ann. Rev. Fluid Mech.*, 16, 365–424, 1984.

Holley, E. R., J. Siemons, and G. Abraham, Some aspects of analyzing transverse diffusion in rivers, *J. Hydraul. Res.*, 10, 27–57, 1972.

Ikeda, S., and M. Kanazawa, Three-dimensional organized vortices above flexible water plants, *J. Hydraul. Eng.*, 122, 634–640, 1996.

Järvelä, J., Flow resistance of flexible and stiff vegetation: A flume study with natural plants, *J. Hydrol.*, 269, 44–54, 2002.

Keller, E. A., Pools, riffles, and channelization, *Environ. Geol.*, 2, 119–127, 1978.

Kiya, M., H. Tamura, and M. Arie, Vortex shedding from a circular cylinder in moderate-Reynolds-number shear flow, *J. Fluid Mech.*, 101, 721–735, 1980.

Kouwen, N., and T. E. Unny, Flexible roughness in open channels, *J. Hydraul. Div., Am. Soc. Civ. Eng.*, 99, 713–728, 1973.

Levi, E., Vortices in hydraulics, *J. Hydraul. Eng.*, 117, 399–413, 1991.

Li, R. M., and H. W. Shen, Effect of tall vegetations on flow and sediment, *J. Hydraul. Div., Am. Soc. Civ. Eng.*, 99, 793–814, 1973.

López., F., and M. García, Synchronized measurement of bed-shear stresses and flow velocity in open channels with simulated vegetation, in *Proceedings of the North American Water and Environment Congress*, (CD-ROM), *Am. Soc. Civ. Eng.*, New York, 1996.

López, F., and M. García, Open channel flow through simulated vegetation: Turbulence modeling and sediment transport, *Wetlands Research Program Technical Report WRP-CP-10*, 106 pp., Waterways Experiment Station, Vicksburg, 1997.

López., F., and M. García, Open-channel flow through simulated vegetation: Suspended sediment transport modeling, *Water Resourc. Res.*, 34, 2341–2352, 1998.

López., F., and M. García, Mean flow and turbulence structure of open-channel flow through non-emergent vegetation, *J. Hydraul. Eng.*, 127, 392–402, 2001.

Nepf, H. M., Drag, turbulence, and diffusion in flow through emergent vegetation, *Water Resourc. Res.*, 35, 479–489, 1999.

Nepf, H. M., J. A. Sullivan, and R. A. Zavistoski, A model for diffusion within emergent vegetation, *Limnol. Oceanogr.*, 42, 1735–1745, 1997.

Nezu, I., and K. Onitsuka, Turbulent structures in partly vegetated open-channel flows with LDA and PIV measurements, *J. Hydraul. Res.*, 39, 629–642, 2001.

Nunnally, N. R., Stream renovation: An alternative to channelization, *Environ. Management*, 2, 403–411, 1978.

Okabe, T., T. Yuuki, and M. Kojima, Bed-load rate on movable beds covered by vegetation, in *Proceedings, Environmental and Coastal Hydraulics: Protecting the Aquatic Habitat, The 27th Congress of the Int. Assoc. Hydraul. Res.*, vol. 2, pp. 1397–1401, 1997.

Raffel, M., C. Willert, and J. Kompenhans, *Particle Image Velocimetry: A Practical Guide*, 253 pp., Springer, New York, 1998.

Richards, K., *Rivers, Forms and Processes in Alluvial Channels*, 358 pp., Methuen, London, 1982.

Righetti, M., and A. Armanini, Flow resistance in open channel flows with sparsely distributed bushes, *J. Hydrol.*, 269, 55–64, 2002.

Rogers, M. M., and R. D. Moser, The three-dimensional evolution of a plane mixing layer: The Kelvin-Helmholtz rollup, *J. Fluid Mech.*, 243, 183–226, 1992.

Rutherford, J. C., *River Mixing*, 347 pp., John Wiley & Sons, Chichester, 1994.

Schmidt, J. C., D. M. Rubin, and H. Ikeda, Flume simulation of recirculating flow and sedimentation, *Water Resourc. Res.*, 29, 2925–2939, 1993.

Schumm, S. A., M. D. Harvey, and C. C. Watson, *Incised Channels: Morphology, Dynamics and Control*, 200 pp., Water Resources Publ., Littleton, 1984.

Shen, H. S., Flow resistance over short simulated vegetation and various tall simulated vegetation groupings on flow resistance and sediment yields, in *Environmental Impacts on Rivers (River Mechanics III)*, edited by H. S. Shen, pp. 3-1 to 3-51, Colorado State University, Fort Collins, 1973.

Shields, F. D., Jr., Hydraulic and hydrologic stability, in *River Channel Restoration*, edited by A. Brookes and F. D. Shields, Jr., pp. 23–74, John Wiley and Sons, Chichester, 1996.

Shields, F. D., Jr., A. J. Bowie, and C. M. Cooper, Control of streambank erosion due to bed degradation with vegetation and structure, *Water Resourc. Bull.*, 31, 475–489, 1995a.

Shields, F. D., Jr., S. S. Knight, and C. M. Cooper, Use of the index of biotic integrity to assess physical habitat degradation in warmwater streams, *Hydrobiol.*, 312, 191–208, 1995b.

Shields, F. D., Jr., S. S. Knight, and C. M. Cooper, Rehabilitation of watersheds with incising channels, *Water Resourc. Bull.*, 31, 971–982, 1995c.

Simon, A., and M. Rinaldi, Channel instability in the loess area of the Midwestern United States, *J. Am. Water Resourc. Assoc.*, 36, 133–150, 2000.

Simpson, R. L., Turbulent boundary-layer separation, *Ann. Rev. Fluid Mech.*, 21, 205–243, 1989.

Stephan, U., and D. Gutknecht, Hydraulic resistance of submerged flexible vegetation, *J. Hydrol.*, 269, 27–43, 2002.

Stone, B. M., and H. T. Shen, Hydraulic resistance of flow in channels with cylindrical roughness, *J. Hydraul. Eng.*, 128, 500–506, 2002.

Thompson, G. T., and J. A. Roberson, A theory of flow resistance for vegetated channels, *Trans. ASAE*, 19, 288–293, 1976.

Tollner, E. W., B. J. Barfield, C. T. Haan, and T. Y. Kao, Suspended sediment filtration capacity of simulated vegetation, *Trans. ASAE*, 19, 678–682, 1976.

Tsujimoto, T., Coherent fluctuations in a vegetated zone of open-channel flow: Cause of bedload lateral transport and sorting, in *Coherent Flow Structures in Open Channels*, edited by P. J. Ashworth, S. J. Bennett, J. L. Best, and S. J. McLelland, pp. 375–396, John Wiley & Sons, Chichester, 1996.

Tsujimoto, T., Fluvial processes in streams with vegetation, *J. Hydraul. Res.*, 37, 789–803, 1999.

U.S. Department of Agriculture, Natural Resources Conservation Service (USDA-NRCS), *National Engineering Handbook Series*, Part 650, Chapter 16, Streambank and Shoreline Protection, U.S. Government Printing Office, Washington, D.C., 1996.

U.S. Environmental Protection Agency, National Water Quality Inventory: 1998 Report to Congress, *Report EPA841-R-00-001*, Washington, D.C., 1998.

U.S. Environmental Protection Agency, The Quality of Our Nation's Waters, A Summary of the National Water Quality Inventory: 1998 Report to Congress, *Report EPA841-S-00-001*, Washington, D.C., 2000.

van Rijn, L. C., 1994. *Principles of Fluid Flow and Surface Waves in Rivers, Estuaries, Seas, and Oceans*, 395 pp., Aqua Publications, Amsterdam, 1994.

Watson, C. C., S. R. Abt, and D. Derrick, Willow posts bank stabilization, *J. Am. Water Resourc. Assoc.*, 33, 293–300, 1997.

Wu., F.-C., H. S. Shen, and Y.-J. Chou, Variation of roughness coefficients for unsubmerged and submerged vegetation, *J. Hydraul. Eng.*, 125, 934–942, 1999.

Sean J. Bennett, U.S. Department of Agriculture-Agricultural Research Service, National Sedimentation Laboratory, P.O. Box 1157, Oxford, MS; currently at: Department of Geography, University of Buffalo, Buffalo, NY 14261-0055.

Riparian Vegetation as a Primary Control on Channel Characteristics in Multi-thread Rivers

Michal Tal[1], Karen Gran[2], A. Brad Murray[3], Chris Paola[1], and D. Murray Hicks[4]

While many previous studies have explored the effects of vegetation on single-thread rivers, systematic studies on multi-thread rivers are scarce. Our approach is to synthesize data and ideas from a field-based study of the Waitaki River in New Zealand with results from laboratory flume experiments and a cellular numerical model for which variables can be controlled. The combination of the results from the three approaches suggests that vegetation affects channel planform mainly through reductions in the total channel width, braiding index, and relative mobility of channels. A major driver is the effect of vegetation on flow dynamics and the apparent cohesion of channel bed and banks. By choking off weaker channels, vegetation corrals the flow into fewer, stronger channels with more uniform higher velocities; by strengthening the banks and exposed bars, vegetation reduces channel migration rates and limits bed-material exchange with islands and bars. At present there is no indication that the maximum shear stress is increased by vegetation, only that the low-stress tail of the total stress distribution is cut off. Thus total flow width appears to be relatively sensitive to changes in vegetation intensity while maximum velocity and width of the region of active sediment transport are less so. In natural channel systems, vegetation reduces the total channel width by occupying freshly exposed areas of bare sediment. Though this happens naturally, it is often enhanced by changes in the flow regime that may be due, for example, to climate change or damming. Colonization by vegetation is not easily reversible, and therefore typically has long-term effects on the system. We investigate historical channel changes on the Platte River in central Nebraska to separate out the effects of discharge reduction and vegetation expansion on channel width, and find that discharge reduction alone cannot account for the current reduced width on the Platte today.

[1]Department of Geology and Geophysics and St. Anthony Falls Laboratory, University of Minnesota Twin Cities, Minneapolis, Minnesota

[2]Department of Earth and Space Sciences, University of Washington, Seattle, Washington

[3]Division of Earth and Ocean Sciences, Nicholas School of the Environment and Earth Sciences, Duke University, Durham, North Carolina

[4]National Institute of Water & Atmospheric Research Ltd, Christchurch, New Zealand

Riparian Vegetation and Fluvial Geomorphology
Water Science and Application 8
Copyright 2004 by the American Geophysical Union
10.1029/008WSA04

1. INTRODUCTION

Vegetation has been recognized as a primary control on river planform, particularly as a determinant of whether a river will adopt a braided or single-thread pattern (e.g. Millar [2000]). Studies have shown that overall behavior of the system correlates with vegetation type or density, shifting between a single-thread channel and a multi-thread system as vegetation changes [*Mackin*, 1956; *Brice*, 1964; *Nevins*, 1969; *Goodwin*, 1996; *Ward and Tockner*, 2000]. *Murray and Paola* [1994] concluded that braiding is the main mode of instability for unconstrained flow over a noncohesive bed. In other words, in the absence of cohesion to stabilize the banks and/or discourage formation of new channels, the flow tends to create new channels until a braided system develops. Therefore, braiding represents the "default" type of channel instability for rivers in non-cohesive sediment without vegetation. In natural systems the cohesion necessary to stabilize banks can be derived either from finegrained sediment (silt, clay) or from vegetation. Seeds are transported and dispersed readily by wind and water, and opportunistically colonize areas of the channel abandoned or exposed at low flows [*Johnson*, 1994, 1997, 2000]. Vegetation increases bank stability through root binding of the sediment and increases the threshold shear stress needed to erode the sediment. In addition, vegetation offers local resistance to flow by increasing drag and reducing velocity, thus decreasing the shear stress available for erosion and transport [*Thorne*, 1990; *Carollo*, 2002]. Vegetation that is not removed while young, when the plants can be uprooted or buried by even minor flows, becomes stronger and increasingly resistant to erosion and removal by the flow. Thus, at least initially, both bank strength and flow resistance increase with time, though this may be reversed for mature plant systems [*Johnson*, 1994, 2000].

The physical details of vegetation effects on river channels are complex. Increased vegetation density is typically linked to a decrease in bank erosion and lateral migration rates [*Smith*, 1976; *Beeson and Doyle*, 1995]. Roots add strength by physically binding particles to roots. Soil without roots has high compressional strength, but little tensile strength. Roots add tensile strength and elasticity, which helps to distribute stresses, thus enhancing the bulk shear strength of the soil [*Vidal*, 1969; *Thorne*, 1990; *Simon and Collison*, 2002]. Roots thus function like the bars in reinforced concrete or the fibers in a carbon-fiber composite material. Several experiments and field studies have documented the connection between increased density of roots and increased soil strength (for example Ziemer [1981], Gray and MacDonald [1989], and Simon and Collison [2002]). Vegetation can also contribute to bank stability through canopy interception and evapotranspiration. These effects lead to drier, better drained banks with reduced bulk unit weight, as well as lower positive pore pressures [*Simon and Collison*, 2002].

Vegetation along banks affects flow dynamics as well. A number of studies have linked physical properties such as width, depth, and velocity to vegetation density in the riparian zone (e.g. *Hadley* [1961], *Brice* [1964], *Zimmerman et al.* [1967], *Charlton et al.* [1978], *Andrews* [1984], *Hey and Thorne* [1986], *Huang and Nanson* [1997], and *Rowntree and Dollar* [1999]). *Vegetation along the banks usually increases roughness, decreasing local velocity and inducing deposition of fines* [*Thorne*, 1990; Hupp and Simon, 1991; *Shimizu and Tsujimoto*, 1994; *Stone and Shen*, 2002]. The addition of fine materials to the banks due to baffling by plants may increase bank cohesion. Increased roughness can reduce conveyance over parts of the channel and force the flow into a smaller area [*Johnson*, 1994]. The resulting higher flow depths can lead to greater flood potential as well as increased bed degradation [*Tsujimoto and Kitamura*, 1996].

1.1. Combined human-vegetation effects

Human development of rivers often alters the natural flow regime and diminishes the threshold flows that would otherwise flush out vegetation in its early stages [*Eschner et al.*, 1983; *Johnson*, 2000]. As a result, vegetation colonizes large areas of the bed and plays an active role in determining the channel morphology. The case is similar for periods of drought. In many river systems new plant species were artificially introduced to stabilize the banks to support agriculture in the river valley. These new species often have characteristics adapted to highly variable environments: abundant seeds during a short dispersal period characterized by rapid germination and root and shoot growth *Graf*, 1978; *Johnson*, 1994; *Pettit et al.*, 2001]. Thus they are often more aggressive colonizers than native equivalents. With the addition of river development (e.g. dams, irrigation) the combination of introduced species and altered flow regime can lead to dramatic changes in river planform. Islands and banks become stabilized and the native species are eliminated.

Studies thus far suggest that the cumulative effects of riparian vegetation on river morphology can be either beneficial or detrimental to the system. In addition to its natural and aesthetic appeal, its effectiveness at offering bank stability makes vegetation an attractive method for restoring degraded streams. For instance, vegetation is being used to alter the stream flow direction and induce meandering in straight degraded stream channels Nevins, 1969; Rowntree and Dollar, 1999; *Bennet et al.*, 2002]. The vegetation around a bend effectively reduces erosion and induces bank accretion and lateral migration *Beeson and Doyle*, 1995]. One of the most striking changes that occur with increasing vegetation is a substantial reduction in the channel width which can reduce the channel capacity and increases the risk of flooding [*Eschner et. al*, 1983]. The channel and near-channel areas of unvegetated systems are typically characterized by unstable sediment deposits that are mobilized during periods of high discharge. With the onset of vegetation the potential for flooding increases due to bars that are immobile and inflexible *Graf*, 1978]. The reduction in the near-bank velocities from increased resistance greatly promotes sediment deposition and bank and bar accretion. The vegetation is also effective in trapping fine suspended material. Over time these processes decrease the median grain size in the system, which can substantially affect the adequacy for spawning by, for example, salmon and trout [*Kondolf and Wolman*, 1993]. The altered systems are characterized by a decrease in the number of active channels, a decrease in the total wetted width, and a decrease in channel mobility; the riverbed has been "pinned" into a well defined fairway that is substantially narrower than the original width. An important biological effect of the planform changes is that critical resting grounds for migratory birds (e.g. sand hill and whooping cranes) are reduced or eliminated [*Eschner*, 1983; *Currier et al*, 1985]. This effect has motivated programs to remove vegetation (e.g. on the Platte River in central Nebraska). Unfortunately we still lack a physically-based understanding of what might be termed "river biophysics" that could aid in developing optimal strategies for adding and removing vegetation.

Much of the research thus far on the interactions between vegetation and channel flow dynamics has been focused on single-thread channels. Research on vegetation effects in multi-thread channels is sparse in comparison. *Johnson* [1994] suggested that the combined effects of flow alterations and vegetation differ considerably between braided and meandering rivers. In meandering rivers floods are needed to overtop the normal bars and produce high elevation point bars through sedimentation in order to enable recruitment. Relative to a braided river, in a meandering river depth is more sensitive and width less sensitive to changes in discharge. In meandering rivers a decrease in flood peaks reduces meandering rates, leading to a reduction in production of new colonizeable area. The opposite is true for braided rivers. The area of active channel exposed is very sensitive to small changes in the flow so that discharge reduction increases the colonizable area.

In this paper our goal is to synthesize what is known about vegetation effects on multi-thread channels in non-cohesive sediments, i.e. rivers where vegetation is the primary source of cohesion in the system. We focus on rivers in humid-temperate climates with well-defined channels and perennial flows. Thus we will consider mainly bank vegetation and not channel-bottom vegetation. We begin with a field-based case study of the Waitaki River in New Zealand. The work is based largely on analysis of aerial photos. As has occurred in many rivers, a change in the natural flow regime due to dam control and irrigation brought about a change in the vegetation cover in the river. While such cases offer us opportunities to use the field as a natural laboratory, it is often difficult to separate the effects of vegetation on the river from the effects of other parameters with which vegetation co-varies (e.g. a change in climate, discharge, sediment type or supply). To isolate individual parameters we turn to laboratory flume experiments and a cellular numerical model as ways of observing the whole channel system under different vegetation conditions with other parameters held constant. Unfortunately, because these studies have not been coordinated, they use different means of measuring vegetation, which complicates comparing them. Thus our approach here is to seek common vegetation effects that are consistent across all three approaches and strong enough to be clearly visible despite the measurement differ-

ences. Finally, we use two additional field studies, the Platte River in Nebraska and an alluvial fan at a tailings mine in northern Minnesota, to consider the implications of a long-term change in discharge in a field situation.

2. BACKGROUND AND METHODS

2.1. Waitaki River: Description

The Waitaki River is located on the South Island of New Zealand. It is the country's largest braided river by discharge (mean discharge ~ 358 m^3/s) and a major source for hydroelectric power. Hydropower works include three dams along the middle, gorged section of the Waitaki Valley and a network of canals, control structures, and power stations that utilize the storage from three natural lakes in the upper basin. The Lower Waitaki River is the portion of the Waitaki that flows ~ 70 km from the furthest downstream dam (Waitaki Dam) to the sea (Figure 1). It is braided for all but the first few km downstream of the dam and has a sandy gravel bed-material with a median size of approximately 30 mm.

Since the completion of the Waitaki Dam in 1937, the discharges to the Lower Waitaki have been controlled, resulting in a damped flood regime and generally steadier river flow, and there has been no bed-material input from the upper catchment. By comparing the measured discharge record with a simulation of the natural, unregulated record, the mean annual flood discharge of the lower river is estimated to have been reduced from 1434 m^3/s to 1171 m^3/s. Tributary sediment budget estimates plus reservoir surveys indicate that the dams have reduced the bed-material supply to the Lower Waitaki River

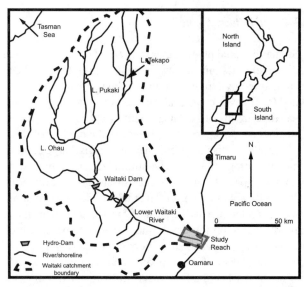

Figure 1. The Waitaki River basin, South Island, New Zealand.

by approximately 50% [*Hicks et al.*, 2002]. The remaining bed-material supply is sourced from tributaries and reworking of the Pleistocene valley-fill. While some of the supply deficit appears to have been compensated by degradation within a few km downstream of the Waitaki Dam, degradation along the braided reach is not obvious; indeed, it is likely that the effect of reduced sediment supply has been moderated by the reduced transport capacity of the flow regime.

The pre-dam riverbed, almost 2 km wide, was characterized by sparse willow trees, temporary islands vegetated mainly by native tussock and scrub, and shifting gravel bars and channels (Figure 2a). The appearance was described as "bare and windswept" [*Thompson et al.*, 1997]. The onset of flow regulation was followed by an invasion of the riverbed by exotic vegetation (notably willow, broom, and gorse) (Figure 2b). This was able to establish because the new flow regime lacked the extreme seasonal variations necessary to flush seedlings and saplings out of the bed or allow grazing of riverbed sites during prolonged spells of lower winter flows [*Hall*, 1984]. In consequence, the less resilient native vegetation was displaced and islands and bars became choked with exotic vegetation and tended to stabilize, while flood break-outs along the riverbed margins became a hazard. Although a policy of devegetating a central "fairway" or braidplain (i.e., area of non- or sparsely vegetated, active riverbed) with spraying and machinery has been implemented since the 1960s, a net increase in vegetation cover remains traceable from aerial photographs. By 2001, the river's braidplain had been reduced to an average width of about 0.5 km (Figure 2b). The total area of riverbed (vegetated and braidplain) has also decreased by conversion of marginal riverbed to pastureland.

2.2. Waitaki River: Methods

Studies of aerial photographs (e.g. *Hall* [1984]) have noted qualitatively that this narrowing of the braidplain has been accompanied by a reduction in braiding activity and a tendency for flows to congregate in one or two principal braids. Clear time-trends of braiding intensity change are difficult to distinguish owing to a dependence of the number of flowing channels on discharge and the varying discharge at times of aerial photography. Here we minimize this complication by analyzing results at a reference discharge. To utilize the only aerial photograph of the river taken prior to regulation, we set this reference discharge at 152 m³/s to match the mean daily discharge on the day of the 1936 photograph, which covered an 11 km reach adjacent to the coast (Figure 2a). A near identical discharge was photographed during a low-flow trial in 2001, when the flow was kept steady at 150 m³/s for 24 hours (Figure 2b).

We used these 1936 and 2001 photographs to measure the number and total width of flowing channels and also the ground cover along 21 valley-normal cross-section lines spaced at 500 m intervals along the coastal reach. Because of a lack of information on channel depth and discharge, all the flowing channels that connected into the braiding network were counted (with backwaters and groundwater-fed channels ignored). The analysis was undertaken using GIS. The 2001 photographs were color digital ortho-images with 0.5 m pixel resolution, georeferenced to the New Zealand Map Grid with a horizontal accuracy of 1 m. The 1936 photographs (black and white) were scanned at 1 m pixel resolution, then rectified and georeferenced using a polynominal fit to approximately 12 control points per frame that were common to both the 2001 and 1936 imagery. While this resulted in absolute positioning errors of up to 5 m on the 1936 imagery, there was negligible distortion of the scaling (i.e., channel widths were accurately measured to within 1-2 m, even if their absolute location was uncertain to ~ 5 m). Ground cover type was classified by eye and digitized. The classes included fenced-off pasture, tall vegetation (mainly willow trees), low vegetation (bushes, typically willow and broom), berm or island grass/tussock, sparse vegetation (isolated bushes on gravel bars), bare gravel, flowing water, and standing water. Total width of riverbed (braidplain width) spans the space of "wildland" between pasture borders. We defined vegetation density as the percentage of riverbed area covered in continuous tall and low vegetation (i.e., trees and bushes). We excluded the native grasses established on transient islands from this definition since they are relatively easily scoured and provide little flow resistance compared with the exotic trees and bushes. In plotting the data, we denoted 1936 as a vegetation density cover of zero, and other years represent the change (positive or negative) from 1936. Changes between 1936 and 2001 over part of this study reach may be compared in Figure 2.

While no intervening historical photographs were available at or near this discharge, we used reach-averaged hydraulic geometry-type relationships determined by *MWD* [1982] to estimate number of channels and total width of flowing channel at 150 m³/s, matching these with ground-cover measurements made from air-photographs taken in 1985. These relations are $N = 1.17\,Q^{0.28}$ and $W = 24Q^{0.50}$, where N is number of flowing channels, W is total width of flowing channels normal to the valley slope (m), and Q is the total discharge (m³/s). We note that for the width relation, we have increased the coefficient by a factor of 1.14 to account for the MWD (1982) widths being measured normal to individual channels, not the valley axis. The 1.14 value is the average channel sinuosity. The 1985 photographs (color) were processed and analyzed in the same fashion as the 1936 photographs. The

A.

B.

Figure 2. The Lower Waitaki River ~ 5 km upstream from the coast, photographed in 1936 (a) and 2001 (b) when the river discharge was ~ 150 m³/s. Flow is left to right. Frame bases span 3.7 km; scales are identical.

ground-cover analysis was also repeated for other photography epochs for the same reach.

2.3. Flume Experiments: Methods

A series of flume experiments was run at the St. Anthony Falls Laboratory to investigate the effects of vegetation density on channel geometry and flow dynamics in multi-thread channels formed in non-cohesive sediment. The details of the flume experiments are described in *Gran and Paola* [2001].

Here we offer a brief description of the experimental procedures pertaining to the results used in this paper. Experiments were conducted in a 2 m by 9 m flume with a slope of 0.014. The experiments consisted of five different runs, four of which are presented here: one with no vegetation and three with varying densities (stems per unit area) of alfalfa (*Medicago sativa*), used for vegetation in the flume [*Ziemer*, 1981; *Gray and MacDonald*, 1989]. Water entered the upstream end of the flume at a constant discharge, and sediment (well sorted quartz sand, D_{50} = 0.5 mm) was fed at a constant rate. An initial

A.   B.

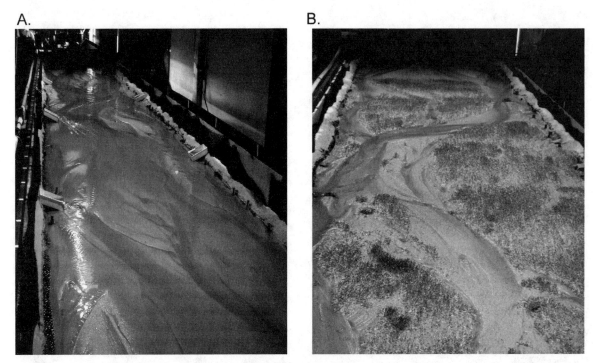

Figure 3. Photographs of the experimental flume at St. Anthony Falls Laboratory during the unvegetated fully braided stage (a) and after the vegetation has been established and only two main channels exist (b).

straight channel widened, and a braided channel system was allowed to develop fully before any vegetation was added (Figure 3a). After the braided channel system developed, the discharge was reduced to half the original discharge, and the sediment feed was shut off during seed dispersal. Seeds were dispersed manually over the entire area of the flume. Some seeds were deposited directly, and some were removed and reworked through the channel and along bars and banks by the flow, thus simulating natural dispersion of seeds by wind and water. The water was shut off for 10–14 days while the plants grew. Once the vegetation was fully established (Figure 3b), the discharge was returned to its original value, and the run continued. Vegetation density was the only variable that changed between runs, varying from 0-9.2 stems/cm². Data from four different runs are presented in this paper, with each run representing a different density.

Width, depth, and number of active channels were measured at 5 equally spaced transects along the study reach every 2 hours for the duration of each run [*Gran and Paola*, 2001]. Measurements were taken at the same place for the unvegetated and vegetated beds. Depths were mapped using an image-based dye density technique. Active channels were defined as those that had a minimum depth of 0.002 m, which corresponds to the threshold shear stress for bedload transport. Summary statistics were compiled on the average num-

ber of active channels, or braiding intensity (*BI*), and the average total width (*w*) of active channels [*Gran and Paola*, 2001]. Summary statistics represent compilations of all cross-sections taken at each time slice for the run (after reaching steady state). Bed topography was measured along all of the same 5 cross-sections every 5–7 hours using a point gauge. This time scale allowed for measurable changes in the bed topography to occur, without sacrificing continuity between sequential profiles.

Gran and Paola [2001] used bed topography data to quantify channel mobility rates by treating the cross-section data as a form of time series data and computing a correlation coefficient (r_0) at each cross-section between sequential profiles (η_1 and η_2)

$$r_0 = \frac{\mathrm{cov}(\eta_1, \eta_2)}{\sqrt{\mathrm{var}(\eta_1)\,\mathrm{var}(\eta_2)}} \qquad (1)$$

Here cov and var refer to the standard statistical definitions of covariance and variance. This correlation coefficient is similar to an auto-correlation coefficient, and in this context gives an index of the coherence of the section geometry with time. The same analysis was done with bed topography data from the cellular model.

2.4. Cellular Model: Methods

The goal of the original cellular model [*Murray and Paola*, 1994] was to include only a small number of processes, represented via simple, abstracted rules, to determine what aspects of the interactions and resulting feedbacks that we see in the field and in the laboratory are sufficient to produce an ongoing, dynamic, multiple-channel pattern. It was found that a nonlinear relationship between sediment flux and local flow strength robustly leads to a braided pattern, but that the gravity-driven component of sediment transport on lateral slopes is necessary to maintain the dynamics indefinitely [*Murray and Paola*, 1994, 1997]. In a similar spirit, *Murray and Paola* [2003] added simplified representations of some of the main effects vegetation has on sediment transport, to investigate what plant effects might be responsible for producing single-thread patterns in bedload-dominated rivers. We stress that, in the cellular approach, some processes are omitted and others represented in simplified form. The goal is thus not to pursue detailed matches with natural examples, but rather to identify model behaviors that are robust and insensitive to specific parameterizations [*Murray and Paola*, 2003]. For example, trends that persist despite changes in parameter values should reflect the basic interactions included in the model. The algorithms of the basic model and the treatment of plant effects are described in detail by Murray and Paola [1994, 1997, 2003]. Here we briefly recap the main points, and describe the changes to the algorithm used in experiments reported in this paper.

2.4.1 Basic algorithm.
A lattice of cells represents the braid plain, with average elevations decreasing longitudinally, creating an overall slope. In the model runs reported in this paper, the lattice was 500 cells long in the downstream direction. The initial elevations decreased linearly in the longitudinal direction, with small-amplitude, white noise perturbations added independently in each cell. Low terraces (three cells wide) along the sides minimized interaction between the flow and the inerodible high sidewalls that confined the flow. Except where noted, the discharge was introduced into the middle 36 cells in the first row, and the braid plain was 48 cells wide.

An iteration begins with the introduction of water into cells at the upslope end. From each of these cells, the water moves into any of the three immediate neighbors in the overall downhill direction that have lower elevations (positive slopes). The amounts of water going to each of these cells are determined by the slopes to those cells. The water routing rule is designed not to represent actual flow with maximal accuracy, but to capture in a simple way the tendency for more water to flow where slopes are steeper.

Discharge is expressed in arbitrary units. One of the experiments reported below involves changing the total discharge from one run to another. Changes in total discharge are represented by altering the number of first-row cells that receive discharge, which is analogous to changing the width of flow entering a stream while holding the discharge per width constant. In these variable-discharge runs, the braid plain was always 12 cells wider than the discharge introduction in the first row.

The amount of sediment transported from cell to cell is related nonlinearly to the stream power (discharge times slope). Runs reported in this paper used Q_s rule 6 from *Murray and Paola* [1997], in which the stream power immediately upstream of a cell, weighted by a factor $å$, is added to the local stream power. In runs reported here, $?å = 0.25$, and the coefficient relating the adjusted stream power to sediment transport, K, = 5 x 10^{-23}. We have performed experiments using different parameter values and different sediment-transport rules [*Murray and Paola*, 1997], and have found that the results reported here do not depend sensitively on the exact form of the rule used.

With noncohesive material, wherever the flow is causing sediment movement, gravity causes a downslope component of sediment transport. Near a channel bank, this component of the transport will be in a direction lateral to the flow direction. This lateral transport moves bank material toward the lower parts of a channel, tending to widen the flow. In the model, if a lateral neighbor cell has a higher elevation than the cell in question, a sediment discharge, Q_{sl}, is transported down the lateral slope, S_l, according to:

$$Q_{sl} = K_l S_l Q_{s0} \qquad (2)$$

where K_l is a constant, adjusted so that Q_{sl} is a few percent of the sediment transport in the cell in question, Q_{s0}, roughly consistent with Parker [1984].

The water routing and sediment transport rules are applied row by row until the water reaches the downslope end of the lattice. Then the elevation in each cell is adjusted according to the difference between the amounts of sediment entering and leaving that cell, conserving sediment mass. The elevation of the cells at the upstream and downstream ends of the lattice remained fixed during each run.

Nothing in the model constrains what scale of stream is simulated; the length represented by a cell, and the time represented by an iteration, are not determined [*Murray and Paola*, 1997]. Using the parameter values above, it takes approximately 100,000 iterations for the discharge pattern to change sufficiently that the correlation coefficient of the elevation pattern at the two times, as defined in the experimental section above, falls below 0.75. This provides a characteristic pattern-change time scale, T_{ch}, for the model.

2.4.2 Simulation of vegetation. The rules that have been added to simulate plant growth and plant effects are designed to be the simplest representations of what we hypothesize to be the main effects in the context of plant/stream interactions [*Murray and Paola*, 2003]. *Murray and Paola* [2003] found that the most important of the plant effects included is an increase in bank strength (a decrease in lateral sediment transport, Q_{sl}), representing the development of root networks. The plant effects increase in any cell, as described below, as long as conditions are conducive to plant growth, up to a limiting time after which the vegetation is assumed to be fully developed. This plant-development time scale, T_{pd}, is equal to T_{ch} (100,000 iterations) in the runs reported here. Any plants in a cell are assumed to be destroyed if either of two conditions is met: 1) the rate of deposition of sediment rises above a cutoff value (either continuously or with interruptions lasting less than a specified plant-resurrection time) for more than a threshold time, effectively burying the plants under too much sediment for recovery; or 2) the rate of erosion rises above a cutoff value (either continuously or with interruptions lasting less than the plant-resurrection time) for longer than a threshold time. When deposition and erosion rates in a cell both fall below the cutoff for longer than the plant-resurrection time, plant growth begins again. The values used here for T_{pd} and the cutoffs and thresholds are consistent with those used previously [*Murray and Paola*, 2003]. Additional experiments in which these values were changed by an order of magnitude produced qualitatively the same results as those presented in *Murray and Paola* [2003].

The effect of plant roots on bank erosion is simulated by decreasing the magnitude of the lateral sediment transport (Q_{sl}) out of a vegetated cell. In this way, if plants are growing in a cell next to a channel (on top of a bank), a steeper slope can develop and be maintained longer between the vegetated cell and the adjacent channel cell. The coefficient in the lat- eral transport rule, K_l, decreases linearly from a value appropriate for noncohesive sediment [*Murray and Paola*, 1997] to a minimum value over the plant growth time scale. We treat the minimum value, K_{lmin}, which is inversely related to plant-enhanced bank strength, as an independent variable in some of the experiments reported below. We vary K_{lmin} by two orders of magnitude; thus, for fully developed vegetation in runs with the "strongest" plants, the bank erosion rate is reduced by two orders of magnitude. This magnitude is conservatively consistent with field experiments that have shown that banks protected by roots can have erosion resistances several orders of magnitude greater than those without protective vegetation [*Smith*, 1976].

The number and total width of active channels, as well as bed topography, were measured along 250 cross-sections every 100,000 iterations for the duration of the run after a statistically steady state was achieved (after approximately 1,000,000 iterations). A discharge threshold of approximately 10% of the typical discharge for a cell in a channel during a run without vegetation was used to delineate the active channels.

3. RESULTS

3.1. Vegetation Parameters

Each of the three approaches described above measures the relative importance of vegetation to the stream system in a different way. We stress that the three parameters used are significantly different from one another, and do not wish to imply that in any sense they are interchangeable.

In the Waitaki River study the vegetation parameter is the fractional areal cover of vegetation over the braid plain, referenced to the vegetation cover in 1936 (defined as a vegetation intensity of zero). For the flume experiments the vegetation parameter is the density of alfalfa stems, from zero to a den-

Table 1. Average number of flowing channels and total width of flowing channels at ~ 150 m³/s, plus average widths of riverbed ground cover for the 11 km reach of the Waitaki River adjacent to the coast, as measured in 1936, 1985, and 2001.

	1936	1985	2001
Number of channels	11.6	4.9	6.8
Total flowing width (m)	416	294	243
Width of tall vegetation (m)	27	272	352
Width of low vegetation (m)	85	181	328
Width of grassed island/berm (m)	298	42	1
Braidplain width (m)	1223	758	549
Total width of riverbed (m)	1632	1254	1229
Vegetation density (% of total width)	6.9	36	55
Normalized braiding index	1	0.42	0.59
Normalized total flowing width	1	0.71	0.58

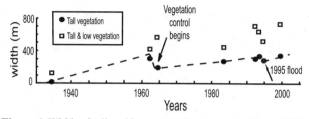

Figure 4. Width of tall and low vegetation across the Lower Waitaki riverbed averaged along the reach within 11 km of the coast. Broken line suggests the trend followed by tall vegetation. Vegetation control began in the early 1960's. A 100-year return period flood occurred in 1995.

sity of 9.2 stems/cm² [*Gran and Paola*, 2001]. Finally, vegetation in the cellular model was varied by changing the plant-effect ratio, which is the vegetation strength measure for that model [*Murray and Paola*, 2003]. The plant-effect ratio is a ratio between the erodibility of a bank with no vegetation and a bank with fully grown vegetation. A plant-effect ratio of 0.1 means that a bank with fully developed vegetation is 1/10 as erodible as one without any vegetation. The vegetation parameter in the model is thus the plant-effect ratio.

3.2. Waitaki River

Results from the Waitaki River (Table 1 and Figure 4) show an increase in riverbed cover by trees and bushes (i.e., tall and low vegetation) from 7 to 55% of riverbed area between 1936 and 2001. This occurred at the expense of the braidplain, while the original areas of grassy berm were effectively reclaimed to pastureland. This overall trend occured despite the substantial phase of vegetation removal by machinery in the early 1960s, ongoing spraying since then, and scour during a 100-year return period flood in December 1995 (Figure 4). Both braiding index and total flowing channel width (at ~150 m³/s) decreased by ~ 40% between 1936 and 2001. The low braiding index in 1985 suggests that the reduction in braiding occurred in the earlier decades. While this result is estimated rather than measured, it confirms previous qualitative reports.

The Lower Waitaki is a typical field situation in that the role of vegetation in driving geomorphic change cannot be isolated from the effects of other controls, such as flow regime change and reduced bed-material supply. Indeed, it appears to have been flow regulation that catalyzed the invasion of the riverbed by the exotic trees and bushes. However, it is clear from our present understanding of the influence of these types of plants on flow dynamics, sedimentation processes, and effective bank strength that, once established, they play a significant role in effecting the morphological evolution toward a less braided state. The role of the reduced upstream supply of bed-material to the Lower Waitaki by entrapment in the

hydro-lakes is less certain. While abundant bed-material supply is a driver of the braiding process (e.g. *Carson* [1984]), it is not clear how much of the Lower Waitaki River's supply deficit has been recovered from storage in its own braidplain. Certainly, any upstream supply deficit effect would be delayed while dispersing to the coastal reach, so we consider it unlikely that this was a dominant driver of the morphological changes that we have measured there.

3.3. Comparison of Results from the Waitaki River, Experimental Flume, and Cellular Model

Despite different approaches to investigating the effects of vegetation on multi-thread channels in the field, experimental, and cellular-model studies, we found similar trends relating riparian vegetation and braiding intensity, channel geometry, and the mobility of channels (Figure 5a,b,c). Increasing vegetation decreases the size and number of active channels (braiding intensity) in all three study systems (Figure 5a,b). Channel width is the sum of the widths of all active (as defined for each study) channels along a cross section. Channel statistics represent compilations of all cross-sections over the study reach for the Waitaki River, and all cross sections through time in each run for the experimental flume and the cellular model. As the riparian vegetation cover increased in the Waitaki River, channels became narrower, with fewer active channels along each cross-section. These trends mirror those found in the experimental flume and the cellular model. Channel widths and braiding intensity in both decreased with an increase in the plant density (experimental flume) and plant-effect ratio (cellular model), both proxies for the stabilizing effect of plant roots.

The field and experimental results reflect the effect of vegetation at a fixed discharge. One advantage of the cellular model is that the discharge can be easily varied without the problem of discharge affecting the vegetation parameter. To test the effect of initial discharge, we carried out a series of runs with constant discharge, but different vegetation strength as measured by the plant-effect parameter. This was repeated five times, for five different discharge values. In all cases, the width decreased with increasing vegetation (Figure 6). Although the fractional width decrease varied between a plant effect ratio of 1 and 0.1 for the different runs, all runs began to converge on the same decreased normalized width value (approximately 0.25) for a plant effect ratio of 0.05.

Because the measures of vegetative influence differ for the three approaches, and because the flume and the cellular model are highly simplified systems, we do not expect quantitative agreement between the results. However, we can compare general behaviors. All the approaches we have compared in this paper show that the effect of vegetation on channel pattern and dynamics is quite strong. In the flume and the cellular model the braiding intensity was reduced by half as the vegetation influence was increased from minimum to maximum (Figure 5a). The channel width was reduced by approximately 80% for the same increase. The Waitaki River had a decrease of approximately 40% in both parameters (Figure 5a,b). The channel mobility as measured by r_0 for the flume and the cellular model decreased by approximately 25% (Figure 5c).

In both the flume and the cellular model we measured channel mobility indirectly by computing correlation coefficients (r_0) on topographic cross-sections between adjacent time steps. The r_0 values for the experimental flume runs with varying vegetation densities and the cellular model runs with varying plant effect ratios are shown in Figure 5c. An $r_0 = 1$ rep-

resents perfect correlation between sequential profiles, with lower r_0 indicating higher channel mobility rates. In both cases, r_0 increased with increasing plant density or strength, indicating lower mobility rates with increasing vegetation intensity. The results of the cellular model from data collected every 100,000 iterations showed similar trends to those collected every 200,000 iterations, showing that the results are not sensitive to a factor of two change in the sampling interval. For simplicity, we have plotted only the results from the 100,000 iteration interval in Figure 5c.

Another common aspect of the effect of vegetation for all three approaches we studied is that the changes in channel characteristics were strongly nonlinear for all measures of vegetation influence. The channel system is most sensitive to initial changes in vegetation and becomes less sensitive as the vegetation continues to increase. For the cellular model, for instance, the changes in channel width, braiding intensity, and channel mobility were strong initially and then weakened, stabilizing around a minimum value (Figure 5a,b,c and Figure 6). The strongest decreases in braiding intensity and channel width corresponded to an initial 30% increase in vegetation; for the channel mobility, the rapid increase in bed

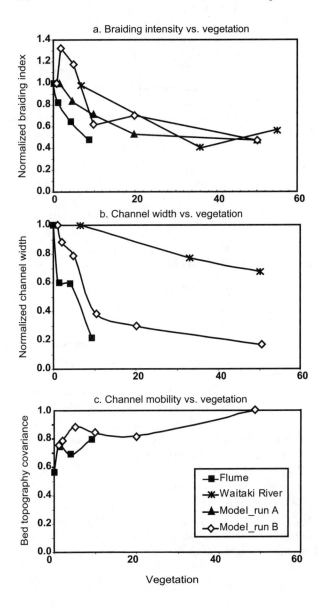

Figure 5. Channel characteristics from an experimental flume [*Gran and Paola*, 2001], cellular model [*Murray and Paola*, 2003], and Waitaki River, NZ [*Hicks et al.*, 2002] are plotted against vegetation. The vegetation parameter in the flume varied as different densities of alfalfa sprout stems, from 0–9.2 stems/cm². Vegetation in the model is represented by the plant effect ratio, the ratio between the lateral transport coefficients in the absence of cohesion to the coefficient value for a fully vegetated bank under varying plant strength conditions. Lateral transport is inversely related to plant enhanced bank strength. Therefore, stronger plants are simulated by decreasing the magnitude of the lateral sediment transport (erodibility) out of a vegetated cell. Model runs A and B use the same parameter for bank strength, however, run A uses a different set of sediment transport rules [*Murray and Paola*, 2003] than the ones described in this paper. Both methods maintain the same trend. Vegetation in the field is represented by a change in the total width of vegetation cover along a cross-section, referenced as a percent of the total river bed width.

a. Braiding intensity is the number of active channels normalized to the # of active channels for the minimum vegetation value.

b. Channel width is the sum of the widths of all active channels along a cross-section, normalized to the width under the minimum vegetation value. Active channels are those above a certain threshold discharge or depth.

c. The bed topography correlation coefficient is a measure of channel mobility rate. Successive bed topography data were measured at set intervals (5–7 hours in the flume, and every 100,000 iterations in the model) and treated as a form of time series data for which a correlation coefficient was calculated.

topography correlation corresponded to an initial 8% change in vegetation. Because we did not continue increasing the vegetation in the flume past a certain density, it is not possible to see the diminishing effect of the vegetation clearly. The same is true for the Waitaki River, due to the small number of data points. We believe however, that given more data, the nonlinearity would hold true in both studies as well, making it independent of how vegetation is measured.

The vegetation caused a reduction in braiding intensity in all three studies (Figure 5a). The systems were transformed from multiple channels to one or two dominant channels. This reduction in the number of active channels corresponds to a transition in planform from a multi-thread system (braided) toward a single-thread one (Figure 3a,b). A true meandering river as defined by *Leopold and Wolman* [1957] is characterized by curves with consistent wavelengths. This was not true for the dominant channels that developed in our studies, although they did follow sinuous paths. Therefore the systems that emerged as a result of an increase in vegetation should be considered wandering or irregularly sinuous streams [*Church and Rood*, 1983].

4. DISCUSSION

Vegetation on river channel banks and bars constrains the flow of the river by stabilizing banks through root reinforcement and by offering resistance to flow. It also induces deposition of fine-grained, cohesive sediments, increasing the overall bank strength. Work done to date in multichannel rivers has established that increasing vegetation density decreases braiding intensity, channel mobility, and total channel width. However, because vegetation in natural rivers often changes along with other parameters such as discharge regime, it is difficult to tease out the effects of these various interactions from field observations alone. The cellular model is especially useful in clarifying the effect of increased bank strength due to vegetation, since increased bank strength is the main way in which vegetation effects are represented in the model. The inhibition of lateral transport due to increased bank strength leads to deeper, narrower channels [*Murray and Paola*, 2003], and a decrease in lateral mobility (Figure 5b,c and Figure 6). These trends are also clearly seen in field and experimental data (Figure 5b,c). The consistency in these trends suggests that the simplified treatment of vegetation in the cellular model does indeed capture a major element of vegetation effects on channel dynamics and that bank strength alone is sufficient to produce many of the changes seen in natural channels.

The trends in decreasing lateral mobility and channel widths as vegetation intensity increases for the multi-thread rivers considered here are consistent with those observed for natu-

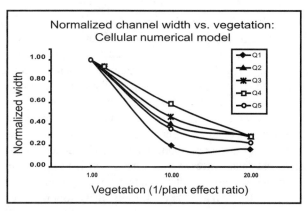

Figure 6. Total channel widths normalized to the width for the unvegetated run are plotted against increasing plant strength. Total width in the model is the number of cells along a cross section that have a discharge above a certain cutoff. The plant effect ratio is a measure of the ratio between the erodibility of the bank with no vegetation and a fully vegetated bank under varying plant strength conditions. Five different discharge values (increasing increments of 12 cells) were simulated for three different plant strength conditions. Each line corresponds to a different discharge. The results from the model correspond to the trend seen in Figure 8 for the Platte River. Although discharge remains the same, there is a reduction in width as the vegetation effect increases.

ral single-thread channels [*Hadley*, 1961; Charlton et al., 1978; Andrews, 1984; Hey and Thorne, 1986; Beeson and Doyle, 1995; Huang and Nanson, 1997; Rowntree and Dollar, 1999]. Furthermore, the runs with highest vegetation density in the model and in the flume produced channels with characteristics that were more similar to those of single-thread rivers (i.e. lower width to depth ratios and lower mobility rates), suggesting that vegetation does indeed play a dominant role in determining whether a river will be single or multi-thread.

So far we have analyzed the effect of plants on river channels in terms of local effects such as increasing bank strength and flow resistance. The common trends in channel behavior we have identified across the three approaches considered here suggest two more global vegetation effects on channel systems: (1) selective colonization, and (2) the vegetation "ratchet" effect. Selective colonization refers to the tendency of plants to selectively occupy channels with relatively low discharge, in effect "corralling" the flow into a few larger channels. In this way even modest amounts of plant growth can quickly reduce total channel width (as noted in all three study approaches) without dramatically affecting flow in the dominant channels. In this view, the effect of the vegetation is mainly to organize the flow rather than to strengthen it. It is noteworthy that, at least according to the laboratory experiments of *Gran and Paola* [2001] the open channels remaining after vegetation establishment do not exhibit larger maximum velocities than the equivalent unvegetated systems; they are simply able to maintain more consistently high velocities. We will return to this point below. The "ratchet" effect refers simply to the observation that once vegetation has colonized a location on the bed, it is relatively difficult to remove. In an unvegetated channel network, a channel may be abandoned and then readily reoccupied due to a slight change in flow conditions elsewhere; with vegetation, colonization of an abandoned channel can quickly make it very hard for the flow to reoccupy. Channel abandonment in a system with active vegetation is much less reversible than in a system without. In the following section, we explore these two ideas in more detail using additional field examples.

4.1. Field Studies: Platte River, Nebraska and Alluvial Fan, Northern Minnesota Tailings Mine

The effects of the trends we have discussed thus far are nicely illustrated in the post-colonial history of the Platte River (central Nebraska, USA) (Figure 7), as presented in Eschner [1983], *Eschner et al.* [1983], *Kircher and Karlinger* [1983]. Moreover, the Platte River presents an opportunity to examine a case where, unlike the experimental and cellular model examples that we have presented, plant colonization accompanies reduction in discharge. In natural rivers, reduction in discharge often sets the stage for vegetation to colonize newly exposed areas of channel resulting in an overall width decrease. Is the vegetation passively colonizing what the reduced discharge offers, or is it also playing an active role in reducing the width?

Historical documents and aerial photographs through the 1900s show that vegetation cover along the river is much higher today than in the past [*Eschner et al.*, 1983]. This increase in vegetation coverage has been accompanied by a decrease in overall width [*Eschner et al.*, 1983; *Kircher and Karlinger*, 1983]. However, during the same period discharge

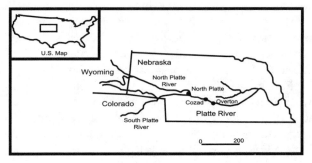

Figure 7. Map of the Platte River, Nebraska showing the approximate location of the 3 gauging stations from which data was used in this paper.

also decreased substantially from pre-settlement values as a result of upstream dams, irrigation withdrawals, and periods of widespread drought [Eschner, 1983; *Eschner et al.*, 1983; *Kircher and Karlinger*, 1983]. While the dams have led to an altered flow regime, the changes in channel morphology resulting from a reduction in the sediment supply have been very small [*Simons and Simons*, 1994]. Was the decrease in channel width the result of lower discharges, increased vegetation density, or both? As local groups remove forest vegetation in an attempt to restore the riparian system to its former state, it is important to know how much of the width change is due to increased vegetation and how much is due simply to reduced discharge and lower peak flows.

To isolate the width reduction due to vegetation from that due to discharge reductions [*Eschner et al.*, 1983; *Kircher and Karlinger*, 1983], we calculated the expected width decrease from lower discharges using the standard hydraulic geometry relationship $w = aQ^b$ [*Leopold and Maddock*, 1953], where w is the width, Q is the discharge, and a and b are constants (Figure 8a,b,c). We used a range of values, $0.3 - 0.7$, for the width exponent b [*Leopold et al.*, 1964; *Richards*, 1982]. We plotted discharge data only for years in which width data was available. To smooth short-term fluctuations in annual peak discharges, we plotted the maximum of annual at-a-station peak discharges for the 5 years prior to the year for which width data was available (Figure 8a,b,c).

Figure 8a,b shows that at North Platte and Cozad, the observed width reductions cannot be explained by discharge reductions alone. In fact, at Cozad, the discharge actually increases from 1936 to 1950, while the width drops sharply. Lack of data during the 1940s makes it difficult to determine exactly what drove the width decrease. The behavior at Overton is also complicated (Figure 8c). Initially, width decreases more or less as expected from the reduction in discharge. However, after 1960, when discharge rises again, the width remains low. This same trend is also apparent at Cozad after 1960 (Figure 8b).

Despite the limitations of the Platte River data, two important insights emerge from the analysis. One is the interplay of the two effects ("ratchet" and selective colonization) discussed above. The stabilizing of the banks and narrowing of channels occurs readily as discharge decreases. The vegetation quickly occupies the exposed banks and stabilizes them; it also chokes off the smaller and weaker channels and corrals the flow into several dominant channels. However, this narrowing process is much harder to reverse. Although the discharge shows a substantial increase in the late 1960s at Cozad and Overton, the width of the channel is unable to recover after the vegetation has taken hold (Figure 8b,c). The second main observation is that channel width does not decrease indefinitely. We observe a strong initial decrease in width which then seems to stabilize

Normalized discharge and channel width vs. time

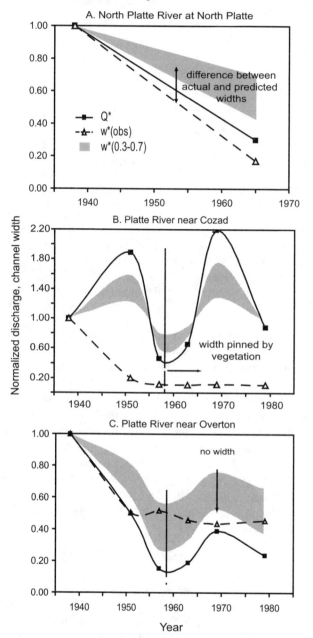

Figure 8. At-a-station discharge (Q^*) and channel width (w^*obs) data normalized to the values of the earliest recorded year for both are plotted against time for three different stations along the North Platte and Platte Rivers. An expected channel width (w^*) is calculated using the formula ($w = aQ^b$) [*Leopold and Maddock*, 1953] for a range (0.3–0.7) of b values. The predicted range of widths shown in gray is plotted along with the observed width in order to highlight the effect of increasing vegetation cover in further reducing channel width from an expected reduction attributed solely to a decrease in discharge. The graph also illustrates how the vegetation pins the river at the banks so that it cannot recover its previous width when the discharge is raised.

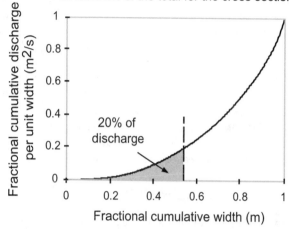

Figure 9. Fractional cumulative discharge is plotted against fractional cumulative width, both as fractions of the total for the cross section, for a sandy, braided alluvial fan in northern Minnesota [*Paola et al.*, 1999]. Note how a 20% reduction in the discharge would lead to a 50% reduction in channel width.

around a minimum value. We observe this trend for the Platte River (Figure 8) as well the cellular model (Figure 6). Although we did not continue the flume experiments to extremely high vegetation density, the observed trends suggest that here too the widths would not decrease indefinitely (Figure 5b). Apparently, as the channel becomes narrower and the velocity becomes consistently high across the channel, near-bank deposition and production of colonizable area are reduced. Although shear stress is not explicitly represented in the cellular model, a comparable effect emerges: as channels narrow with increasing vegetation influence, the stream power in cells next to the banks increases. As a result, erosion and deposition in these cells also increases which prevents plants from taking hold. Johnson [1997] points out that the open channel area on the Platte River reached a general equilibrium due to the fact that as portions of the river were taken over by new woodland and rendered inactive, water depth and coverage had to increase in the remaining channels.

In a braided river, most of the discharge is accounted for by a relatively small fraction of the width. This is in part because braided rivers tend to have high width/depth ratios, and low mean topographic relief. Thus, the flow is free to spread opportunistically over a variety of flow paths that are distinguished by only slight topographic differences. A study of a well-controlled alluvial fan at a tailings basin in northern Minnesota [*Paola et al.* 1999] shows how the discharge is distributed over the width of a braided river reach, and illustrates how blocking of relatively low-discharge parts of the braided system could lead to substantial reductions in width. Eliminating flow width representing the lowest 20% of the discharge would result in a loss of over 50% of the total channel width (Figure 9). Thus, on the Platte River, we believe that the low discharge values in the late 1950s allowed a large portion of the braidplain to become exposed and then colonized by vegetation. This vegetation then kept the width pinned at the lower values, even though the discharge subsequently increased. Therefore, the initial width reduction may be primarily related to the discharge reduction, but the maintenance of the reduced width is driven by vegetation. Thus vegetation acts to amplify the effect of a reduction in discharge.

4.2. Bank Strength and Sediment Flux

Bed topography data in the flume experiments [*Gran and Paola*, 2001] show a decrease in lateral mobility as the vegetation density increased (correlation between sequential bed topography profiles increased; Figure 5c), meaning that the banks were eroding or aggrading more slowly. In addition, these channels became narrower (Figure 5b) and deeper. These tendencies offer an additional insight: as the banks become stabilized by the vegetation, the bank-attached bars no longer serve as storage and source areas for bed-material, as in the classic braiding process. Thus, as the flow converges into a deepening main channel, bedload should move through the system more continuously as temporary storage and release in bars is reduced. This hypothesis is currently being tested in a new set of experiments.

One might conclude that if river banks are strengthened (made more cohesive) with vegetation, they should be able to resist higher shear stresses in the channel without eroding. Therefore, one might expect a higher unit sediment flux in a river with vegetated banks. We do not have data that allow systematic comparison of shear stresses between vegetated and unvegetated rivers. However, comparison of dimensionless Shields stresses in sand-bed rivers with cohesive and noncohesive banks (Figure 10) shows no systematic difference between the two cases. In addition, Gran and Paola [2001] found no correlation between vegetation density and mean velocity. What did change was the *variability* in velocity, which was reduced: "corralling" the flow into a smaller number of well-defined channels led to a reduction in velocity variability by eliminating small ineffective side channels.

Based on this, there is no indication that vegetation would increase overall bedload transport capacity. Based on the arguments given in Paola [1996] on the effect of stress fluctuations on total sediment flux, it is even possible that bank stabilization could reduce total sediment flux by reducing the frequency of formation of ephemeral high-stress zones typi-

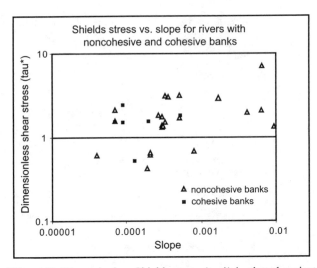

Figure 10. Dimensionless Shields stress (tau*) is plotted against slope for a variety of rivers with noncohesive and cohesive banks. The data are from a compilation by *Church and Rood* [1983]. Note that the data do not show any significant difference between the two systems.

cal of active braiding. Overall, at this point the indications are that vegetation may not have much effect on overall sediment transport efficiency, if it acts mainly to eliminate small side channels that do not move much sediment anyway. It may also be that the width of natural channels is effectively limited by transient deposits of relatively fresh, unvegetated sediment along the insides of the longer-lived, vegetated banks, which could produce the same result. The overall picture seems to be that plants can dramatically reduce the total width of the wetted surface and organize the flow by corralling it into fewer stronger channels, but their effect on bedload transport capacity and the width of the 'effective' sediment-carrying channels is much weaker.

5. CONCLUSIONS

The common trends that represent the robust effects of vegetation on multi-thread rivers include:

1. A decrease in channel lateral mobility; channels become more stable and have lower lateral migration rates.
2. A decrease in the braiding intensity and the total wetted width; the number of active channels is reduced and the channels are narrower and deeper.
3. A nonlinear change in channel parameters with increasing vegetation intensity. The effects of the vegetation are strong initially, and then weaken as easily occupied channels are eliminated.

In addition, we conclude that:

4. While in natural rivers an increase in vegetation often coincides with a decrease in discharge, vegetation reduces

channel widths beyond the width reduction expected solely from the change in discharge.

5. Vegetation effects on overall channel dynamics take two main forms: selective colonization of small, low-discharge channels; and a "ratchet" effect such that once the vegetation has taken hold (e.g. due to a reduction in discharge), the effects are not easily reversed even if the discharge is returned to its previous value.

6. Vegetation leads to a reduction in velocity variability as smaller, weaker channels are choked off and the flow is forced into fewer channels with more consistently high velocities.

7. Based on evidence thus far, increased bank strength does not appear to lead to a consistent increase in shear stress in the main (sediment-carrying) channels and therefore there is no reason to expect major increases in total sediment transport capacity.

Acknowledgments. This material is based upon work supported in part by the STC Program of the National Science Foundation under Agreement Number EAR-0120914, NSF grant No. EAR-0207556, and NSF grant No. EAR-9628393. DMH was supported in part by the New Zealand Foundation for Research, Science and Technology under Contract C01X0023. ABM received support from the Andrew W. Mellon Foundation. MT was supported in part by a Graduate Assistance in Areas of National Need Fellowship from the Department of Education. We thank Gary Parker and Efi Foufoula for helpful discussions, the staff, fellow students, and undergraduate interns at the St. Anthony Falls Laboratory for help with experiments, and Ude Shankar at NIWA for GIS work on the Waitaki River. We appreciate the comments from two reviewers.

REFERENCES

Andrews, E. D., Bed-material entrainment and hydraulic geometry of gravel-bed rivers in Colorado, *Geological Society of America Bulletin*, 95, 371–378, 1984.

Beeson, C. E., and P. F. Doyle, Comparison of bank erosion at vegetated and non-vegetated channel bends, *Water Resources Bulletin*, 31, 983–990, 1995.

Bennet, S. J., T. Pirim, and B. D. Barkdoll, Using simulated emergent vegetation to alter stream flow direction within a straight experimental channel, *Geomorphology*, 44, 113–126, 2002.

Brice, J. C., Channel patterns and terraces of the Loup Rivers in Nebraska, pp. 41 pp., U.S. Geological Survey, Washington, D.C., 1964.

Carollo, F. G., V. Ferro, and D. Termini, Flow velocity measurements in vegetated channels, *Journal of Hydraulic Engineering*, 128, 664–673, 2002.

Carson, M. A., Observations on the meandering-braided river transition, Canterbury Plains, New Zealand, *New Zealand Geographer*, 40, 89–99, 1984.

Charlton, F. G., P. M. Brown, and R. W. Benson, The hydraulic geometry of some gravel rivers in Britain, pp. 48, Hydraul. Res. Stat., Wallingford, England, 1978.

Church, M., and K. M. Rood, Catalogue of alluvial river channel regime data, Department of Geography, University of British Columbia, Vancouver, 1983.

Currier, P. J., C. R. Lingle, and J. G. Vanderwalker, Migratory bird habitat on the Platte and North Platte Rivers in Nebraska, The Platte River Whooping Crane Critical Habitat Maintenance Trust, Grand Island, Nebraska, 1985.

Eschner, T. R., Hydraulic geometry of the Platte River near Overton, south-central Nebraska, in *Hydrologic and Geomorphic Studies of the Platte River Basin*, pp. C1–C32, U.S. Geological Survey, Washington, D.C., 1983.

Eschner, T. R., R. F. Hadley, and K. D. Crowley, Hydrologic and morphologic changes in channels of the Platte River basin in Colorado, Wyoming, and Nebraska: A historical perspective, pp. A1–A39, U.S. Geological Survey, 1983.

Goodwin, C. N., Channel widening and bank erosion processes on a cobble-bed river, *Geol. Soc. Am. Abstr. Programs*, 28, 262, 1996.

Graf, W. L., Fluvial adjustments to the spread of tamarisk in the Colorado Plateau region, *Geological Society of America Bulletin*, 89, 1491–1501, 1978.

Gran, K., and C. Paola, Riparian vegetation controls on braided stream dynamics, *Water Resources Research*, 37, 3275–3283, 2001.

Gray, D. H., and A. MacDonald, The role of vegetation in riverbank erosion, in *Proceedings of the National Conference on Hydraulic Engineering*, edited by M.A. Ports, pp. 218–223, Am. Soc. of Civ. Eng., New York, 1989.

Hadley, R. F., Influence of riparian vegetation on channel shape, northeastern Arizona, pp. 30–31, U.S. Geological Survey, 1961.

Hall, R. J., Lower Waitaki River: management strategy, Waitaki Catchment Commission and Regional Water Board, Timaru, New Zealand, 1984.

Hey, R. D., and C.R. Thorne, Stable channels with mobile gravel beds, *Journal of Hydraulic Engineering*, 112, 671–689, 1986.

Hicks, D. M., M. J. Duncan, U. Shankar, M. Wild, and J. R. Walsh, Project Aqua: Lower Waitaki River geomorphology and sediment transport, National Institute of Water & Atmospheric Research Ltd, Christchurch, 2002.

Huang, H. Q., and G. C. Nanson, Vegetation and channel variation; a case study of four small streams in southeastern Australia, *Geomorphology*, 18, 237–249, 1997.

Hupp, C. R., and A. Simon, Bank accretion and the development of vegetated depositional surfaces along modified alluvial channels, *Geomorphology*, 4, 111–124, 1991.

Johnson, W. C., Woodland expansion in the Platter River, Nebraska: Patterns and causes, *Ecological Monographs*, 64, 45–84, 1994.

Johnson, W. C., Equilibrium response of riparian vegetation to flow regulation in the Platte River, Nebraska, *Regulated Rivers: Research and Management*, 13, 403–415, 1997.

Johnson, W. C., Tree recruitment and survival in rivers: influence of hydrological processes, *Hydrol. Process.*, 14, 3051–3074, 2000.

Kircher, J. E., and M. R. Karlinger, Effects of water development on surface-water hydrology, Platte River basin in Colorado,

Wyoming, and Nebraska upstream from Duncan, Nebraska, in *Hydrologic and Geomorphic Studies of the Platte River Basin*, pp. B1-B49, U.S. Geological Survey, Washington D.C., 1983.

Kondolf, G. M., and M. G. Wolman, The sizes of salmonid spawning gravels, *Water Resources Research*, 29, 2275–2285, 1993.

Leopold, L. B., and T. J. Maddock, The hydraulic geometry of stream channels and some physiographic implications, pp. 1–57, U.S. Geological Survey, 1953.

Leopold, L. B., and M. G. Wolman, River channel patterns: braided, meandering, and straight, pp. 39–73, United States Geological Survey, Washington, 1957.

Leopold, L. B., M. G. Wolman, and J. P. Miller, *Fluvial Processes in Geomorphology*, 522 pp., W.H. Freeman and Company, San Francisco, 1964.

Mackin, J. H., Cause of braiding by a graded river, *Geological Society of America Bulletin*, 67, 1717–1718, 1956.

Millar, R. G., Influence of bank vegetation on alluvial channel patterns, *Water Resources Research*, 36, 1109–1118, 2000.

Murray, A. B., and C. Paola, A cellular model of braided rivers, *Nature (London)*, 371 (6492), 54–57, 1994.

Murray, A. B., and C. Paola, Properties of a cellular braided-stream model, *Earth Surface processes and landforms*, 22, 1001–1025, 1997.

Murray, A. B., and C. Paola, Modelling the effects of vegetation on channel pattern in bedload rivers, *Earth Surface processes and landforms*, 28, 131–143, 2003.

MWD, Lower Waitaki River hydro-electric pwer investigations report, Part I, Power Directorate, Ministry of Works and Development, Wellington, New Zealand, 1982.

Nevins, T. H. F., River training—The single-thread channel, *New Zealand Engineering*, 367–373, 1969.

Paola, C., Incoherent structure; Turbulence as a metaphor for stream braiding, in *Coherent Flow Structures in Open Channels*, edited by P. J. Ashworth, S. J. Bennett, J. L. Best, and S. J. McLelland, pp. 705–723, John Wiley & Sons, Ltd., 1996.

Paola, C., G. Parker, D. C. Mohrig, and K. X. Whipple, The influence of transport fluctuations on spatially averaged topography on a sandy, braided fluvial fan, in *Numerical Experiments in Stratigraphy; Recent Advances in Stratigraphic and Sedimentologic Computer Simulations*, edited by J. W. Harbaugh, W. L. Watney, E. C. Rankey, R. Slingerland, R. H. Goldstein, and E. K. Franseen, pp. 211–218, SEPM, Lawrence, 1999.

Parker, G., Lateral bedload transport on side slopes, *Journal of Hydraulic Engineering*, 110, 197–199, 1984.

Pettit, N. E., R. H. Froend, and P. M. Davies, Identifying the natural flow regime and the relationship with riparian vegetation for two contrasting western australian rivers, *Regulated Rivers: Research & Management*, 17, 201–215, 2001.

Richards, K., *Rivers; Form and Process in Alluvial Channels*, 358 pp., Methuen & Co., New York, 1982.

Rowntree, K. M., and E. S. J. Dollar, Vegetation controls on channel stability in the Bell River, Eastern Cape, South Africa, *Earth Surface Processes and Landforms*, 24, 127–134, 1999.

Shimizu, Y., and T. Tsujimoto, Numerical analysis of turbulent open-channel flow over a vegetation layer using a k-e turbulence model, *J. Hydroscience and Hydraul. Engrg., JSCE*, 11, 57–67, 1994.

Simon, A., and A. J. C. Collison, Quantifying the mechanical and hydrologic effects of riparian vegetation on streambank stability, *Earth Surface processes and landforms*, 27, 527–546, 2002.

Simons, R. K., and D. B. Simons, An analysis of Platter River channel changes, in *The Variability of Large Alluvial Rivers*, edited by S. A. Schumm, and B. R. Winkley, pp. 341–361, ASCE Press, New York, 1994.

Smith, C. E., Modeling high sinuosity meanders in a small flume, *Geomorphology*, 25, 19–30, 1998.

Smith, D. G., Effect of vegetation on lateral migration of anastomosed channels of a glacier meltwater river, *Geological Society of America Bulletin*, 87, 857–860, 1976.

Stone, B. M., and H. T. Shen, Hydraulic resistance of flow in channels with cylindrical roughness, *Journal of Hydraulic Engineering*, 128, 500–506, 2002.

Thompson, S. M., I. G. Jowett, and M. P. Mosely, Morphology of the Lower Waitaki River, National Institute of Water and Atmospheric Research, Wellington, New Zealand, 1997.

Thorne, C. R., Effects of vegetation on riverbank erosion and stability, in *Vegetation and Erosion*, edited by J. B. Thornes, pp. 125–144, John Wiley & Sons, Ltd., 1990.

Tsujimoto, T., and T. Kitamura, Rotational degradation and growth of vegetation along a stream, in *International Conference on New/Emerging Concepts for Rivers*, pp. 632–657, Rivertech 96, Chicago, Illinois, 1996.

Vidal, H., The principle of reinforced earth, *Highway Research Record*, 282, 1–16, 1969.

Ward, J. V., and K. Tockner, Linking ecology and hydrology in alluvial flood plains, in *European Geophysical Society, 25th general assembly*, Nice, France, 2000.

Ziemer, R. R., Roots and the stability of forested slopes, *IAHS Publ.*, 132, 343–361, 1981.

Zimmerman, R. C., J. C. Goodlett, and G. H. Comer, The influence of vegetation on channel form of small streams, in *Symposium on River Morphology*, pp. 255–275, Int. Assoc. Sci. Hydrol. Publ., 1967.

Karen Gran, Dept. of Earth & Space Sciences, University of Washington, Seattle, WA 98195, U.S.A.

D. Murray Hicks, NIWA, PO Box 8602, Christchurch, NZ.

A. Brad Murray, Div. of Earth and Ocean Science/Center for Nonlinear and Complex Systems, Duke University, Box 90230, Durham, NC 27708-0230, U.S.A.

Chris Paola, St Anthony Falls Laboratory, University of Minnesota, 2 - 3rd Avenue SE, Minneapolis, MN 55414, U.S.A.

Michal Tal, St Anthony Falls Laboratory, University of Minnesota, 2 - 3rd Avenue SE, Minneapolis, MN 55414, U.S.A.

Transport Mechanics of Stream-Borne Logs

Carlos V. Alonso

U.S. Department of Agriculture-Agricultural Research Service, National Sedimentation Laboratory, Oxford, Mississippi.

Large woody debris is increasingly regarded as an integral component of stream stabilization and restoration programs. Unravelling the dynamics of complex interaction of multiple logs among themselves and with the stream environs must start with a characterization of the transport mechanics of individual logs. This paper presents a generalized modelling concept of log motion, and examines in detail available information on the influence of log roughness, orientation, Reynolds number, proximity to the streambed, interaction with the free surface, and flow unsteadiness on the mean hydrodynamic forces acting on submerged cylindrical logs. It is shown that (1) drag and inertial forces acting on logs transported by unsteady streams can be approximated with sufficient accuracy with drag and inertia coefficients developed for steady flows, (2) lift can influence the entrainment of logs sufficiently close to the bed, (3) log inertia is not a significant factor, and (4) drag and buoyancy are the main mobilizing forces.

1. INTRODUCTION

Large woody debris (LWD) resulting from trees falling into streams and rivers is commonly defined as log pieces larger than about 0.12 m in diameter and longer than about 2 m [*Cherry and Beschta*, 1989]. These logs frequently move as part of debris flows in steep-grade, upland streams, or they are entrained and transported by fluvial currents in large, low-land rivers [*Braudrick et al.*, 1997]. Fluvial transport of LWD can pose potential flooding risks and their instream recruitment is known to have an important influence on channel morphology and flow resistance [*Cherry and Beschta*, 1989; *Wallerstein et al.*, 2001; *Young*, 1991]. The ecological value of LWD has long been recognized [*Harmon et al.*, 1986], and river engineers are beginning to examine the use of LWD as an integral component of stream stabilization and river restoration programs [*Abbe et al.*, 1997].

The complex interaction of multiple logs among themselves and with the stream environs controls in the end the overall behaviour of LWD and determines the hydraulic conditions governing the stability and entrainment of individual logs. These elements can be arbitrarily oriented with respect to the stream channel and the fields of flow velocities and accelerations. Thus, unravelling the dynamics of that interaction must start with a complete characterization of the mechanics of fluvial transport of single logs [*Braudrick and Grant*, 2000].

The first part of this paper presents a modelling concept based on the general equations governing the motion of cylindrical logs and incorporating the effect of transient flow conditions, log orientation, and interaction with the free surface and the streambanks. To be sure, an exhaustive characterization of log transport requires a correct specification of all the forces acting on logs as well as on the root wads and branches; however, such an undertaking is beyond the bounds of this paper. Hence, the second part of the study is limited to examine some properties of steady and unsteady mean hydrodynamic forces on cylindrical bodies that are relevant to log transport mechanics. It should be noted that although this chapter uses S.I. units, all the equations and formulas presented below are homogeneous relationships valid in any consistent system of units.

Riparian Vegetation and Fluvial Geomorphology
Water Science and Application 8
10.1029/008WSA05

2. GENERAL EQUATIONS OF LOG MOTION

Figure 1 depicts a log of length L and diameter D, arbitrarily oriented and partially submerged in a time-varying streamflow of instantaneous depth h. The shape of the log is approximated as a circular cylinder to take advantage of the wealth of knowledge accumulated about hydrodynamic forces on cylindrical bodies. The presence of roots and branches can inhibit the entrainment of the log and, thus, the frictional/anchoring effects of these end pieces are implicitly accounted for in the present analysis through the reactive forces, $\vec{R}_i$, that develop at the points where the log comes in contact with the stream bed and banks (Figure 1).

During transport, the instantaneous motion of the log can be considered as the combination of a translation of its center of mass, C, and a rotation of the log about C. Let $\vec{x}_C(t)$ and $\vec{x}(t)$ represent the locations the center of mass and any arbitrary point along the log axis, respectively, occupied at time t in relation to a fixed frame of reference, and let $\vec{r} = s\,\vec{s}$ denote the directed distance along the log axis between C and any other cross section. Similarly, let m_L denote the total mass of the log and dm the mass of a differential log segment. Then, the general equations of motion governing log transport are given by the kinematic equations:

$$\vec{x} = \vec{x}_C + \vec{r}$$
$$\vec{V} = \vec{V}_C + \vec{\omega}_C \times \vec{r} \tag{1}$$
$$\dot{\vec{V}} = \dot{\vec{V}}_C + \dot{\vec{\omega}}_C \times \vec{r} + \vec{\omega}_C \times \left(\vec{\omega}_C \times \vec{r}\right)$$

and the linear and angular momentum equations:

$$\int_L d\vec{F} = m_L\,\dot{\vec{V}}_C$$
$$\int_L d\vec{F} \times \vec{r} = \int_L \vec{r} \times \left(\omega_C \times \vec{r}\right) dm \tag{2}$$

where $\vec{v}$ is the instantaneous velocity of any point on the log's axis, $\vec{v}_C$ is the instantaneous velocity of the log's center of mass, and $\vec{\omega}_c$ is the log angular velocity. The force acting on a differential length of the log is given by:

$$d\vec{F} = C_D \frac{\rho_w}{2} dA_S \left[\vec{n} \cdot \left(\vec{U} - \vec{V}\right)\right]^2 \left(\vec{s} \times \vec{e}\right)$$
$$+ \; C_M \rho_w \, d\forall_S \left(\dot{\vec{U}} - \dot{\vec{V}}\right)$$
$$+ \; C_L \frac{\rho_w}{2} dA_S \left[\vec{n} \cdot \left(\vec{U} - \vec{V}\right)\right]^2 \vec{e} + \rho_\ell d\forall \, \vec{g} \tag{3}$$
$$- \; \rho_w d\forall_S \, \vec{g} + \vec{R} \, \delta_{\tilde{R}}\left(\vec{x} - \vec{x}_{\tilde{R}}\right)$$

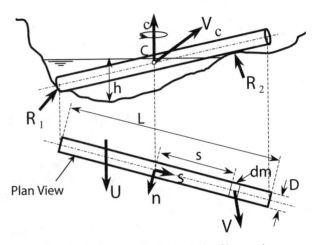

Figure 1. Free-body diagram used in analysis of log motion.

where A_s and $\forall_s$ represent the segment's submerged area normal to the flow and the segment's submerged volume, $\forall$ is the total volume of the log segment, C_D, C_L, and C_M are the hydrodynamic drag, lift and inertia coefficients, $\vec{g}$ is the gravitational acceleration field, $\vec{n}$ is the unit vector normal to the log's longitudinal axis, t represents time, $\vec{U}$ is the instantaneous flow velocity in the absence of the log, ρ_ℓ is the density of the log, ρ_w is the density of water, $\vec{e} = \{(\vec{U} - \vec{V})/|\vec{U} - \vec{V}|\} \times \vec{s}$ is the unit vector normal to the plane formed by the cylinder's axis and the instantaneous velocity vector relative to the log, and $\delta_{\tilde{R}}$ is a Kronecker-delta definition. The first term on the right-hand side of equation (3) represents the in-line (parallel to incident flow) drag; the second is the apparent-mass force due to acceleration of the relative velocity; the third term is the cross-current (normal to incident flow or "lift") force; the fourth and fifth are the true weight and buoyant force, respectively, and the last term represents the bed reaction acting on the log segment. The inertia coefficient is a correction factor that accounts for both the force that must be applied to accelerate the mass of the log and the added force needed to accelerate the mass of water displaced by the log. The buoyant force is considered separately from the drag and lift forces because this model assumes that the relative flow does not alter the hydrostatic pressure distribution about the log. It should also be noted that this model ignores log spinning about its axis but incorporates log pivoting (ω_c), and that the first, third, and fourth terms in this equation vanish for any portion of the log above the free surface.

The instantaneous flow velocity and acceleration fields ($\vec{U}, \dot{\vec{U}}$) and the bed reactions ($\vec{R}_i$) are assumed known at all times and generated by ancillary models of channel evolution [e.g., *Langendoen*, 2000] and log-bed interactions not considered here. Hence, Equations (1)–(3) yield a determined system of six vector equations in the six unknown vector quantities $d\vec{F}$, $\vec{x}_C$, $\vec{V}_C$, $\dot{\vec{V}}_C$, $\vec{\omega}_c$, and $\dot{\vec{\omega}}_c$. Clearly, the physical process represented by these equations behaves uniquely. Hence, a well posed numerical solution to the set of Equations (1)–(3) and the ancillary models will define the position and state of motion of the log at all times $t > 0$ once the initial conditions for the process are specified at $t = 0$. In other words, the model can be regarded as an initial-boundary value problem where knowledge of the state of the physical system at some initial time is used to obtain its state at a later time using a forward-marching numerical scheme together with boundary conditions imposed by the presence of the channel boundaries and the stream free surface. An initial condition typical of ephemeral streams in Mississippi is a log resting on the dry channel bed ($\vec{x}(0) = \vec{x}_c$, $V(0) = \vec{\omega}_c(0) = 0$, $\vec{R}_i(0) \neq 0$) and subjected to an advancing flood wave ($\vec{U}(0) = 0$). Similarly, the case of a log falling into a perennial stream ($\vec{U}(0) > 0$) with a certain orientation and velocity ($\vec{x}(0) = \vec{x}_0$, $\vec{V}(0) = \vec{V}_0$, $\vec{\omega}_c(0) = \vec{R}_i(0) = 0$) describes an initial condition commonly encountered in western rivers. In both instances, the log will first undergo a complex trajectory within the stream and, if unimpeded by interactions with the channel boundary, it will eventually rise to the stream surface and continue its trajectory as a floating body largely driven by gravity, buoyancy, and drag forces. Obviously, the proposed model requires solving a large set of coupled, nonlinear equations and, thus, its actual implementation is only feasible through computational solvers.

2.1. Complex Dependence of Hydrodynamic Forces on Log Geometry and Flow Conditions

The accurate evaluation of Equation (3) under general flow conditions is complicated by the following factors. Drag, lift, and inertial forces vary with time, the cylinder's roughness, slenderness (L/D), its proximity to the channel boundary and the free surface, and its orientation in relation to the incident flow (yaw). In addition, when a cylinder is in close proximity to other cylinders they interfere with the flow resistance of each other [Zdravkovich, 1977; Roshko et al., 1975]. Other than for the cases of (a) a stationary cylinder submerged in an unbounded flow with constant unidirectional acceleration [Sarpkaya and Garrison, 1963], and (b) harmonic-wave loading on offshore structures [Lighthill, 1986] there is no general proof that the terms of Equation (3) can be treated as additive or that the drag and lift components can be expressed in terms of the square of the relative flow velocity.

Moreover, vortex shedding due to flow separation from the cylinder results in considerable wake unsteadiness, independent of the time-dependence of the mean ambient flow [Abernathy and Kronauer, 1962; Gerrard, 1961; Perry et al., 1982]. This is important because the vortex-shedding frequency locks on to the natural frequency of the cylinder and this, in turn, develops in-line and cross-current oscillations

by extracting energy from the unsteady wake. Under resonant conditions, this phenomenon may lead to severe flow-induced body vibrations [Sainsbury and King, 1971; Schewe, 1983; Wootton et al., 1974] that can accelerate log mobilization. Another source of drag fluctuations are the buffeting forces induced by the turbulence of the free stream [So and Savkar, 1981]. Because of space limitations, these important hydrodynamic aspects are not addressed in the present paper and the treatment is limited to characterizations of time-mean forces.

The following sections briefly review the dependence of the force coefficients on dimensionless dynamic and geometric parameters governing bluff bodies and fluid flow interactions [Barenblatt, 1987]. Specifically, the cylinder Froude and Reynolds numbers referred to in the paper are defined as $\boldsymbol{F}_r = U/(gD)^{1/2}$ and $\boldsymbol{R}_e = UD/\nu$, respectively, where $g = |\vec{g}|$, $U = |\vec{U}|$, and ν is the kinematic viscosity of water. The functional dependence of force coefficients on these parameters reduces to particular forms in specific cases. For instance, the drag of cylinders interacting with the free surface is affected by viscous and gravitational forces and thus $C_D = C_D$ (geometry, $\boldsymbol{F}_r$, $\boldsymbol{R}_e$). On the hand, the gravitational effect decreases as the depth of submergence increases and the drag coefficient converges to the form $C_D = C_D$ (geometry, $\boldsymbol{R}_e$) valid for fluids of large extent. However, the drag coefficient depends continuously on the governing parameters and for this reason drag is represented by a single coefficient in Equation (3) regardless of any log-boundary interaction. Similar considerations apply to the other force coefficients.

Much research is still needed to completely delineate the effect of the above factors, particularly in the range of cylinder Froude and Reynolds numbers prevalent in log transport. This information will have to be obtained from carefully planned experimental and computer-simulation studies. The following sections summarize information available on the in-line and cross-current mean forces on circular cylinders normal to uniform flows characterized by a unidirectional velocity $U(t)$.

2.2. Steady Drag of Submerged Circular Cylinders at High Reynolds Number

Flow separation and vortex shedding control the shape and flow pattern in the near wake of cylinders submerged in steady, unbounded, uniform flows. This phenomenon depends on the cylinder Reynolds number and as this parameter increases concurrent changes take place in the pressure distribution around the cylinder and, thus, in the associated mean drag. The first parametric study on the effects of Reynolds number on the mean drag of smooth circular cylinders normal to uniform, steady turbulent flows was carried out by Wieselsberger

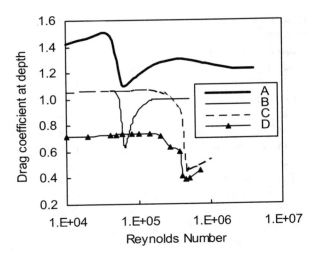

Figure 2. Drag coefficient of submerged circular cylinders normal to flow. (A) $L/D = \infty$, $k/D = 0.009$ [*Achenbach and Heinecke,* 1981]; (B) $L/D = \infty$, $k/D = 0.0045$ [*Achenbach,* 1971]; (C) $L/D = \infty$, $k/D = 0$ [*Wieselsberger,* 1923]; (D) $L/D = 5$, $k/D = 0$ [*Wieselsberger,* 1923].

[1923] and subsequently reported in the English literature by *Prandtl and Tietjens* [1934]. *Fage and Warsap* [1930] were the first to investigate the influence of cylinder roughness on drag up to cylinder Reynolds numbers $R_e = 2.5 \times 10^5$. *Achenbach* [1971] and later *Achenbach and Heinecke* [1981] extended these measurements to $R_e = 4 \times 10^6$ and relative surface roughness up to $k/D = 0.03$, where k represents the cylinder's effective roughness.

Figure 2 presents a few selected runs from these studies to illustrate the influence of slenderness and relative roughness within the range of Reynolds numbers relevant to log transport. Infinitely large values of L/D are used herein to denote tests where the ambient flow past the cylinder was two-dimensional. Although there exist some experimental differences among the results from these studies and those carried by subsequent investigators, the existing body of data is consistent with the trends depicted in Figure 2 and shows that (1) the drag decreases as the cylinder's length is reduced because fluid at free-stream pressure leaks around the ends of the cylinder into the low-pressure wake region, (2) the drag drop at the critical Reynolds number shifts to lower R_e values with increasing roughness and (3) beyond some R_e larger than the critical threshold the drag coefficient depends only on the cylinder relative roughness. For the typical log populations encountered in central Mississippi having an average diameter $D = 0.20$ m and length $L = 15$ m and experiencing peak-flow velocities of about $U = 1.6$ m s^{-1}, the average Reynolds number and slenderness are approximately $R_e = 3 \times 10^5$ and $L/D = 83$. For these parameters and the large roughness associated with natural logs, the data reported by *Achenbach and*

Heinecke [1981] suggests that the mean drag coefficient for logs can reach values as high as $C_D = 1.25$ or greater.

2.3. Steady Drag and Lift of Submerged Circular Cylinders Near a Flat Stream Bed

Roshko et al. [1975] measured the drag coefficient of a cylinder positioned normal to the flow and parallel to a flat wall at a Reynolds number of 2×10^4 and for gap ratios G/D ranging from 0 to 6, where G is the separation between the wall and the cylinder lower edge. *Göktun* [1975] conducted similar measurements for $9 \times 10^4 \leq R_e \leq 2.5 \times 10^5$ and $0 \leq G/D < 3$. *Bearman and Zdravkovich* [1978] reported pressure distributions about a circular cylinder for $R_e = 4.5 \times 10^4$ and $0 \leq G/D \leq 3.5$ from which the writer computed the pressure drag coefficients. The results from these investigations are shown in Figure 3 and the differences among the sets of measurements are suspected to be the result of uncorrected differences in the thickness of the wall boundary layer. These differences notwithstanding, it is clear that far from the wall the drag coefficient approaches asymptotically the value for unbounded flows, and as the gap between the cylinder and the wall closes the drag coefficient decreases rapidly to an intermediate value. To be sure, this value depends on characteristics of the wall boundary layer and the separation zones that form immediately upstream and downstream of the cylinder when the cylinder touches the wall as reported by the fore-

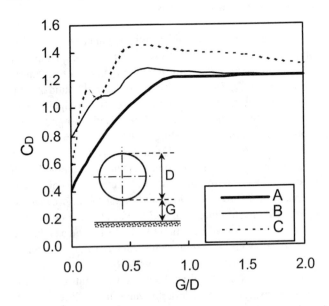

Figure 3. Effect of wall proximity on drag coefficient of circular cylinders, $L/D = \infty$, $k/D = 0$, (A) $R_e = 45,000$ [*Bearman and Zdravkovich,* 1978]; (B) $R_e = 20,000$ [*Rosko et al.,* 1975]; (C) $R_e = 153,000$ [*Götkun,* 1975]. The parameter k/D represents the relative roughness of the log surface.

going investigators. The rise in drag observed in the *Rosko* and *Götkun* datasets around $G/D = 0.6$ are most likely due to uncorrected blockage conditions in their tests.

Flow separation also plays a key role in the development of hydrodynamic lift forces acting on well-submerged cylinders. Within the range of Reynolds numbers giving rise to alternating vortex-shedding, the separating streamline surfaces are asymmetric with respect to the plane normal to $\vec{e}$ (Equation 3). This flow asymmetry gives rise to a pressure gradient normal to the plane and directed away from the last detached vortex which results in an oscillating lift force with zero mean [*Schewe*, 1983]. In the background of this entire process, the far-field flow away from the cylinder remains undisturbed and independent of the body Reynolds number. However, this picture changes drastically when the cylinder is close to a flat bed. For one thing, the proximity of the bed affects the far-field flow around the cylinder by inhibiting velocities normal to the boundary, enables the formation and growth of a boundary layer along the bed, interferes with the vortex shedding process, and triggers flow separation on the bed when the cylinder moves very close to the boundary. Secondly, the flow asymmetry created when the cylinder is placed near a flat boundary generates a net pressure gradient (lift) force normal to the boundary that is flow velocity dependent [*Taneda*, 1965]. Figure 4 shows the effect of wall proximity on the lift coefficient of circular cylinders measured by *Knoblock and Troller* [1935] for $R_e = 1.5 \times 10^5$ and $0 \le G/D \le 1.25$, and by *Roshko et al.* [1975] for $R_e = 2 \times 10^4$ and $0 \le G/D < 2$. It is apparent that lift development follows the reverse trend of drag, reaches a maximum value of about 0.5 when the cylinder is in contact with the wall and essentially vanishes when $G/D > 0.5$.

These drag and lift measurements where conducted with fully submerged, smooth cylinders normal to the approaching flow. Dependence on Reynolds number was most likely non-existent because the tests were all conducted in the subcritical regime. On the other hand, there does not seem to be any information on the relation between these coefficients, G, and R_e for different cylinder roughness and orientation values. Furthermore, drag and lift forces on logs near erodible beds can be expected to decrease rapidly with time as scour reshapes the bed topography in the proximity of the log [*Wallerstein et al.*, 2001].

2.4. Drag of Circular Cylinders Close to a Free Surface

A horizontal cylinder positioned beneath a free surface creates differences in the water-surface level around the body resulting in the formation of a standing surface wave. The experiments reported by *Sheridan et al.* [1997] demonstrate that the standing wave is associated with the generation of a vorticity layer issuing from the free surface due to localized separation, similar to a breaking wave front. This wave state

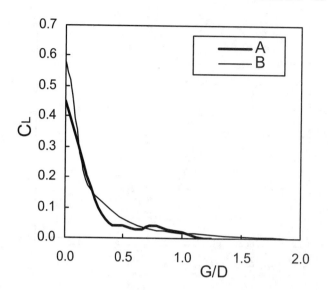

Figure 4. Effect of wall proximity on lift coefficient of circular cylinders, $L/D = \infty$, $k/D = 0$, (A) $R_e = 150,000$ [*Knoblock and Troller*, 1935]; (B) $R_e = 20,000$ [*Rosko et al.*, 1975].

gives rise to shear and normal-stress distributions on the cylinder's surface distinctly different from those on deeply submerged cylinders. Hence, the drag created by the standing wave can be regarded as an added form-resistance. *Lamb* [1945] presented the first analysis of this phenomenon for the case of a two-dimensional, circular, stationary cylinder beneath the free surface of a steady, uniform, potential flow. He developed a linearized, small amplitude wave solution valid for cylinders with $D < z$, where z is the distance from the cylinder's axis to the undisturbed free surface. He obtained solutions for the free-surface displacement $\eta(\zeta, t)$ about the undisturbed free surface, $\eta(\zeta, 0) = 0$ where ζ is the streamwise distance from the vertical passing through the cylinder's axis, and for the wave drag force f_W acting on the cylinder per unit length and for all $t > 0$. His solutions are rewritten below in dimensionless form:

$$\eta^0 = \frac{1}{2}\frac{z^0}{(\zeta^0)^2 + (z^0)^2} - \left[4\pi F_r^{-2}\exp\left(-z^0 F_r^{-2}\right)\sin\left(\zeta^0 F_r^{-2}\right)\right]\delta_\zeta\left(\zeta - 0^+\right) \quad (4)$$

$$f_w^0 = \frac{f_w L}{\rho_w \forall} = \pi F_r^{-4}\exp\left(-2z^0/F_r^{-2}\right) \quad (5)$$

where $\zeta^0 = \zeta/D$, $z^0 = z/D$, $\eta^0 = \eta/D$, $\delta_\zeta = \{0;1 : \zeta = 0; > 0\}$ is a Kronecker-delta definition, and $\rho_w \forall/L$ is the weight of water displaced per unit length of log. The stated approximations

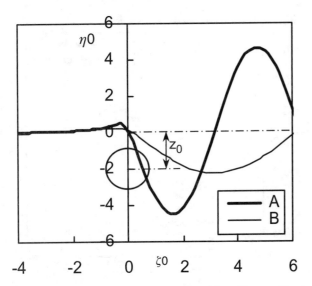

Figure 5. Standing surface waves produced by submerged horizontal circular cylinders [*Lamb*, 1945]. Flow direction is from left to right. (A) $z^0 = 1.0$, $F_r = 0.50$; (B) $z^0 = 2.0$, $F_r = 0.70$.

notwithstanding, Lamb's analysis provides insight on the force that arises from the interaction between the body and the free surface. Equation (4) shows that there is a local free-surface disturbance immediately upstream of the cylinder followed by a wave train that remains stationary with respect to the cylinder and decreases in amplitude as the cylinder's depth increases (Figure 5). Similarly, Equation (5) shows that wave drag, in addition to increasing as the submergence ($z^0 = z/D$) decreases, vanishes at $F_r = 0$, goes through a maximum at $z^0 = 4F_r^2$, and decreases rapidly for $F_r \gg 1$ (Figure 6). In particular, wave drag peaks as $z \to D$. Subsequent studies on wave drag have been reported for submerged prolate-spheroids [*Wigley*, 1953] and for submerged hemispherical bodies resting on a channel bed [*Flammer et al.*, 1970]. In both instances, wave drag dependence on body Froude number and submergence is consistent with the behavior predicted by Lamb's solution.

Wallerstein et al. [2002] reported measurements of wave drag for horizontal cylinders normal to the flow conducted in a laboratory flume for a range of cylinder submergence and slenderness values. These experiments were carried out at $F_r = 0.5$ to maximize the observed wave drag forces, and the results are plotted in Figure 7. In this plot the percent drag increase due to surface wave generation is expressed as $100 (C_w - C_{D\infty}) / C_{D\infty}$, where C_w is the drag coefficient near the free surface and $C_{D\infty}$ is the drag coefficient at depth obtained from Figure 2C. It should be noted that C_w and $C_{D\infty}$ are symbols used solely to characterize the limits the coefficient C_D approaches at low and high submergence, respectively. *Hygelund and Manga* [2003] measured drag forces on large-

scale smooth-log models in two Oregon streams. Reanalysis of their data for a log normal to the flow with $L/D = 15.8$, $R_e = 1.8 \times 10^4$, and $F_r = 0.15$ yielded the wave-drag component shown in Figure 7D. The difference between this data and the drag increase observed by *Wallerstein et al.* [2002] for the same slenderness most likely reflects the different measuring techniques and flow conditions used in those studies. It can be seen that drag at depths less than eight cylinder diameters is consistently higher than at larger depths and decreases as slenderness increases.

2.5. Effect of Cylinder Orientation on Steady Drag

For fully submerged cylinders at rest and oriented perpendicular to a steady uniform flow with velocity U, the first term on the right-hand-side of Equation (3) reduces to the well-known form of the drag force per unit length:

$$f_D = C_{D\infty} D \frac{\rho_w U^2}{2} \qquad (6)$$

This section examines the applicability of this relationship to yawed cylinders, that is, cylinders inclined at an angle with the flow direction. The yaw angle, θ, is defined here as the angle between the incident velocity and the normal to the log axis (Figure 1). *Hoerner* [1965] proposed that the flow pattern and pressure distribution on a plane normal to the cylinder are independent of the flow component parallel to the cylinder for subcritical Reynolds numbers. Based on this hypothesis he approximated the normal component of velocity as

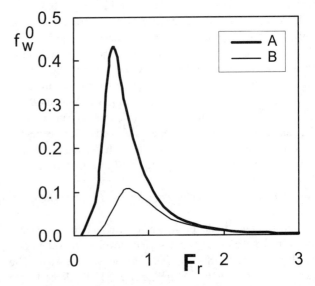

Figure 6. Relationship between the wave drag of a submerged horizontal, circular cylinder and its Froude number and submergence [*Lamb*, 1945]. (A) $z^0 = 1.0$; (B) $z^0 = 2.0$.

$U_n = U \cos \theta$ to compute the force normal to the cylinder, f_{Dn}, using the drag coefficient reported in Figure 2. Accordingly, *Horner* rewrote Equation (6) as:

$$f_D = f_{Dn} \cos\theta = C_{D\infty} D \frac{\rho_w U_n^2}{2} \cos\theta$$
$$= C_{D\infty} D \frac{\rho_w (U \cos\theta)^2}{2} \cos\theta = C_{Dn} D \frac{\rho_w U^2}{2} \quad (7)$$

where $C_{Dn} = C_{D\infty} \cos^3 \theta$, and then compared this equation with C_{Dn} values evaluated for a range of yaw angles using Equation (7) and $[f_D, U]$ data reported by other investigators. *Horner* presented the agreement between the measured and computed values of $C_{Dn}(\theta)$ as proof of his hypothesis, but overlooked the fact that his scaling did not result in a unique relationship between C_{Dn} and $R_{en} = DU_n / \nu$ independently of the yaw angle. Indeed, *Bursnall and Loftin* [1951] have shown this to be the case and, in particular, that the critical R_{en} strongly depends on the yaw angle (Figure 8).

So far it has not been possible to correlate the normal drag force with a single Reynolds number. Part of the problem is that the cylinder's inclination introduces three-dimensional effects that play a major role in the distribution of vorticity in all directions. Under these circumstances, using the cosine approximation may be a gross simplification of both the behaviour of flow in the near wake region and the cylinder's drag, as well as of its lift.

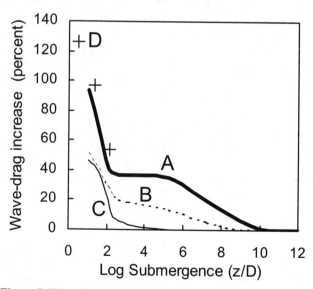

Figure 7. Wave-drag increase for circular cylinders as a function of slenderness and submergence. $R_e = 4,000$, $F_r = 0.5$, $k/D = 0$, (A) $L/D = 5.3$, (B) $L/D = 7.9$, (C) $L/D = 15.9$ [*Wallerstein et al.*, 2002]; (D) $R_e = 18,000$, $F_r = 0.15$, $k/D = 0$, $L/D = 15.8$, adapted from *Hygelund and Manga* [2003].

2.6. Drag and Inertial Resistance in Unsteady Flows

Any characterization of the mechanics of fluvial log transport must account for the unsteady nature of stream flows typical of upland tributaries. Obviously, the complexity and infinite variety of possible unsteady flow motions around cylinders oriented at random renders impractical any attempt to a wholesale characterization of hydrodynamic forces under unsteady conditions. However, resorting to existing idealized analyses of the relation of transient loadings to flow unsteadiness can provide some simplifying insights.

In the context of the present study, let us imagine a horizontal cylindrical log submerged in the path of a stream flood wave and oriented normal to the streamwise direction. One can conceptualize the rising limb of the velocity hydrograph, $U(t)$, as a sequence of linear velocity transitions between consecutive break points, each transition having a different constant acceleration, $\dot{U}(t)$. *Sarpkaya and Garrison* [1963] studied the hydrodynamics of this type of uniformly accelerated flow around stationary cylinders in unbounded fluids. They expressed the drag and inertial coefficients as a function of position, velocity, and circulatory strength of the vortices growing in the near wake of the cylinder, and on the relative displacement of the ambient flow. These vortical parameters were evaluated experimentally for cylinder Reynolds numbers up to a maximum $R_e = 5.2 \times 10^5$ and constant flow accelerations ranging from 1.5 to 9.7 m s^{-2}. The drag and inertial coefficients computed in this manner are shown in Figure 9, where $Ut/D = (\dot{U} t/2) t/D = Ut^2/2D$ represents the ambient flow displacement and t is the time elapsed from the onset of the flow acceleration. It should be noted that the cylinder's Reynolds number and roughness do not appear explicitly in this relationship because the theory is based on the irrotational flow assumption; however, these parameters are expected to affect the onset of flow separation in viscous fluids. Figure 9 shows that the drag coefficient increases to a maximum of 1.25 while simultaneously the inertial coefficient decreases from 2 to 1.2 by the time $Ut/D \sim 2.5$. *Sarpkaya and Garrison* [1963] also showed that beyond this point C_D and C_M fluctuate between smaller limits with ever-decreasing amplitudes, each fluctuation coinciding with a shedded vortex.

It is important to note that the relationships shown in Figure 9 do not depend parametrically on the magnitude of the acceleration and, hence, the drag and inertial coefficients will exhibit the same dependence on flow displacement for any hydrograph transition. Quasi-constant flow accelerations of about 0.001 m s^{-2} sustained for roughly an hour are commonly observed in ephemeral streams of northern Mississippi. According to the *Sarpkaya and Garrison* [1963] relationships (Figure 9), the drag and inertial coefficients for 0.30-m diameter logs submerged in a flow undergoing such acceleration rate will approach their oscillating asymptotic limits in less than a minute. Hence, the evolutions undergone by the drag and inertial coefficients following the onset of flow acceleration are short enough that these coefficients can be approximated by their steady-state asymptotic values. Moreover, it is logical to extend this conclusion to lift coefficients given that lift forces have the same vortical origin as drag forces.

3. APPLICATION

This section presents a hypothetical example that uses the preceding information to illustrate the relative order of magnitude of transient forces acting on logs exposed to ephemeral flow events. The input parameters selected for this example are drawn from conditions typically encountered in northern Mississippi streams. The present scenario considers a stream reach with a fairly flat bed and assumes a log positioned parallel to the bed, oriented normal to the streamflow, and with its longitudinal axis 0.25-m above the bed. To keep the presentation within the bounds of the available space, the example further assumes that the log somehow remains anchored at both ends during a selected runoff event and it is thus prevented from being entrained by the flow.

The hydrograph selected for this example is taken from a runoff event recorded at Station W-5 of the now deactivated ARS Pidgeon Roost Experimental Watershed during a storm on February 21, 1971 [Alonso et al., 1978]. The measured hydrograph is approximated with a synthetic hydrograph pat-

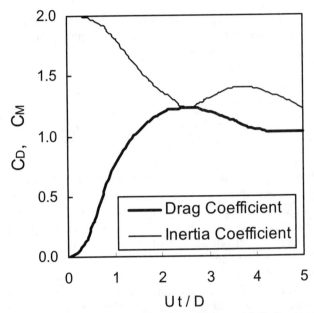

Figure 9. Variation of drag and inertia coefficients with displacement of uniform flow with constant acceleration past a circular cylinder [after *Sarpkaya and Garrison*, 1963]

terned after the SCS Unit Hydrograph [*USDA Soil Conservation Service*, 1985]. In 1971, the approaching channel had a fairly trapezoidal cross-section with a 4.5-m wide sand bed, vegetated banks with roughly 1:2 slopes, and an average thalweg slope of 0.0045 [*Bowie et al.*, 1975]. With this channel data, assuming the channel was running at bank-full capacity during peak flow, and a Manning roughness coefficient estimated at 0.05, the discharge hydrograph was converted to the velocity hydrograph shown in Figure 10. The peak sediment discharge measured during the selected runoff event was about 11,700 ppm. This rate is considerably less than the peak transport capacity of 45,760 ppm estimated with the Engelund-Hansen formula [*Alonso et al.*, 1981] and for a mean sand size of about 0.3mm typical of Northern Mississippi streams. For the purpose of the present example, the excess flow transport capacity is assumed in balance with the upstream sediment supply, and the channel bed is treated as if operating in equilibrium with the flow capacity.

Under these conditions Equations (1) reduce to $\vec{x} = [0; 0.25; 0]$, $\vec{V}(\vec{x}) = 0$, $\vec{\omega}_C = 0$, and from Equations (2) and (3) we arrive at the following equilibrium of forces per unit length of log in terms of their horizontal and vertical components:

$$f_D + f_M - 2R_H/L = 0$$
$$f_W + f_L - 2R_V/L = 0 \qquad (8)$$

where R_H and R_V are the components of the equal reactive forces acting on the ends of the log, and f_D, f_L, f_M, f_W, repre-

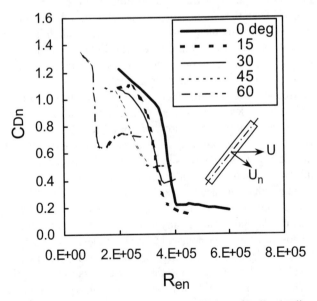

Figure 8. Variation of the normal drag coefficient of inclined cylinders with the normal Reynolds number and the yaw angle shown in the inset, $L/D = \infty$, $k/D = 0$ [*Bursnall and Loftin*, 1951]

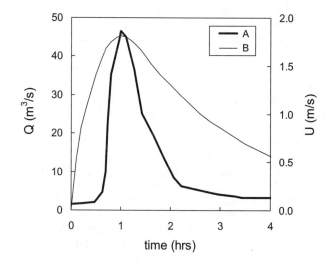

Figure 10. (A) Discharge hydrograph recorded at the station W-5 of the ARS Pigeon Roost Watershed, Mississippi, during the runoff event of February 21, 1971. The origin of the time scale corresponds to the local time 6:32 pm; (B) Velocity hydrograph developed from the SCS Unit hydrograph and theoretical rating curve.

sent the drag, lift, inertial, and effective-weight forces per unit length of log, respectively. The anchored-log condition is achieved by requiring that R_H and R_V satisfy Equations (8) regardless of the values taken by the hydrodynamic forces. Similarly, the opposite case of a log free to move is obtained by setting $R_H = R_V = 0$.

By further neglecting the vertical flow velocities and accelerations *vis-à-vis* their horizontal counterparts, yields the following expressions for these forces:

$$f_D = C_D \frac{\rho_w}{2} DU^2 \qquad (9)$$

$$f_L = C_L \frac{\rho_w}{2} DU^2 \qquad (10)$$

$$f_M = C_M \rho_w \frac{\pi D^2}{4} \frac{dU}{dt} \qquad (11)$$

$$f_W = g \frac{\pi D^2}{4} (\rho_\ell - \rho_w) \qquad (12)$$

Equations (9)-(12) were evaluated for a log with average diameter $D = 0.40$ m and mass density $\rho_\ell = 600$ kg m^{-3}. These parameters yield a gap ratio $G/D = 0.125$, a submerged weight $f_W = -493.1$ N m^{-1}, and cylinder Reynolds-numbers in the range $R_e = 1.5 \times 10^{-5}$ to 7.2×10^{-5}. The steady drag coefficient

at depth was applied to transient conditions and computed from the *Achenbach and Heinicke*'s [1981] dataset (Figure 2A) using the regression fit $C_{D\infty} = 1.062 + 2 \times 10^{-6} R_e - 3 \times 10^{-12} R_e{}^2 + 2 \times 10^{-18} R_e{}^3$. Similarly, the datasets of *Rosko et al.* [1975] plotted in Figures 3B and 4B were approximated as $C_D = C_{D\infty} [1 - 0.35 \exp(-4G/D)]$ and $C_L = 0.5 \exp(-4G/D)$, respectively. Drag increase due to surface waves was computed by adding to C_D the values of C_w tabulated for $L/D = 15.9$ (Figure 7). The inertial coefficient was fixed at $C_M = 2.0$ based on the values reported by *Wright and Yamamoto* [1979] for cylinders near plane walls.

The drag, lift, and inertial forces per unit length of log computed for the displayed hydrograph period are plotted on Figure 11. The inertial force is plotted at 10^3 times its actual value to enable displaying its evolution. This figure reveals some interesting points. Inertia does not appear to have a significant influence on the transport of logs in ephemeral streams; however, it may still turn out to be important in the case of an assembly of logs (i.e., LWD jams) because of their large combined added mass. Lift is not negligible at small values of G/D contrary to common assumptions, and it can have considerable influence in the entrainment of logs situated close to the bed. Drag remains the main mobilizing force and, in the present example, the contribution of wave drag is evident in the brief increase in drag during the rising and falling stages.

4. CONCLUSIONS

A generalized framework is proposed to model entrainment and transport of single water-borne logs. In its more comprehensive form this modeling concept must be implemented in conjunction with computational stream evolution models. In addition to accounting for gravitational and mechanical-contact forces, the model incorporates all the hydrodynamic forces known to act on submerged bodies. Because of space limitations, the formulation is restricted to time-mean forces and the important issue of in-line and cross-current force fluctuations is not considered.

The degree to which a model of this type will agree with field tests depends on accurate characterization of all the hydrodynamic forces. A brief review of these forces and the available experimental data needed for their characterization is presented. Hydrodynamic coefficients obtained for steady flows can be used in unsteady streams because these coefficients experience short-lived fluctuations as the stream velocity varies. Some issues related to the characterization of hydrodynamic forces acting on logs in a fluvial environment still require attention. Among these are the effect of log roughness, orientation, and interaction with other logs and with the channel boundaries.

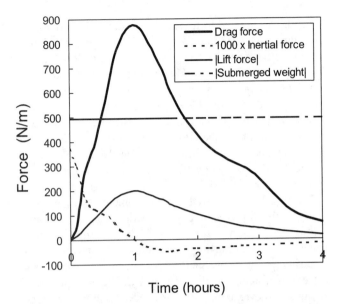

Figure 11. Comparison of theoretical drag, inertial, lift, and submerged-weight forces per unit length acting on an immobilized circular log during the storm event shown in Figure 10. $D = 0.40$ m, $G/D = 0.13$, $\rho_\ell = 600$ kg m^{-3}, $f_W = -493.1$ N m^{-1}.

The simple case of a stationary cylindrical log exposed to an actual stream flood event is used to highlight the relative influence of drag, lift, and inertial forces. This example shows that lift can influence the entrainment of logs sufficiently close to the bed. Drag is the main mobilizing force, and the inertia of single logs is not a significant factor.

Acknowledgements. The support of this work by the Agricultural Research Service of the U.S. Department of Agriculture is gratefully acknowledged. Reviews provided by C.A. Braudrick and F.D. Shields, Jr., were very helpful.

REFERENCES

Abbe, T. B., D. R. Montgomery, and C. Petroff, Design of stable in-channel wood debris structures for bank protection and habitat restoration: An example from the Cowlitz River, WA, in *Management of Landscapes Disturbed by Channel Incision: Stabilization, Rehabilitation, Restoration*, edited by S. S. Y. Wang, E. J. Langendoen, and F. D. Shields, Jr., pp. 809–815, University of Mississippi Press, University, 1997.

Abernathy, F. H., and R. E. Kronauer, The formation of vortex streets, *Journal of Fuid Mechanics*, 13, 1–20, 1962.

Achenbach, E., Influence of surface roughness on the cross-flow around a circular cylinder, *Journal of Fuid Mechanics*, 46, 321–335, 1971.

Achenbach, E., and E. Heinecke, On vortex shedding from smooth and rough cylinders in the range of Reynolds numbers 6 x 10^3 to 5 x 10^6, *Journal of Fluid Mechanics*, 109, 239–251, 1981.

Alonso, C. V., D. G. DeCoursey, S. N. Prasad, and A. J. Bowie, Field test of a distributed sediment yield model, in *Proceedings of the 26th ASCE Hydraulics Division Specialty Conference*, pp. 671–678, 1978.

Alonso, C. V., W. H. Neibling, and G. R. Foster, Estimating sediment transport capacity in watershed modeling, *Transactions of the ASAE*, 24(5), 1211–1220, and 1226, 1981.

Barenblatt, G. I., *Dimensional Analysis*, English translation by P. Maniken, 135 pp., Gordon and Breach Science Publishers, New York, 1987.

Bowie, A. J., G. C. Bolton, and J. A. Spraberry, Sediment yields related to characteristics of two adjacent watersheds, *Report ARS-S-40*, USDA Agricultural Research Service, 89–99, 1975.

Braudrick, C. A., and G. E. Grant, When do logs move in rivers? *Water Resources Research*, 36(2), 571–583, 2000.

Braudrick, C. A., G. E. Grant, Y. Ishikawa, and H. Ikeda, Dynamics of wood transport in streams: A flume experiment, *Earth Surface Processes and Landforms*, 22, 669–683, 1997.

Bursnall, W. J., and L. K. Loftin, Experimental investigation of the pressure distribution about a yawed circular cylinder in the critical Reynolds number range, *NASA Technical Note* 2463, 34 pp, 1951.

Cherry, J., and R. L. Beschta, Coarse woody debris and channel morphology: A flume study, *Water Resources Bulletin*, 25(5), 1031–1036, 1989.

Fage, A., and J. H. Warsap, The effects of turbulence and surface roughness on the drag of circular cylinders, *Aeronautical Research Council*, Reports & Memoranda No. 1283, London, 1930.

Flammer, G. H., J. P. Tullis, and E. S. Mason. Free surface, velocity gradient flow past hemisphere, *J. Hydraulics Division, ASCE*, 96(HY7), 1485–1502, 1970.

Gerrard, J. H., An experimental investigation of the oscillating lift and drag of a circular cylinder shedding turbulent vortices, *Journal of Fluid Mechanics*, 11, 244–256, 1961.

Göktun, S., *The drag and lift characteristics of a cylinder placed near a plane surface*, unpublished M.Sc. thesis, Naval Postgraduate School, Monterrey, California, 1975.

Harmon, M. E., J. F. Franklin, F. J. Swanson, P. Sollins, S. V. Gregory, J. D. Lattin, N. H. Anderson, S. P. Cline, N. G. Aumen, J. R. Sedell, G. W. Lienkaemper, K. Cromack, Jr., and K. W. Cummins, Ecology of coarse woody debris in temperate ecosystems, *Advances in Ecological Research*, 15, 133–302, 1986.

Hoerner, S. F., *Fluid-Dynamic Drag*, 3rd. Edition, published by the author, New Jersey, 1965.

Hygelund, B., and M. Manga, Field measurements of drag coefficients for large woody debris, *Geomorphology*, 51, 175–185, 2003.

Knoblock, F. D., and T. Troller, Tests on the effect of sidewind on the ground handling of airships, *Publication No. 2*, The Daniel Guggenheim Airship Institute, pp. 53–57, 1935.

Lamb, H., *Hydrodynamics*, Dover Publications, 1945.

Langendoen, E. J., CONCEPTS-Conservational Channel Evolution and Transport System, *Research Report No. 16*, USDA-ARS National Sedimentation Laboratory, 160 pp., 2000.

Lighthill, J., Fundamentals concerning wave loading on offshore structures, *Journal of Fluid Mechanics*, 173, 667–681, 1986.

Perry, A. E., M. S. Chong, and T. T. Lim, The vortex-shedding process behind two-dimensional bluff bodies, *Journal of Fluid Mechanics*, 116, 77–90, 1982.

Prandtl, L., and O. Tietjens, *Applied Hydro- & Aeromechanics*, vol. II, English translation by J.P. den Hartog, 311 pp., Dover Publications, 1934.

Roshko, A., A. Steinolfson, and V. Chattoorgoon, Flow forces on a cylinder near a wall or near another cylinder, in *Proceedings, 2nd U.S. Conference on Wind Engineering Research, Fort Collins, Colorado*, Paper IV-15, 1975.

Sainsbury, R. N., and D. King, The flow induced oscillation of marine structures, *Proceedings of the Institution of Civil Engineers*, 49, 269–301, 1971.

Sarpkaya, T., and C. J. Garrison, Vortex formation and resistance in unsteady flow, *Journal of Applied Mechanics*, 30, Series E, No.1, 16–24, 1963.

Schewe, G., On the force fluctuations acting on a circular cylinder in crossflow from subcritical up to transcritical Reynolds numbers, *Journal of Fluid Mechanics*, 133, 265–285, 1983.

Sheridan, J., J. C. Lin, and D. Rockwell, Flow past a cylinder close to a free surface, *Journal of Fluid Mechanics*, 330, 1-30, 1997.

So, R. M. C., and S. D. Savkar, Buffeting forces on rigid circular cylinders in cross flows, *Journal of Fluid Mechanics*, 105, 397–425, 1981.

Taneda, S., Experimental investigation of vortex streets, *J. Phys. Soc. Japan*, 20, 1714–1721, 1965.

USDA Soil Conservation Service, *SCS National Engineering Hanbook, Section 4: Hydrology*, Washington, D.C., 1985.

Wallerstein, N. P., C. V. Alonso, S. J. Bennett, and C. R. Thorne, Distorted Froude-scaled flume analysis of large woody debris, *Earth Surface Processes and Landforms*, 26, 1265–1283, 2001.

Wallerstein, N.P., C. V. Alonso, S. J. Bennett, and C. R. Thorne, Surface wave forces acting on submerged logs. *J. of Hydraulic Engineering*, 128(3), 349–353, 2002.

Wieselsberger, C., Versuche über den Luftwiderstand gerundeter und kantiger Körper, *Ergebnisse Aerodyn. Versuchsanstalt Göttingen*, edited by L. Prandtl, II. Lieferung, 23 pp., 1923.

Wigley, W. C. S., Water forces on submerged bodies in motion, *Trans. Inst. Naval Arch.*, 95, 268–279, 1953.

Wootton, L. R., M. H. Warner, and D. H. Cooper, Some aspects of the oscillations of full-scale piles, in *IUTAM-IAHR Symposium on Flow-Induced Structural Vibrations*, edited by E. Naudascher, pp. 587-601, Springer-Verlag, New York, 1974.

Wright, J. C., and T. Yamamoto, Wave forces on cylinders near plane boundaries, *J. Waterway, Port, Coastal and Ocean Div.*, 105, 1–13, 1979.

Young, W. J., Flume study of the hydraulic effects of large woody debris in lowland rivers, *Regulated Rivers: Research & Management*, 6, 203–211, 1991.

Zdravkovich, M. M., Review of flow interference between two cylinders in various arrangements, *Journal of Fluids Engineering*, 99, Series 1, 618–633, 1977.

Carlos V. Alonso, U.S. Department of Agriculture-Agricultural Research Service, National Sedimentation Laboratory, P.O. Box 1157, Oxford, MS, 38655.

The Role of Riparian Shrubs in Preventing Floodplain Unraveling along the Clark Fork of the Columbia River in the Deer Lodge Valley, Montana

J. Dungan Smith

U. S. Geological Survey, Boulder, Colorado

Intensive land use along streams in the semi-arid western United States has resulted in reduced riparian-shrub densities. Loss of shrubs in the meander belt of a steep, highly sinuous, single-threaded river can put the adjacent floodplain in jeopardy of catastrophically eroding (unraveling) during a large overbank-flow and, thereby, altering the single-threaded morphology to a multiple-threaded one. A new model for flow over floodplains covered with woody vegetation is used to evaluate (1) the characteristics of a nearly 300-year-recurrence-interval flood on Clark Fork of the Columbia River in the Deer Lodge Valley, Montana, in 1908 and (2) the present vulnerability to unraveling of this fluvial system. Reconstruction of the hydraulics of the 1908 flood indicates that a dense shrub community in the meander belt was responsible for protecting its single-threaded morphology throughout the Deer Lodge Valley and making possible a relatively thick deposit of mine tailings on its floodplain. In the decades since the flood, the metals-contaminated sediments deposited on the floodplain physiologically stressed and eventually killed much of the woody vegetation near the river. As a consequence, large portions of the meander belt of the Clark Fork through the Deer Lodge Valley have, at present, a very sparse shrub community. Calculations using the overbank-flow model indicate that, as a result of the current impoverished riparian flora, large segments of the meander belt will unravel when subjected to the boundary shear stresses produced by multi-decadal-recurrence-interval floods.

1. INTRODUCTION

As populations and consequent land use increase in the semi-arid western United States, the densities of shrubs on the floodplains of sinuous, single-threaded streams and small rivers in this part of the country are being seriously impacted. Pastures, grass with widely spaced trees, and gravel are intruding into and often replacing the original shrub carrs [*Mitsch and Gosselink*, 2000] and their understories of erosion resist-

Riparian Vegetation and Fluvial Geomorphology
Water Science and Application 8
This paper is not subject to U.S. copyright. Published in 2004 by the American Geophysical Union
10.1029/008WSA06

ant herbaceous plants. In addition to loss of the considerable form drag on the stems of the shrubs comprising the carrs, the new surfaces typically are much smoother than those beneath the original shrub carrs, thereby, producing both less resistance to the inevitable overbank flows that occur during multi-decadal-recurrence-interval floods and much higher local shear stresses on the floodplain sediments. As the riparian vegetation along these streams and small rivers is lost to development, the issue arises of how to determine the robustness of the impacted fluvial systems when subjected to inevitable large floods.

A process-based model for overbank flow through shrub carrs was developed recently in order to investigate the geomorphic vulnerability of fluvial systems to large floods [*Smith*, 2001; *Smith and Griffin*, 2002 and 2003; *Griffin and Smith*, this volume]. This model couples the hydraulics of flow in a sinuous channel with the hydraulics of flow over a grass-, shrub-, or tree-covered floodplain. The resistance of the topographic elements comprising the channel bed and the floodplain surface and that of the woody vegetation on the floodplain is calculated using drag equations with geometrically appropriate, independently determined drag coefficients and the average of the square of the velocity over the locations of the stems and branches of the plants.

A version of the above-mentioned floodplain model was previously applied to a headwater tributary of East Plum Creek, Colorado, during a large flood [*Griffin and Smith*, 2001] in order to test its accuracy [*Smith*, 2001; *Smith and Griffin*, 2003; *Griffin and Smith*, this volume]. In 1965, a large flood caused all of East Plum Creek and then Plum Creek to unravel. For one headwater tributary, post-flood aerial photographs showed the site of initiation of unraveling. The floodplain model was used successfully to predict the location and shrub density at which the reconstructed large flood began to erode the vegetated floodplain. This successful test of the model provides the foundation for its use in other situations. In this paper, the model is used to examine the hydraulics of the Clark Fork of the Columbia River in the Deer Lodge Valley, Montana, first to reconstruct a major flood that occurred in 1908 and then to evaluate the current vulnerability of that fluvial system to unraveling.

The primary advantages of this model for the present application are that (1) it treats the stems of the shrubs and tree trunks as fields of water-surface-penetrating cylinders, the number and mean diameter of which can be measured in the field

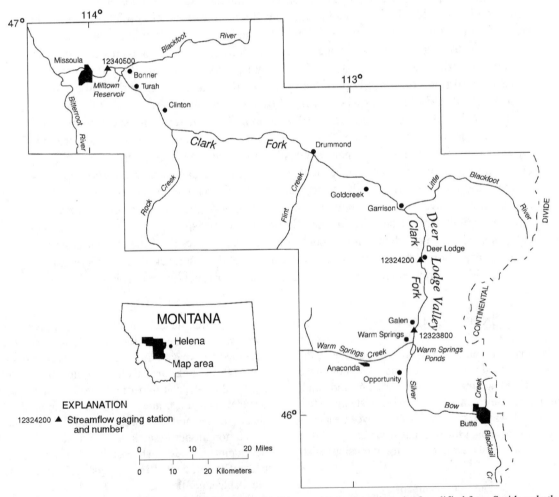

Figure 1. Location of the upper Clark Fork of the Columbia River and its major tributaries [modified from Smith and others, 1998]. The Clark Fork begins just below the Warm Springs Ponds, at the confluence of Silver Bow and Warm Springs Creeks, and flows northward through the Deer Lodge Valley to Garrison. The first major tributary of Clark Fork, the Little Blackfoot River, joins it at Garrison.

or estimated from botanical and land use information and provided as inputs to the overbank-flow model and (2) it needs no calibration because all parameters are specified a priori. Calculations using the model indicate that as the density of the woody riparian vegetation decreases, the flow velocity over the floodplain and the boundary shear stress on the underlying floodplain surface increase substantially [*Smith and Griffin*, 2002 and 2003]. Therefore, it is possible that for a floodplain along a stream or small river, a reduction in shrub density could lead to destabilization and unraveling during a large flood.

Catastrophic geomorphic transformation of a sinuous, single-threaded fluvial system to a multi-threaded one can have serious ecological consequences. By mixing coarse sediment from the original streambed with a large amount of much finer sediment from the floodplain, the median size and sorting of the sedimentary material on the bed of each thread decreases considerably, making the bed material mobile during much lower flows. This makes the benthic microhabitats unstable and, thus, uninhabitable for many of the insect larvae that lived in the stable pebble or cobble armor on the bed of the original single-threaded stream or river. Many shallower channels that are far from the vegetated banks of the system replace a single, relatively narrow and deep channel, typically shaded by shrubs along the banks. The result, over a prolonged distance, is a large increase in stream temperature. Also, with the loss of shrubs at the edges of the threads and the absence of significant stream-centerline curvature, there are few cut banks and little of the topographic complexity that is essential for healthy trout populations. The lack of cover and the shallow water make even warm-water fish easy targets for predatory birds. Finally, if there were contaminated materials on the floodplain surface before unraveling, then these materials would become incorporated into the active streambed and the braid bars during unraveling. This would become an especially serious problem if the contaminated materials were mobilized by oxidation during low-flow conditions, because movement of the sediment comprising the bars separating the threads would then permit dissolved metals to enter the river during floods.

The Clark Fork begins at the confluence of Warm Springs Creek and Silver Bow Creek near Warm Springs, Montana, and flows northward through the Deer Lodge Valley to Garrison (Figure 1). It then flows northwestward to Missoula, Montana. *Smith and Griffin* [2002] quote reports from the mid 1800's indicating that there were dense shrub carrs along the Clark Fork in the Deer Lodge Valley at that time, and they argue that these shrub carrs persisted, in large measure, into the 1900's. During the flood of record in 1908, the Clark Fork upstream of Garrison appears to have been a robust meandering river with a floodplain covered by large willow and water birch shrubs. The flood brought tailings down Silver

Bow Creek from the Butte mining district and into the Deer Lodge Valley. A thick deposit of mine tailings was left on the floodplain of the Clark Fork throughout the Deer Lodge Valley [*Smith and others*, 1998].

Tailings maps by *Nimick* [1990] and by *Schafer and Associates* [1997] have been used here to delineate the path, stage, and boundaries of the flood. Subsequent to the flood, the tailings deposits physiologically stressed the shrubs and initiated a thinning of the shrub communities that appears to have occurred over many subsequent decades and that has resulted in areas now devoid of vegetation, locally called slickens. Fortunately, beneath some of these barren areas, the stems and roots of the dead shrubs can still be found making it possible to confirm that the pre-1908-flood stem fields are similar in size and density to those found at the few nearby sites where there presently are small but dense stands of willows and water birch. Shrubs in both communities appear to have emanated from similar very large and possibly centuries old root balls.

The primary issues to be addressed with the floodplain model are: (1) Why did the 1908 flood not significantly alter the geomorphology of the Clark Fork in the Deer Lodge Valley? (2) How was it possible to deposit several decimeters of tailings over a large area during this five-day flood? (3) In light of the impoverished riparian carrs, is the meander belt of the Clark Fork in the Deer Lodge Valley protected from unraveling during large floods as it was in 1908?

2. METHODS

The theoretical approach used in this model is conceptually similar to that introduced by *Smith* [2001], who considered a floodplain that was populated by sandbar willow and confined by high terraces. The narrowness of the floodplain in that model application caused much deeper flows than are possible in the Clark Fork. In addition, sandbar willows bend over when subjected to deep overbank flows, making a multi-layer model essential. In contrast, the flows of concern in this report do not exceed two meters in depth and are through stands of much taller, rigid types of willows. Consequently, the three-layer model that was developed by *Smith* [2001] is unnecessary for the floodplain of the Clark Fork upstream of Garrison, and a simpler, one-layer model is used.

2.1 Floodplain Characteristics Affecting Overbank Flow in the Deer Lodge Valley

Throughout most of the Deer Lodge Valley, the Clark Fork is highly sinuous and can be represented by a sine-generated centerline trace [*Langbein and Leopold*, 1966; *Smith and McLean*, 1984] that oscillates back and forth across the axis of the valley. The sinuosity through this reach is 1.82, the average width of the river is 20.0 ± 2.9 m, the average width of the meander belt is 138 ± 56 m, the average meander wavelength is 508 ± 168 m, the average crossing angle for the sine-generated pattern is $77.6 \pm 21.3°$, and the average cross-tab distance in the down-valley direction is 139 ± 76 m [*Griffin and Smith*, 2002]. Therefore, the average floodplain tab width is 118m and the ratio of average down-valley to average cross-valley tab lengths is 1.2. The simple sine-generated pattern, the small ratio of down-valley to cross-valley tab lengths, the large areas of the protrusions of floodplain toward the point bars (denoted here as floodplain tabs) relative to those of the river at the tab margins, and the large crossing-angles imply that, except through partitioning of the total discharge between the river and the floodplain, the overbank flow down the meander belt is largely unaffected by the flow in the river. That is, the flow across the floodplain tabs is driven primarily by the specific weight of the water times the down-valley slope of the meander belt.

The willows and water birch on the floodplain of the Clark Fork in the Deer Lodge Valley have few branches that would be submerged by a one-to-two-meter overbank flow. Therefore, the shrub carrs would have interacted with the 1908 flow as approximately uniformly spaced clusters of stems, and each shrub or stem cluster would have interacted with the flow as a group of nearly vertical, surface-penetrating cylinders. In addition, the areas of the floodplain tabs are large relative to the horizontal scale area for the canopy of a single shrub, indicating that thousands of stem groups or shrubs would have been interacting with the flow on each floodplain tab in 1908. At 80% canopy coverage of the floodplain in the meander belt, there would have been more than 3000 shrubs with over 100,000 stems on an average tab. As a result, the detailed distributions of the stems in the shrubs or the shrubs on the floodplain are of secondary importance with respect to the overall flow characteristics.

At present, the low density of shrubs on the floodplain of the Clark Fork in the Deer Lodge Valley [*Griffin and Smith*, 2002] makes grouping of the shrubs more important from a hydrodynamic point of view. When shrub canopies cover less than 50% of the floodplain area and the shrubs are distributed across the floodplain in patches, the flow is retarded within the patches and is increased in the broad downstream-oriented paths between them. This results in a boundary-shear-stress field that is much greater in magnitude beneath the open-flow paths between the shrub patches than would occur on a floodplain for which the shrubs were uniformly or randomly distributed. Consequently, if the threshold shear stress for floodplain erosion were exceeded when the shrubs were uniformly distributed, then it also would be exceeded beneath the open flow paths on a floodplain for which the shrubs are in patches on the tabs. The issue of concern for the calculation is whether or not the present floodplain of the Clark Fork will unravel. The situation where the shrubs are uniformly distributed across the tabs provides a minimum estimate of the boundary shear stresses to which the surfaces beneath preferred flow paths across that tab are subjected without having to know the spatial configurations of the patches of shrubs. A uniform distribution of shrubs, therefore, is used in the model.

2.2 Fluid-Mechanical Considerations for a Shrub-Covered Floodplain Model

The drag force on the stem field per unit area of bed is the variable of primary fluid-mechanical importance. In the equation for fluid momentum, this reduces to the drag force on the average stem per unit volume of fluid. The volume of flow ($\lambda^2 h$) affected by each stem is the bed area affected by that stem (λ^2) times the flow depth (h). The drag force per unit volume acts like a hydrostatic pressure gradient opposing the flow and, thereby, reduces both the velocity and shear stress fields. The drag force is calculated by assuming the stems to be circular cylinders that extend from the floodplain surface through the free surface of the flow, and by using the local vertically averaged flow velocity that results from the calculation as the reference velocity in the drag equation.

The reference velocity in the drag equation is precisely defined and cannot be chosen arbitrarily. The expression for form drag at high Reynolds number can be derived from the Navier-Stokes equation, but in doing so a boundary-layer approximation is required so that Bernoulli's equation can be used to relate the local pressure field to the local, dominantly inviscid velocity field [*Batchelor*, 1967]. In perusing the derivation of the familiar equation for form drag in this manner, it becomes obvious that, for the drag coefficient to be constant in the Reynolds similarity regime, the square of the reference velocity in the drag equation must be the average square of the velocity evaluated on the surface that ultimately surrounds the object, but with the object removed. This is equivalent to the average over the cross-sectional area of the undisturbed-flow-velocity squared. If there is only one object in a uniform flow, then it can be represented by the square of the velocity at infinity, but if there are many obstacles in the domain of interest, then they all may affect the flow at the site of interest. This situation frequently arises in sediment transport problems, and it is the case for shrubs on a floodplain.

2.3 Equations for Flow Over a Simple Shrub-Covered Floodplain

The goal of the Clark Fork investigation has been to develop a process-based model that quantifies the dominant features associated with flow, sediment transport, and erosion or deposition on shrub-covered floodplains. This is done without incorporating in the model any parameters that are not fully specified for a wide variety of situations in the fluid-mechanical literature. In order to accomplish this goal for the complex fluid mechanical situations that exist on most floodplains, only what the author perceives to be the essential zero-order processes are included in the model. In particular the turbulence field is characterized in the simplest possible manner. The test of how well this goal has been accomplished lies in the accuracy with which the model predicts comprehensive field data pertaining to the large-scale response of a wide variety of floodplains to large floods.

Based on the above considerations, the approach taken in this paper is (1) to treat the shrubs in the meander belt of the Clark Fork in the Deer Lodge Valley as cylinders that extend throughout the water column, (2) to assume that the shrub carrs in this meander belt are statistically uniform over an area sufficiently large to smooth out the variations in shrub density within the carrs, (3) to assume that the flow is quasi-uniform in the horizontal over each floodplain tab, (4) to drive the overbank flow down the meander belt with the down-valley slope and (5) to treat the interaction between the floodplain and the channel by maintaining the same overbank flow depth for each. These criteria produce a reach averaged overbank-flow model. The flow in the channel also is calculated using a reach-averaged model, but in this case it driven by the channel centerline slope at bankfull flow and employs the flow depth that satisfies the discharge constraint.

Under these assumptions, the force balance per unit volume over the floodplain is

$$\frac{\partial \tau_{zx}}{\partial z} = \rho g(\sin \alpha) + \sigma_D \frac{\tau_b}{h} \qquad (1a)$$

where

$$\sigma_D = \frac{F_D}{\lambda^2 h \left(\frac{\tau_b}{h}\right)} = \frac{C_D}{2k^2} \left((\ln(\frac{h}{z_0}) - 0.74\right)^2 \frac{hD_S}{\lambda^2} \qquad (1b)$$

In equation (1a), τ_{zx} is the streamwise shear stress component on planes parallel to the average bed of the flow, z is the distance above the average bed, ρ is the density of the fluid, g is the acceleration due to gravity, and α is the slope of the bed relative to the horizontal. The conventional mathematical definition of slope is used here, so that a negative value of α gives a downward-sloping bed. In equation (1b), $\tau_b = (\tau_{zx})_b$ is the shear stress on the average bed, z = s is the equation for the free surface, z = η is the equation for the bed, h = s-η is the thickness of the water layer penetrated by the stems of shrubs, F_D is the drag force on the woody vegetation, λ^2 is the mean area of the bed affected by an average-sized stem of

diameter D_S, C_D is the drag coefficient for a stem or branch, k = 0.408 is von Karman's constant, and z_0 is the roughness parameter for the mean bed.

In equation (1b), the non-dimensional parameter $D_S h/\lambda^2$, the ratio of the cross-sectional area of the stem to the area of the bed affected by that stem, governs the relative importance of the drag on the floodplain shrubs. Therefore, floodplain tabs that have the same basic topography and the same cross-tab slopes and that have shrubs with the same values of the dimensional parameter D_S/λ^2 have the same net resistances to flow across them and will have the same flow velocities. For example, 10 mm diameter stems spaced 100 mm apart have the same affect on the flow as 40 mm diameter stems spaced 200 mm apart. The ratio of canopy diameter to the basal diameter of multi-stemmed shrubs is small compared to the ratio of canopy diameter to trunk diameter for trees, resulting in large values of D_S/λ^2. Canopy diameters for trees increase approximately linearly with trunk diameters, and trees typically are spaced not much closer than one canopy diameter. Therefore, D_S/λ^2 decreases inversely with canopy diameter for these large woody plants. This argument demonstrates that it is not the large riparian trees, such as cottonwoods, but their shrub understories that reduce the flow velocities and shear stresses in overbank flows through forested floodplains, thereby, protecting these forested floodplains from deep overbank flows produced by large floods. Small coniferous trees with numerous lower branches act more like shrubs.

Neither of the terms on the right side of (1a) are functions of z; therefore, integration of (1a) from z = 0 to the free surface where z = s and τ_{zx} = 0 yields an expression for τ_{zx} that is linear in z. Evaluating this expression for τ_{zx} at z = 0, where $\tau_{zx} = \tau_b$ and solving for τ_b gives

$$\tau_b = \frac{(-\rho gh(\sin \alpha))}{1 + \sigma_D}. \qquad (2)$$

Writing the linear expression for τ_{zx} in terms of τ_b then gives the familiar equation $\tau_{zx} = \tau_b(1-z/h)$. Also, τ_{zx} can be equated to the density of the water (ρ), times a scalar eddy viscosity (K), times the rate of shearing in the flow ($\partial u/\partial z$) because the flow is assumed to be steady and quasi uniform in the horizontal. This gives equation (3), which relates the boundary shear stress to the velocity.

$$\tau_{zx} = \tau_b(1 - \frac{z}{h}) = \rho K \frac{\partial u}{\partial z} \qquad (3)$$

Following *Smith* [2001], the kinematic eddy viscosity (K) can be written in two parts. Near the bottom of the flow, the turbulent eddies increase in scale with distance from the boundary [*Schlichting*, 1979], and dimensional analysis requires that the eddy viscosity vary linearly with z. Further from the boundary, the effect of the boundary on the turbulent eddies becomes small and the length scale for the turbulence becomes proportional to the flow depth. In this region, therefore, the eddy viscosity becomes independent of distance from the lower boundary. Using the method of *Rattray and Mitsuda* [1974], the two velocity profiles can be matched at 0.20h. It should be noted, however, that the analysis employed here gives a slightly different and somewhat more general result than that obtained by *Rattray and Mitsuda* [*Smith*, 1999]. The fully specified, two-part eddy viscosity, written in terms of the shear velocity ($u_* = \tau_b/\rho$), is as follows:

$$K = ku_* z(1 - \frac{z}{h}), \text{ when } z \leq z_m = 0.20h \quad (4a)$$

and

$$K = \frac{ku_* h}{\beta}, \text{ where } \beta = 6.24, \text{ when } z \geq z_m = 0.20h \quad (4b)$$

Substituting (4a) into (3) and integrating from $z = z_0$ to $z \leq z_m$ gives

$$u = \frac{u_*}{k}(\ln \frac{z}{z_0}), \qquad \text{for } z \leq z_m \quad (5a)$$

Similarly, substituting (4b) into (3) and integrating from z_m to $z \leq s$ gives

$$u = \frac{u_*}{k}\left(\beta\left(\xi - \frac{1}{2}\xi^2 \right) - \left(\xi_m - \frac{1}{2}\xi_m^2 \right) + \left(\ln \frac{\xi_m}{\xi_0} \right) \right) \quad (5b)$$

for $z \geq z_m$. In equation (5b), $\xi = z/h$, $\xi_m = z_m/h$, and $\xi_0 = z_0/h$.

The average of the velocity field given by (5), from z_0 to s, is

$$u = \frac{u_*}{k}((\ln \frac{h}{z_0}) - 0.74) \quad (6)$$

When $\ln(h/z_0) >> 1$, the square of the average velocity is approximately the average of the velocity squared. This approximation has been used together with (6) in (1b).

In equation (6), the shear velocity carries all of the information on what type of flow is being considered. If the velocity field represented by (6) were for a steady, horizontally uniform undisturbed flow, then the shear velocity would be obtained from the boundary shear stress calculated using the depth-slope product or determined by setting $\sigma_D = 0$ in equation (2). In contrast, for the overbank-flow problem of concern in this report, the appropriate shear velocity is obtained from the boundary shear stress given by equation (2) when $\sigma_D > 0$.

The process-based modeling approach taken in this paper requires information on the characteristics of the shrubs as input. For the yellow, Bebbs, and Geyer willows and the water birch on the floodplain of the Clark Fork in the Deer Lodge

Valley, the typical allometry of the shrubs was determined at several sites by *Smith and Griffin* [2002]. The characteristics of the shrubs all appear to scale either with shrub or stem diameter. For the willows and water birch along the Clark Fork upstream of Garrison (Figure 1), a typical shrub has a basal (D_G) diameter of about 0.5 m and contains about 36 stems. The average stem diameter is approximately 0.04 m. Average shrub height is about 130 times the average stem diameter. As mentioned above the ratio of canopy diameter to basal shrub diameter is about five, giving a typical canopy diameter of 2.5m.

2.4 Computation of Flow Near an Irregular Floodplain Surface

No floodplain has a perfectly planar surface and there is form drag on the small-scale topographic features of that irregular surface. The effectiveness of a long-crested topographic feature in producing form drag depends primarily on the ratio of its height to its streamwise spacing, and on its drag coefficient, which depends on the ratio of height to streamwise breadth [*Kean*, 1998]. Long, low topographic features, therefore, do not contribute much to the drag stress on the floodplain surface, whereas, short steep ones do. Consequently, a floodplain surface can be modeled effectively by treating it as a mildly distorted plane on which are located protrusions denoted topographic elements. The form drag on these topographic elements, once geometrically characterized, can be calculated using the method of *Smith and McLean* [1977], which was developed to determine the form drag on ripples and dunes. In a ripple or dune field, the streamwise scale of a topographic element is the same as its spacing, but on a floodplain, a topographic element has a height (H_t), a streamwise length scale (σ_t), and a spacing (λ_t). A simple geometric shape that can be used to represent these topographic elements is a long-crested Gaussian form. For these Gaussian topographic features the overall shear stress [τ_T, denoted the total shear stress by *Smith and McLean*, 1977] can be partitioned into a spatially averaged shear stress on the actual surface (τ_{SF}, denoted skin friction by *Smith and McLean*) and a shear stress in the reach averaged flow resulting from the form or pressure drag on the feature divided by the area of the floodplain affected by that feature (τ_D).

The equation derived by Smith and McLean still applies to this more general formulation of the stress-partitioning problem. That is,

$$\tau_T = \tau_{SF} + \tau_D = \tau_{SF} + \frac{1}{2}C_{Dt}\rho U_{rt}^2(\frac{H_t}{\lambda_t}) \quad (7)$$

Here, C_{Dt} is the drag coefficient for the topographic element, which varies with H_t/σ_t, and U_{rt} is the reference velocity for the form drag on the Gaussian element. Substituting the undisturbed velocity next to the boundary (which has a shear velocity of $(u_*)_{SF}$ and a roughness length of $(z_0)_{SF}$), for the reference velocity, averaging this velocity from z_0 to H_t, and rearranging gives

$$\frac{\tau_T}{\tau_{SF}} = 1 + (\frac{C_D}{2k^2})(\frac{H_t}{\lambda_t})(\ln(\frac{H_t}{(z_0)_{SF}}) - 1)^2 \qquad (8)$$

On floodplains with few shrubs, the natural topographic elements are associated with the activities of animals and with features produced by old flows, whereas, on shrub-covered floodplains the topographic elements associated with the woody plants dominate. In both cases, those features resulting from old floodplain channels typically parallel the old overbank-flow directions and do not produce much drag. In the case of the meander belt of the Clark Fork in the Deer Lodge Valley, the mounds beneath the shrubs have heights of $0.2D_G$, streamwise breadths about D_G, and streamwise spacings comparable to the streamwise spacings of the shrubs. In the direction perpendicular to the flow paths, barriers with long undulating crests can be used to approximate these topographic elements. Employing the measurements of *Hopson* [1999] and the theory of *Kean* [1998] for Gaussian shapes, the drag coefficient (C_{DM}) for these barriers is estimated to be about 0.8. In addition to the topographic elements that scale with D_G, a large background roughness was added. This was composed of two sets of topographic elements with drag coefficients of 0.8, heights of 0.050 m, and spacings of 0.50 and 4.0 m. This produces an exceptionally high roughness for floodplains with few or no shrubs and, thus, results in an underestimate of the skin friction and of the potential of the flow to erode the floodplain.

Writing the reciprocal of equation (8) for the mounds and defining the ratio τ_{SF}/τ_T as γ_R, gives

$$\gamma_R = (1 + (\frac{C_D}{2k^2})(\frac{H_t}{\lambda_t})(\ln(\frac{H_t}{(z_0)_{SF}}) - 1)^2)^{-1} \qquad (9)$$

The approach of *Smith and McLean* [1977] assumes that there are inner and an outer logarithmic profiles, scaled by $(u_*)_{SF} = (\tau_{SF}/\rho)^{1/2}$ and $(u_*)_T = (\tau_T/\rho)^{1/2}$ respectively and that they match at the average height (z_*) of the growing internal boundary layer. This matching procedure specifies the roughness parameter for the outer layer $(z_0)_T$. They use an equation by *Elliot* [1958] for the height of the internal boundary layer. It gives

$$z_* = a_1((z_0)_{SF})^{4/5}(\frac{\lambda_M}{(z_0)_{SF}}) \qquad (10)$$

and the matching yields

$$(z_0)_T = (z_0)_{SF}(\frac{z_*}{(z_0)_{SF}})^{\gamma_R} \qquad (11)$$

Together with $a_1 = 0.1$, $C_{DM} = 0.8$, $H_M = 0.30D_G$, and $\lambda_M = \lambda$, equations (7) through (9) completely specify the structures of both the internal and external boundary layers.

2.5 Computation of Flow Over the Riverbed

In order to determine the overall discharge of a flood, a channel discharge for the same flood stage must be added to the computed overbank-flow discharge. As mentioned above, a reach-averaged channel-flow model is used for this purpose. The overall channel roughness is caused by drag on the gravel bed, pebble clusters, incipient gravel dunes, and point bars. For the Clark Fork near Galen, the D_{50} and D_{84} of the bed surface are 30 and 54 mm respectively. The D_{50} and D_{84} of the bed sub-surface are approximately 20 and 38 mm. The skin friction z_0, therefore, is about 5.4 mm at present and would have been about 3.8 mm for the active surface during the 1908 flood [*Smith and Griffin*, 2002]. Currently, the pebble clusters are $2D_{84} = 0.11$ m high and have spacings of $40D_{84} = 2.1$ m. They are modeled using the same wavelength-to-height ratio and $3D_{84} = 0.11$ m for the more active surface during the 1908 flood. The incipient gravel dunes are modeled to have a height of $12D_{50} = 0.24$ m and a wavelength-to-height ratio of 50. The pebble clusters and the incipient dunes are modeled using the method of *Smith and McLean* [1977], with their published drag coefficient for round-crested dunes of 0.84. The point bars are modeled for the 1908 flood by the same method, using their stream-centerline half wavelength and their calculated height during the flood of 1.2 m.

2.6 Inputs for the Coupled River and Overbank Flow Calculations

The only area along the Clark Fork where the cross-floodplain topography is known accurately is in the vicinity of Perkins Lane Bridge (located a few meters north of the Galen gage shown on Fig. 1). Here a two-foot contour map for the present floodplain and the tailings-thickness maps of *Nimick* [1990] and *Shaffer and Associates* [1997] are used to determine both the present and the pre-1908 floodplain cross sections of the meander belt. The tailings-thickness maps are particularly accurate in this area, and they permit the margins of floodplain deposition and, thus, flow depth to be ascertained reasonably accurately for the 1908 flood. The bridge was not present in 1908. The maximum overbank flow depth next to the river, determined with the tailings present, is 1.34 m. Both the present and pre-1908 cross sections are shown in Figure 2. The pre-

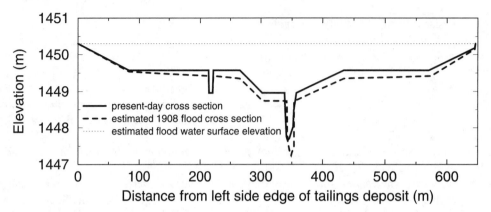

Figure 2. Cross-section of the floodplain of the Clark Fork near the Galen gage constructed from a map with a 2 ft (0.61m) contour interval. The upper limits of the tailings deposits on each side of the valley provide an accurate record of stage during the 1908 flood. The solid line indicates the present cross-floodplain topographic profile and the dashed line indicates the topographic profile as it would have been at the beginning of the 1908 flood, that is, with the tailings removed. Owing to the loss of riparian shrubs along the channel margins, the present channel is broader than would have been the case in 1908; therefore, a channel shape from a heavily vegetated site somewhat further upstream has been used to complete the pre-1908 cross section.

1908 floodplain vegetation, discussed above, appears to have been sufficiently dense to produce a narrower channel than the one near the Perkins Lane Bridge at present. Therefore, the pre-1908 channel cross section is thought to have been more like the one found in a heavily revegetated reach of the Clark Fork a short distance upstream. As a consequence, the 1908 channel shown in Figure 2 has been adjusted in width to match the one at this upstream reach. Owing to the disturbance of the present cross section by the bridge, the channel a short distance upstream of the bridge is used in the cross-section for the modern case.

Discharge for the 1908 flood was estimated to be 130 m³/s at the Perkins Lane site using the flow measured at a gage in Missoula and partitioning it by drainage basin area. This approach requires assuming that the rainfall was uniform over the entire drainage basin upstream of Missoula in 1908, which includes the Blackfoot River. The large area affected by the flood and the long duration of the event make the uniform rainfall assumption plausible as an average value for the five-day period. Combining the estimated flood discharge, the floodplain cross-section, and the maximum stage permits the coupled channel-floodplain hydraulic model to be solved for shrub density.

2.7 Model for the Tailings-Deposit Thickness Profile with Down-Valley Distance

During the five-day 1908 flood, an average of 0.32 m of metals-contaminated mine tailings was deposited within the meander belt where Perkins Lane now crosses the Clark Fork in an approximately cross-valley direction [*Schafer and Asso-*

ciates, 1997]. Two kilometers upstream, the deposit thickness in the meander belt exceeds 0.7 m, but within several kilometers of that location the deposit thickness decreases to 0.2 m as a consequence of the rapid settling of the coarse suspended sediment. The tailings thicknesses require the boundary shear stress on the floodplain surface during the flood to have been below the critical value for entrainment of silt. As a consequence, the local deposition rate during the flood was simply the local sediment concentration in the flow at Galen (C_s) times the settling velocity of the suspended material (w_s). The thickness of the deposit (Δz), therefore, is equal to $C_s w_s T$, where T is the duration of the flood.

The effective settling speed of the silt-sized tailings in 1908 is not known but can be determined from the deposit thickness a few kilometers downstream from Perkins Lane. The equilibrium, bed-supported suspended sediment concentration, calculated for the outlet of Silver Bow Creek during the 1908 flood, exceeds 12%, but that silt concentration includes coarse material that would have settled out in the first several kilometers of the Clark Fork. The mean concentration of suspended sediment that would have remained in the overbank flow throughout the Deer Lodge Valley is about 8%. The cross-stream-averaged deposit thickness at a location approximately two kilometers downstream of Perkins Lane is 0.16 m. Using this deposit thickness and a silt concentration of 8% with the five-day duration of the flood yields a settling velocity of 0.41 m/day for the material settling out of suspension in the Deer Lodge Valley.

The tailings deposition rate as a function of downstream distance (x) can be calculated by noting that the rate of decrease of sediment volume in the flow per unit increment of down-

Table 1. Calculated hydraulic characteristics of the 1908 flood as functions of shrub spacings for an overbank flow depth of 1.34 m next to the river. The discharge over the riverbed was 65.6 m^3/s.

Non-dimensional shrub spacing	Stem spacing (m)	Vertically-averaged velocity (m/s)	Overbank discharge (m^3/s)	Flood discharge (m^3/s)	Flood recurrence interval (years)
1.2	0.100	0.132	64.4	130	274
1.4	0.117	0.154	75.2	141	397
1.6	0.133	0.176	85.9	151	569
1.8	0.150	0.198	96.6	162	811
2.0	0.167	0.220	107	173	1146

stream distance following the flow equals the rate of sediment loss through deposition. That is,

$$\frac{d(C_s bh)}{dt} = U\frac{\partial(C_s bh)}{\partial x} = C_s w_s b \tag{12}$$

where U is the velocity averaged over the floodplain cross section, A = bh is the cross-sectional area of the overbank flow, b is the flow width, and h is the cross-stream averaged depth of the overbank flow. Therefore,

$$\frac{\partial(C_s bh)}{\partial x} = -(C_s b)\frac{w_s}{U} = -\frac{(C_s A)w_s}{hU} \tag{13}$$

In the dense shrub asymptote discussed in a subsequent section, U depends only on shrub spacing, and h depends on shrub spacing and cross-valley width. Assuming that these two variables vary slowly in the down-valley direction, (12) can be integrated to yield

$$C_s A = (C_s A)_0 \exp(-\frac{w_s x}{Uh}). \tag{14}$$

The predicted thickness of the tailings deposit on the floodplain downstream of Perkins Lane, therefore, is

$$\Delta z = (\Delta z)_0 (\frac{A_0}{A}) \exp(-\frac{(\Delta z)_0}{(C_s)_0 UhT}x) \tag{15}$$

where $(\Delta z)_0 = 0.32$ m.

2.8 Tailings Thickness Data

The maps of *Shafer and Associates* [1997] permit the width of the deposition zone, the thickness of the tailings in the meander belt, and the cross-stream averaged tailings thickness to be determined as functions of downstream distance. The

width of the deposit as a function of downstream distance (b = b(x)) is reasonably accurately resolved and is used here to calculate h = h(x) in (14). The tailings thickness is not as accurately resolved. The tailings thickness data were binned in classes that range from 0 to 3, 4 to 6, 7 to 12, and 13 to 24 inches (0 to 0.076, 0.102 to 0.152, 0.178 to 0.305, 0.330 to 0.610 meters). There is also a class for tailings thicknesses in excess of 25 inches (0.635 m), but there are data in this class only for the first 2.0 km along the channel of the Clark Fork. There are no data in this class downstream of the Perkins Lane Bridge. As a result of data inaccuracy and the consequent binning of the tailings thicknesses, the measured downstream profile of tailings thickness is very noisy. To smooth these data for comparison to the sediment deposition model, they were filtered with a 13-point running mean. The degree of smoothing was chosen to yield a wavelength structure comparable to that given by the deposition model.

3. RESULTS

3.1 Reconstruction of the 1908 Flood Near Galen

The shrub densities, floodplain velocities, overbank discharges, and flood recurrence intervals calculated for the 1908 flood are shown for flood discharges ranging from approximately 130 to 170 m^3/s in Table 1. The predicted shrubs densities for the meander belt from these calculations are exceedingly high (Table 1). They result in overbank velocities of only 0.13 to 0.22 m/s, and they certainly justify the assumptions imposed on the model (listed in section 2.3). The calculated close shrub spacings specifically permit the detailed hydraulics of the channel and floodplain to be decoupled. (See the dense shrub cases of Kean and Smith, this volume.)

The cross-sectional geometry and the stage for the Perkins Lane section are reasonably accurate for the 1908 flood because they are constrained by the tailings deposit. The dis-

Table 2. Calculated hydraulic characteristics for the 1908 flood as functions of stem and shrub spacings for a 1.14 m over-bank flow depth next to the river. Discharge over the riverbed was 57.5 m^3/s.

Non-dimensional shrub spacing	Stem spacing (m)	Vertically-averaged velocity (m/s)	Overbank discharge (m^3/s)	Flood discharge (m^3/s)	Flood recurrence interval (years)
1.8	0.150	0.198	71.3	129	261
2.0	0.167	0.220	79.2	137	344
2.2	0.183	0.242	87.0	145	450
2.4	0.200	0.264	94.9	152	586
2.6	0.217	0.286	103	160	760
2.8	0.233	0.307	111	168	981

charge reconstructed for this site probably is less accurate. It is possible that the discharge at the head of the Clark Fork was higher than given by the uniform rainfall assumption. The drainage basin of the Blackfoot River accounts for 38% of the total drainage area upstream of Missoula and is located somewhat to the north of the rest of the system. If it received only half as much rain as originally estimated, then the discharge in 1908 at the present location of Perkins Lane would be 30% higher or approximately 170m^3/s. Typically the Black-foot receives a somewhat larger fraction of total rainfall than does the Clark Fork during major floods, so this enhancement of the discharge near Galen should be considered an upper bound. Non-dimensional shrub spacings less than four result in essentially 100% canopy cover and a non-dimensional shrub spacing of two results in a very dense carr. Actual non-dimensional shrub spacings rarely are less than 2.0 over a wide area, and below 1.8 is unlikely. According to Table 1, it is possible to have had a discharge of 162 m^3/s and a non-dimension shrub spacing of 1.8 or a discharge of 172 m^3/s and a spacing of 2.0, but these values are at the margin of acceptability.

It is possible that the overbank flow depth next to the river, averaged over the duration of the flood, was somewhat below 1.34 m. Table 2 shows the hydraulic characteristics of the 1908 flood at the Perkins Lane location for an overbank flow depth 0.2m lower than the 1.34 m used to produce Table 1. This is considered to be an extreme reduction in overbank flow depth, but it does yield somewhat lower discharges and higher non-dimensional shrub spacings. Skin friction shear stresses for all of the 17 cases are an order of magnitude below the critical values for (1) erosion of the non-cohesive silt and sand on the floodplain and (2) resuspension of the tailings. For the cases with reasonable discharges and shrub spacings, the vertically averaged floodplain velocities range between 0.18 and 0.25 m/s. Summarizing the results, the canopy cover in 1908

was approximately 100% and the overbank flow was slowed by the shrubs to less than 0.25 m/s. The most probable scenario has a non-dimensional shrub spacing of about two, an over-bank flow depth next to the river of about 1.2 m, and a discharge of 147 m^3/s. The overbank velocity for this case is 0.22 m/s and the skin friction is 0.034 N/m^2.

3.2 Deposition of Sediment on the Floodplain During the 1908 Flood

The cross-valley average tailings thicknesses are presented as a function of down-valley distance from the origin of the Clark Fork in Figure 3. Included on this figure is the filtered down-valley tailings thickness profile. Figure 3 also shows a tailings thickness profile from (15), and the two-part floodplain velocity profile used in this equation. The magnitudes of the two velocity segments were both adjusted to maximize agreement with the filtered version of the empirical data. A two-part velocity profile was expected because the shrubs would not have been as dense downstream of Deer Lodge. The river is now bordered by cottonwoods from Deer Lodge to Missoula, and a cottonwood forest with a dense shrub understory probably bordered it in 1908. The increase in floodplain velocity is about as expected for going from a dense shrub carr to a dense shrub understory.

The parameter group $(C_s w_s T)_0$ controls the initial deposit thickness, whereas the parameter ratio (w_s/U) controls the deposit thickness profile. Although w_s is not known very accurately, $(Cs)_0$ and the duration of the flood (T) are sufficiently well known to preclude substantially larger values of U. The calculated deposit thickness profile yields a value of velocity for the floodplain near Galen (0.20 m/s) that falls within the range of overbank-flow velocities presented in the previous section, and it is very close to the value for the favored case (0.22

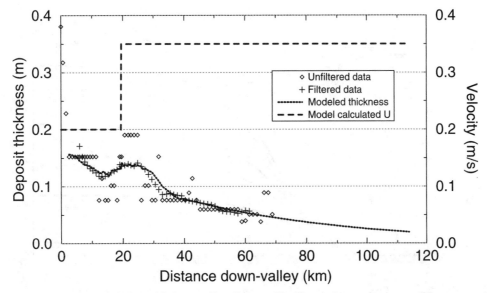

Figure 3. Tailings thickness variation with down-valley distance. The diamonds represent cross-valley averaged tailings thickness at each one-mile interval from the beginning of the Clark Fork to Milltown Reservoir, estimated using the data of Schafer and Associates, 1997. The plus signs represent a filtered version of these data using a 13-point running mean, while the heavy dotted line is the deposit thickness calculated using the two-part velocity profile (dashed line) in equation (15).

m/s). This calculation also confirms that the skin friction shear stress over the floodplain was well below the critical value for re-entrainment of the tailings throughout the flood. Consequently, it supports the conjecture that Clark Fork was densely vegetated with shrubs in the Deer Lodge Valley prior to the 1908 flood. This conclusion is in agreement with the mid-1800's reports of Warren Ferris and Father Jean-Pierre DeSmet [*Smith and Griffin*, 2002; *Smith and others*, 1998]. It also is in agreement with the few slickens beneath which there is good evidence of a dense pre-1908 shrub flora. The results of the model suggests that the reason the Clark Fork in the Deer Lodge Valley retained its meandering morphology through the 1908 flood is that its meander belt was a dense shrub carr.

3.3 Results of Calculations Pertaining to Erosion of the Clark Fork Floodplain

Over the many decades since the 1908 flood, the tailings in the meander belt of the Clark Fork have been killing the willows or forcing them into dormancy, and by the mid 20th century, the dead stems were breaking off. By the 1960's or 1970's, the floodplain in the meander belt of the Clark Fork was losing its protection against large floods. The question, therefore, arises as to whether or not the floodplain of the Clark Fork through the Deer Lodge Valley has become vulnerable to unraveling during floods with multi-decadal recurrence intervals. To address this issue, series of calculations using the

overbank-flow model were made for a range of discharges represented as flood recurrence intervals at the Galen gage. The calculations also were carried out for a wide range of shrub spacings.

Figure 4 shows the skin friction stress as a function of non-dimensional shrub spacing for flood recurrence intervals ranging from 5 to 300 years. Each profile has a similar shape and the curves converge to nearly the same function as the shrub flora becomes dense. At very large shrub spacings the boundary shear stress is controlled entirely by flow depth, whereas for very close shrub spacings it is controlled mostly by shrub density. For non-dimensional shrub spacings of less than eight (where the canopy cover is 20% and the boundary shear stress is approximately 1.7 N/m^2), the curves are essentially the same for recurrence intervals in excess of ten years, whereas, for non-dimensional shrub spacings of less than ten (where the canopy cover is 13% and the boundary shear stress is approximately 2.2 N/m^2), the curves are essentially the same for recurrence intervals in excess of 20 years. A non-dimensional shrub spacing of eight for the above defined characteristic shrub has a shrub spacing of 4.0 m and a mean spacing for uniformly distributed stems of 0.67 m. A non-dimensional shrub spacing of 10 has a shrub spacing of 5.0 m and a mean spacing for uniformly distributed stems of 0.83 m.

Also shown in Figure 4 are the critical shear stresses for the mineral soil on the Clark Fork floodplain (0.3 N/m^2) and for the more cohesive sediment comprising the slickens (1.0 N/m^2). The former was calculated by the method of Shields,

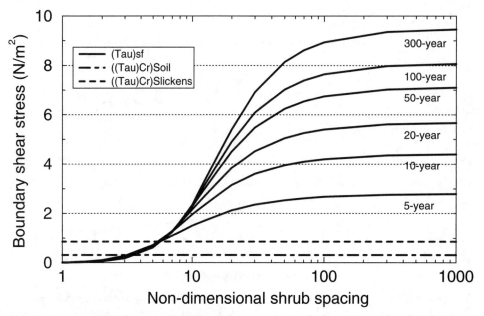

Figure 4. Skin friction as a function of non-dimensional shrub spacing for floods with recurrence intervals ranging from 5 to 300 years. Shrub spacing is non-dimensionalized by the mean shrub diameter in the meander belt. For non-dimensional shrub spacings less than about 8 the skin friction is essentially independent of flow depth. The lower (dot-dashed) horizontal line represents the critical shear stress for the floodplain soil, and the upper (dashed) line represents the critical shear stress for the tailings in the slickens.

as modified by *Wiberg and Smith* [1987] for poorly sorted sediment, and confirmed in the field during an overbank flow through sparse grass [*Smith and Griffin*, 2002]. The value for the tailings was measured in a small flume [*Smith and Griffin*, 2002]. The threshold shear stress for floodplain erosion along the Clark Fork in the Deer Lodge Valley, therefore, is less than 1.0 N/m^2 for bare soil, sparse grass, and the barren slickens. For non-irrigated grass on uncontaminated parts of the floodplain it can be assumed to be less than 2.0 N/m^2 [*Smith and Griffin*, 2003: *Griffin and Smith*, this volume]. Outside the asymptotic regime for which the flow properties are controlled primarily by shrub spacing and there is no significant dependence of boundary shear stress on discharge or flow depth, a boundary shear stress of 2.2 N/m^2 exceeds all three of the relevant threshold shear stresses for floodplain erosion. The boundary shear stress increases rapidly as the stem density decreases and flow depth becomes a more important hydraulic variable. To a first approximation, therefore, floodplain erosion begins at a non-dimensional shrub spacing of ten, when stage becomes an important hydraulic variable for floods with recurrence intervals greater than ten years. That is, once canopy covers drop below 13%, floodplain protection depends on having a thick, dense, continuous grass sod throughout the meander belt. This type of continuous sod is rare outside of wetlands and irrigated fields in the semi-arid western United States, and it does not exist at present on most of the floodplain

of the Clark Fork in the Deer Lodge Valley. According to *Griffin and Smith* [2002] about 20% of the floodplain tabs on the Clark Fork upstream of Deer Lodge (Figure 1) have less than 13% canopy cover.

In reality, much of the Clark Fork floodplain will erode at boundary shear stresses greater than 1.0 N/m^2, and this occurs at non-dimensional shrub spacings of 6 (shrub spacings of 3 m, stem spacings of 0.5 m, and canopy cover of 35%). *Griffin and Smith* [2002] have determined that about 60% of the 178-floodplain tabs on the Clark Fork upstream of Deer Lodge (Figure 1) have less than 35% canopy cover, 50% have less than 30% canopy cover, and 35% have less than 20% canopy cover. Accordingly 20% of the floodplain tabs will certainly erode during a prolonged flood with recurrence intervals of only 20-years or more, and 60% of the 178 tabs probably will erode during such an event. Furthermore, the erosion of only a few tabs will put the rest of the tabs in jeopardy of unraveling. (See *Smith and Griffin*, 2002 and 2003, and *Griffin and Smith*, this volume, for discussions of the mechanism by which floodplains unravel.)

Figure 5 is analogous to Figure 4 except that it is for vertically averaged overbank-flow velocity. In this case, the curves for floods of all recurrence intervals merge in the dense shrub asymptote, and the flow velocity becomes independent of flow depth. As increasing shrub density reduces the flow velocity, the flow depth (Figure 6) increases and the over-

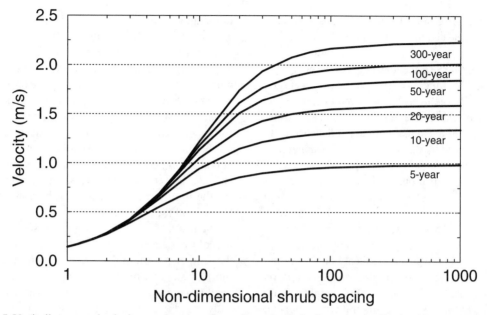

Figure 5. Vertically averaged velocity as a function of non-dimensional shrub spacing for floods with recurrence intervals ranging from 5 to 300 years. The six curves all increase monotonically with increasing non-dimensional spacing of the shrubs. They have the general form of the skin friction component of the boundary shear stress, but they are scaled by τ_B.

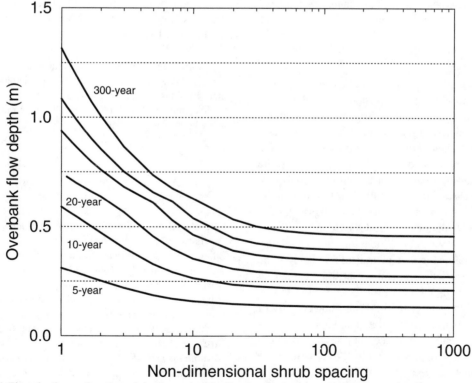

Figure 6. Flow depth as a function of shrub spacing for floods with recurrence intervals ranging from 5 years (bottom) to 300 years (top). The recurrence intervals increase monotonically and are the same as on Figures 4 and 5. The wiggles in the curves for the greater depths result from the cross-sectional structure of the floodplain, which increases in area in a monotonic but non-uniform manner near the Galen gage.

bank-flow discharge decreases. This occurs because the sum of the discharge over the channel and that over the floodplain must remain constant. As the shrubs cause the flow velocity over the floodplain to decrease, the flow depth of the entire system must increase. This increases the vertically averaged velocity and the discharge over the river and decreases the fraction of the total discharge over the floodplain in a manner controlled by the ratio of the total roughness of the channel to that of the floodplain and by the width of the floodplain relative to that of the river.

Of interest are the implications of these results to use of Manning's equation for overbank flows through trees and shrubs during floods. As seen in Figures 5 and 6, the overbank flow velocities, flow depths, and discharges become nearly independent of shrub density at non-dimensional shrub spacings exceeding 30. For relatively frequent floods this transition occurs at smaller shrub spacing because the drag on the stems is proportional to flow depth. On floodplains with lower stem densities than occur at this transition, normal hydraulic relations, such as Manning's equation, are applicable, but on floodplains with greater stem densities, depth dependent hydraulic relations are no longer valid. That is, if the flow depends on both stem spacing and depth or just on stem spacing, parameterizations, such as Manning's equation, that depend on flow depth are no longer valid.

4. SUMMARY AND CONCLUSIONS

Calculations using a cross section near the Galen gage and information from the 1908 tailings deposit in the neighborhood of that site show that a dense shrub community (100% canopy cover) must have existed in the meander belt of the Clark Fork in this part of the Deer Lodge Valley prior to 1908. This result is supported by the above-mentioned reports on the state of the river and its riparian zone from the mid-1800s, the depositional nature of the overbank flow throughout the meander belt, and the down-valley profile of tailings thickness. The sets of model calculations for the 1908 flood results in a reconstruction of the flow and sediment transport characteristics of the event that is highly constrained and likely to be accurate. The reconstruction indicates that the meandering morphology of the Clark Fork in the Deer Lodge Valley remained essentially unaltered by the 1908 flood because of the high density of riparian shrubs that populated the meander belt prior to the flood. Comparison of present and calculated pre-1908 shrub densities on the meander belt of the Clark Fork throughout the Deer Lodge Valley indicates that the riparian shrub community has been severely thinned.

Using the overbank flow model with a full range of shrub densities provides a comprehensive assessment of the role that vegetation plays in protecting the geomorphic integrity of this single-threaded, sinuous fluvial system and provides specific criteria for determining whether or not the floodplain can withstand a severe flood. The results suggest that the Clark Fork upstream of Deer Lodge can no longer retain its meandering morphology in the face of a prolonged flood with a multi-decadal recurrence interval. The floodplain within the meander belt will likely unravel during such a flood, transforming the river to one with large braided sections, an unstable bed, and many shallow threads. Associated with this alteration in geomorphic structure there will be major geochemical and ecological changes. River stability can be assured for the long term only if the shrub community is restored throughout the meander belt.

The results presented in this paper show that a significant loss of riparian shrubs can put a fluvial system in jeopardy in the semi-arid western United States, and suggest that the geomorphic stability of other heavily impacted riparian systems need to be carefully examined with modern fluid-mechanically based river modeling techniques. Of particular concern are the small rivers for which the geomorphology evolved when the riparian wetlands were being actively maintained by unfettered beaver activity. The relatively steep valley profiles and meandering geometries of these fluvial systems exist at present because dense shrub floras protected them during previous large floods. Sufficient protection may not exist during the next large flood.

Acknowledgements. The author has had many discussions concerning this work with his colleagues, particularly Jonathon Friedman, Ellie Griffin, and Jason Kean. Ned Andrews, Larry Benson, Carlos Amos, and Abdul Khan reviewed the manuscript and made many helpful suggestions.

REFERENCES

Batchelor, G. K., *An introduction to fluid dynamics*, Cambridge University Press, London, 615 pp, 1967.

Elliot, W. P., The growth of the atmospheric internal boundary layer, *Trans. Am. Geophys. Union*, 38, 1048–1055, 1958.

Griffin, E. R. and J. Dungan Smith, Computation of bankfull and flood-generated hydraulic geometries in East Plum Creek, Colorado in *Proceedings of the Seventh Federal Interagency Sedimentation Conference*, Reno, Nevada, vol. 1, section II, 50–56, 2001.

Griffin, E. R. and J. Dungan Smith, *State of floodplain vegetation within the meander belt of the Clark Fork of the Columbia River, Deer Lodge Valley, Montana*, U. S. Geological Survey Water-Resources Investigations Report 02-4109, 17pp, 2002.

Hopson, T. H., *The form drag of large natural vegetation along the banks of open channels*, unpublished Master's thesis, University of Colorado, Boulder, 114 pp, 1999.

Kean, J. W., *A model for form drag on channel banks*, unpublished Master's thesis, University of Colorado, Boulder, 56pp, 1998.

Langbein, W. B. and L. B .Leopold, *River meanders—theory of minimum variance*, U. S. Geological Survey Professional Paper 422-H, H1-H15, 1966.

Mitsch, W. J. and J. G. Gosselink, *Wetlands*, Third Edition, John Wiley and Sons, New York, 920pp, 2000.

Nimick, D. A., *Stratigraphy and chemistry of metal-contaminated flood plain sediments, upper Clark Fork River, Montana*, unpublished Master's thesis, University of Montana, Missoula, 118 pp, 1990.

Rattray, M., Jr. and E. Mitsuda, Theoretical analysis of conditions in a salt wedge, *Estuarine and Coastal Marine Sciences*, 2, 373–394, 1974.

Schafer and Associates, *Soil and tailings map of a portion of the Clark Fork River floodplain, Bozeman, Montana*, unpublished report prepared for ARCO, 56 pp, 1997.

Schlichting, H., *Boundary-layer theory*, Seventh Edition, McGraw-Hill Book Company, New York, 817pp, 1979.

Smith, J. Dungan, Flow and sediment transport in the Colorado River near National Canyon in *The Controlled Flood in Grand Canyon* edited by R. H. Webb, J. C. Schmidt, G. R. Marzolf, G. R. and R. A. Valdez, Geophysical Monograph 110, American Geophysical Union , Washington, 99–115, 1999.

Smith, J. Dungan, On quantifying the effects of riparian vegetation in stabilizing single threaded streams, *Proceedings of the Seventh Federal Interagency Sedimentation Conference*, Reno, Nevada, vol. 1, section IV, 22–29, 2001.

Smith, J. Dungan and E. R. Griffin, *Relation between geomorphic stability and the density of large shrubs on the floodplain of the Clark Fork of the Columbia River in the Deer Lodge Valley, Montana*, U. S. Geological Survey Water-Resources Investigations Report 02-4070, 25 pp, 2002.

Smith, J. Dungan and E. R. Griffin, *Quantitative analysis of catastrophic geomorphic transformation from a narrow, sinuous to a broad, straight creek*, U. S. Geological Survey Water-Resources Investigations Report 02-4065, 2003.

Smith, J.D., J. H. Lambing, D. A. Nimick, Charles Parrett, Michael Ramey, and William Schafer, *Geomorphology, flood-plain tailings, and metal transport in the Upper Clark Fork Valley, Montana*, U.S. Geological Survey Water-Resources Investigations Report 98-4170, 56pp, 1998.

Smith, J. Dungan and S. R. McLean, Spatially averaged flow over a wavy surface, *Journal of Geophysical Research*, 82, 1735–1746, 1977.

Smith, J. Dungan, and S. R. McLean, Flow in meandering streams, *Water Resources Research*, 20, 1301–1315, 1984.

Wiberg, P. L. and J. Dungan Smith, Calculations of the critical shear stress for motion of uniform and heterogeneous sediments, *Water Resources Research*, 23, 1471–1480, 1987.

J. Dungan Smith, U.S. Geological Survey, 3215 Marine Street, Suite E-127, Boulder, CO 80303

Spatial Pattern of Turbulence Kinetic Energy and Shear Stress in a Meander Bend with Large Woody Debris

Melinda D. Daniels

Department of Geography, University of Connecticut, Storrs, Connecticut

Bruce L. Rhoads

Department of Geography, University of Illinois, Urbana, Illinois

This paper explores the effect of a partial LWD dam on the spatial pattern of turbulence kinetic energy (TKE) and shear stress within a meander bend of a small stream in East Central Illinois, USA. Field data on three-dimensional velocity components were collected using an acoustic doppler velocimeter (ADV). Results show that upstream of the bend apex the highest values of TKE occur near the base of the outer bank. As flow approaches the LWD obstruction, which is located along the outer bank downstream of the bend apex, a zone of stagnant fluid develops along the outer bank, shifting the region of maximum TKE toward the center of the channel. Shearing of fluid between the stagnation zone and the adjacent streaming flow produces near-bed maximum values of TKE that are more than two times greater than those along the outer bank upstream of the bend apex. Downstream of the LWD obstruction, the zone of maximum TKE is located near the inner bank along a shear layer extending downstream from the margin of the obstruction. The net effect of the LWD obstruction is to locally increase maximum TKE values, but to shift the zone of maximum TKE away from the toe of the outer bank downstream of the bend apex—a location where rates of bank erosion usually are greatest in meander bends.

1. INTRODUCTION

Despite the increasing recognition that turbulence has a profound influence on the distributions of shear stress and sediment transport [*Sukhodolov et al.*, 1998], few field studies have investigated the properties of turbulence in river channels. The lack of such studies is largely due to historical limitations on measurement technology [*Clifford and French*, 1993]. Only recently have instruments been produced that are capable of measuring streamflow at high frequencies required for turbulence studies. Moreover, advances in measurement technology now allow collection of three-dimensional velocity data in field settings [*Lane et al.*, 1998]. The capacity to measure velocity components in three dimensions is critical in meander bends [*Frothingham and Rhoads*, 2003] and confluences [*Rhoads and Sukhodolov*, 2001] where flow patterns often are highly three-dimensional. The exploration of certain turbulence properties, such as turbulence kinetic energy, necessitates three-dimensional measurements because calculations of turbulence kinetic energy (TKE) require information on velocity fluctuations in three dimensions [*Sukhodolov and Rhoads*, 2001].

Research on woody debris in streams, while noting the potential of individual LWD obstructions to strongly influence flow structure, sediment transport, and patterns of erosion and deposition [*Keller and Swanson*, 1979; *Lisle*, 1981;

Riparian Vegetation and Fluvial Geomorphology
Water Science and Application 8

Robinson and Beschta, 1990; *Gregory et al.*, 1994; *Maser and Sedell*, 1994; *Assani and Petit*, 1995; *Fetherston et al.*, 1995; *Gippel*, 1995; *Gurnell et al.*, 1995; *McKenny et al.*, 1995; *Montgomery et al.*, 1995; *Richmond and Fausch*, 1995; *Abbe and Montgomery*, 1996; *Myers and Swanson*, 1997; *Gurnell and Sweet*, 1998], has yet to examine in detail the local effects of LWD on three-dimensional patterns of fluid motion. While field and laboratory studies have shown that LWD directly impacts the hydraulics of in-channel flows [*Gurnell and Sweet*, 1998], most investigations have examined the effects of large obstructions on mean downstream velocities [*Wallace and Benke*, 1984; *Young*, 1991] and on overbank flooding [*Young*, 1991; *Ehrman and Lamberti*, 1992; *Gippel*, 1995; *Gurnell and Gregory*, 1995; *Shields and Gipple*, 1995; *Gippel and Finlayson*, 1996]. Research on the influence of vegetation on turbulence in open channels has concentrated mainly on material with flexible stems, such as grasses [e.g. *Lopez and Garcia*, 2001, *Carollo et al.*, 2002], rather than on sizeable persistent obstacles, such as large woody debris.

Field studies of flow through meander bends have focused predominantly on two-dimensional patterns of mean downstream and cross-stream velocity components in relatively simple bends that do not contain major obstructions [*Hey and Thorne*, 1975; *Dietrich and Smith*, 1983; *Thorne and Rais*, 1984; *Thorne et al.*, 1985; *Markham and Thorne*, 1992]. In reality, some, if not many meander bends flanked by riparian forests are complicated by the occurrence of persistent obsta-cles composed of LWD and living vegetation. The strong influence of LWD on fluvial processes in meander bends was first demonstrated by *Thorne and Furbish* [1995], who found that dense vegetation and tree roots lowered velocities near the outer bank. Recent research based on field measurements of time-averaged downstream, cross-stream and vertical veloc-ity components has shown how LWD not only can increase flow resistance along the outer bank of a bend, but actively can steer flow away from this bank [*Daniels and Rhoads*, 2003]. While many studies have investigated turbulent flows in lab-oratory flumes [e.g. *Nezu and Nakagawa*, 1993; *Bennet and Best*, 1995] few studies have investigated three-dimensional turbulent flow in natural channels, particularly in meandering channels. Only a few studies have examined patterns of tur-bulence kinetic energy in meander bends [*Anwar*, 1986; *Blankaert and Graf*, 2001], and none of these investigations has considered the effects of large obstacles on patterns and intensities of turbulence.

This paper extends the work of *Daniels and Rhoads* [2003] by examining the influence of a LWD obstruction on spa-tial patterns of TKE within a meander bend along a small stream in East Central Illinois, USA. In particular, it explores how patterns of TKE are related to patterns of mean fluid motion and channel bed morphology. It also relates the pat-terns of TKE to the pattern of bed shear stress in the bend. The results illustrate how LWD can strongly affect patterns of turbulence in rivers.

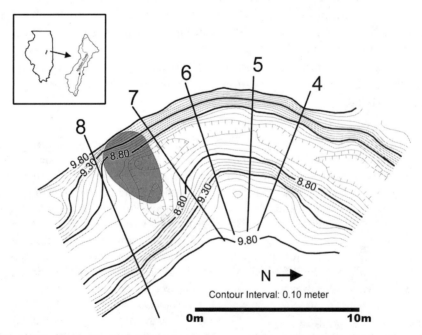

Figure 1. Location of the field site and detailed topographic map of the study site showing location of measurement cross sections (numbered lines) and location of LWD obstruction (shaded area) (arbitrary datum of 10.0 m) (adapted from Daniels and Rhoads, 2003).

Figure 2. Photograph of the study site (adapted from Daniels and Rhoads, 2003).

2. METHODS

2.1 Study site

The study site is a meander bend along Madden Creek—a headwater agricultural stream in East Central Illinois (Figures 1 and 2). This site has been the focus of research on the influence of LWD on the fluvial dynamics of agricultural streams in Illinois [*Daniels and Rhoads, 2003*]. Madden Creek drains a landscape shaped predominately by the Wisconsinan stage of Pleistocene glaciation, has a drainage area of 42-km², and local relief between the valley floor and adjacent uplands is about 25 to 30 m. Although many portions of Madden Creek have been cleared of vegetation and artificially straightened for the purpose of agricultural drainage, the study site lies within an approximately 2.5-km-long meandering section of Madden Creek with a narrow, forested riparian corridor.

The meander bend that constitutes the study site contains a complex LWD obstruction consisting of the trunk and exposed roots of a tree that leans at a low angle into the channel, trapping an accumulation of branches against its upstream side. The obstruction is firmly rooted in the outer bank and extends across the channel immediately downstream of the bend apex, occupying approximately one third of the channel width and extending from the bed of the channel to well above the bankfull channel elevation. As stage changes, the accumulated floating component of the debris obstruction shifts with the changing water surface, resulting in significant obstruction of the flow at all stages (Figures 1 and 2).

The channel of Madden Creek within the bend has an average bankfull width (W_b) of approximately 9 m, a bankfull depth of 1.2 to 1.4 m, and a dimensionless bend curvature (r/W_b), where r is radius of curvature, of approximately 2.0 [*Daniels and Rhoads, 2003*]. A point bar along the inner bank extends gradually into the channel, whereas the outer bank is nearly vertical. The thalweg, or zone of maximum depth, is generally located near the base of the outer bank, but turns abruptly toward the inner bank immediately upstream from the LWD obstruction. Bed material in Madden Creek consists of bimodal sand and gravel [*Daniels and Rhoads, 2003*].

2.2 Field Data Collection, Processing, and Analysis

Field data on 3-D velocity components were collected using an acoustic doppler velocimeter (ADV) on July 9, 1998 at five cross sections aligned orthogonally to the direction of the channel centerline (Figures 1 and 2). The flow stage during the period of measurements was about one-half of the bankfull depth and remained relatively constant, rising gradually by 0.03 m over the 8-10 hour measurement period [*Daniels and Rhoads, 2003*]. Point measurements of downstream (u), cross-stream (v), and vertical (w) velocities were obtained at several locations over the flow depth at seven to eight verticals along each cross section. Velocity components were measured in a < 0.25 cm³ sampling volume located ~ 5 cm away from the probe head at a rate of 25 Hz over an interval of 60 seconds.

Alignment of the sensor is a critical issue when obtaining field measurements of three-dimensional velocity components. The positioning system described by [*Rhoads and Sukhodolov* [2001] was used to ensure that velocities were measured within a fixed frame of reference at each cross section. This system involved mounting the sensor on a custom-built wading rod that in turn is attached to a steel tag line stretched tautly between the end points of the cross section. By using a level to plumb the wading rod, the sensor at each measurement location is aligned in a 3-D frame of reference with an X-axis perpendicular to the cross section, a Y-axis parallel to the cross section and a Z-axis coincident with the vertical plane of the cross section [*Daniels and Rhoads, 2003*]. This positioning system also precludes the need for an operator to stabilize the wading rod, eliminating the possibility of operator-induced unsteadiness during measurements. After properly aligning the wading rod, the operator moved a sufficient distance away from the ADV to avoid disruption of the flow field near the sensor (Figure 2). The entire process of 3D-flow data collection took approximately 2 hours per cross section.

Because the variances of the velocity components, which provide the basis for estimation of values of TKE, are sensitive to extremes in the data, every effort was made to eliminate high-magnitude errors from the time series. During collection of the velocity data occasional interruption of the acoustic signal by floating debris or intermittent intersection of the acoustic beam with the channel bed produced periods of excessive noise or large spikes in the velocity records. These situations were noted and the sampling duration was increased to obtain an uninterrupted 60-second record of clean data. Any remaining spikes resulting from acoustic noise were removed using a 3σ-filter, where σ is the standard deviation of the time series of velocities. The occurrence of such spikes

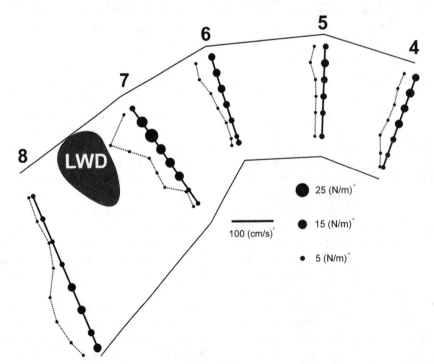

Figure 3. Spatial patterns of depth-averaged values of turbulence kinetic energy and estimates of near-bed shear stress. Distance of the dashed line from the cross section (solid line) indicates the level of turbulence kinetic energy. Estimates of near-bed shear stress are indicated by the size of the filled circles. The shaded area shows the location of the LWD obstruction.

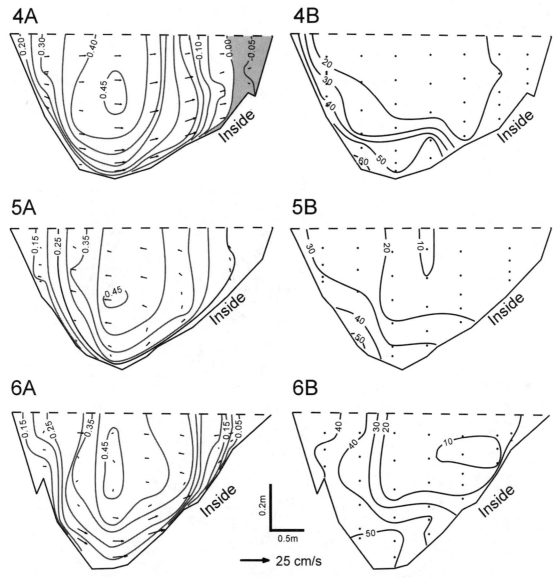

Figure 4. Patterns of mean velocity components (column A) and turbulence kinetic energy (column B) at each cross section (looking upstream). In plots of mean velocity components (column A), the downstream component is defined by contours, and the cross-stream/vertical velocity components are illustrated as vectors. Shaded areas indicate zones of separated or stagnant flow.

was rare. Most data sets contained no spikes and those that did had only one or two spikes. In all cases, application of the 3σ filter removed the spikes, but left the remainder of the data series unaltered.

The kinetic energy of turbulence, TKE, is calculated as

$$K = 0.5 \times (u'u' + v'v' + w'w')$$ (1)

where u′, w′ and v′ are fluctuations in the downstream, cross-stream and vertical velocity directions. Using the TKE method

[*Kim et al.*, 2000], values of near-bed shear stress were calculated as:

$$\tau = CK$$ (2)

where the proportionality constant C is assumed to be 0.21 and K is the kinetic energy of turbulence. The TKE method was selected for two reasons: 1) in a comparative analysis of four methods of estimating near-bed shear stress, *Kim et al.* [2000] found that the TKE method was the most consistent, and

Figure 4. *Continued*

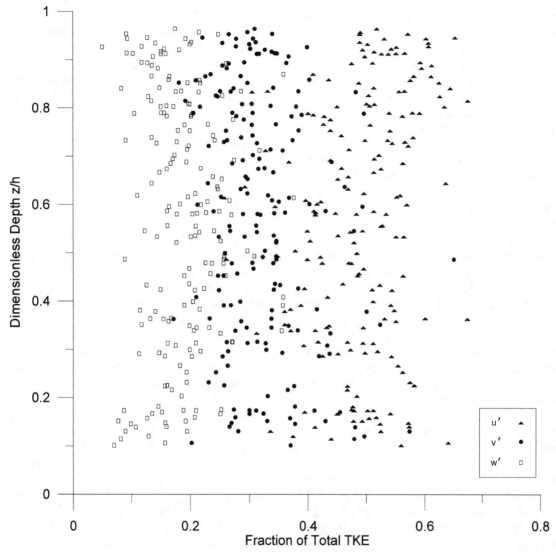

Figure 5. Contributions of u', v', and w' turbulent velocity components to turbulence kinetic energy (K) over flow depth for all data points.

2) because flow near the bed at Madden Creek is highly three-dimensional, the TKE method, which is not affected by extreme deviations between the path of the flow and the orientation of sampling cross sections, should give the best unbiased estimate of bed shear stress.

3. RESULTS

3.1 Spatial Patterns of Turbulence Kinetic Energy

Between the bend entrance (cross-section 4) and a position slightly downstream of the bend apex (cross-section 6) depth-averaged values of turbulence kinetic energy (K_{DA})

are fairly uniform across the channel, but exhibit a slight tendency to increase towards the outer bank (Figure 3). As the flow approaches the LWD obstruction (cross section 7), values of K_{DA} still increase towards the outer bank, but are consistently larger across the stream than at cross sections upstream. Moreover, the maximum value of K_{DA} is much larger than maximums upstream and is positioned immediately upstream of the LWD obstacle. Downstream of the obstacle (cross section 8), values of K_{DA} are greatest along the inner bank, where flow is directed past the woody debris. Values of K_{DA} are smallest in the center of the channel and increase slightly towards the outer bank in the lee of the LWD.

Distributions of point values of turbulence kinetic energy (K) within each cross section reveal a spatial pattern of K that changes dramatically as flow moves through the meander bend (Figure 4). At the bend entrance (cross-section 4), values of K increase towards the bed and banks, with the largest values positioned near the base of the outer bank. Near the bend apex (cross-sections 5 and 6), the intensity of turbulence along the face of the outer bank increases slightly, especially near the water surface. Immediately upstream of the LWD obstacle (cross-section 7), values of K increase dramatically in the outer third of the channel with maximum values aligned vertically from the bed to the surface over the thalweg. This region of vertically oriented high turbulence kinetic energy is indicative of a shear layer [*Sukhodolov and Rhoads*, 2001]—in this case between an area of stagnate fluid along the outer bank and the adjacent streaming flow. The largest values of K within the shear layer occur near the bed at the base of the thalweg. Downstream of the obstruction (cross section 8), a complex spatial pattern of K develops. The highest values of K are located toward the inner bank where turbulent eddies are shed off of the lateral margin of the LWD obstacle near the water surface. Turbulence values are lowest over the outer half of the channel in the lee of the obstacle where downstream velocities range from about 0.15 m s^{-1} to slightly less than 0.05 m s^{-1} (Figure 4).

3.2 Contributions of Velocity Components to Turbulence Kinetic Energy

The contributions of each velocity component to the total turbulence kinetic energy can be quantified as proportions (e.g. $(w'w')/2K$). Within the bend, contributions to K are: 1) u'=22-68%, 2) v'=18-68% and 3) w'=5-40% (Figure 5). Although the components separate roughly into three domains of $u'u'/2K > v'v'/2K > w'w'/2K$, these domains are not as distinct as those for straight sand-bed rivers [*Sukhodolov et al.*, 1998]. Considerable overlap occurs among the three domains due to relatively large ranges of the three proportions. In particular, maximum values of $v'v'/2K$ and $w'w'/2K$ are much greater than corresponding maximums for straight rivers. On the other hand, mean percentages of $u'u'/2K$, $v'v'/2K$, and $w'w'/2K$ of 49%, 32% and 19%, respectively, are consistent with mean percentages of $u'u'/2K \approx 50\%$, $v'v'/2K \approx 30\%$ and $w'w'/2K \approx 20\%$ for straight rivers [*Sukhodolov et al.*, 1998].

3.3 Near-Bed Shear Stress

Estimates of near-bed shear stress (τ_b) largely mirror the pattern of K_{DA} in the bend (Figures 3). From the bend entrance through the bend apex (cross-sections 4–6) values of τ_b are greatest close, but not immediately adjacent to, the outer bank. Immediately upstream of the LWD obstruction (cross-section 7) maximum values of τ_b increase dramatically as near-bed K is enhanced within the shear layer between the stagnation zone along the outer bank and the adjacent streaming flow. Moreover, values of τ_b within the center of the channel exceed maximum values at cross-sections farther upstream. Downstream of the LWD obstruction (cross section 8), values of τ_b are greatest along the inner bank, where flow is directed past the woody debris, and decrease towards the outer bank in the lee of the LWD.

4. DISCUSSION

The most pronounced influence of the LWD obstacle on the spatial pattern of turbulence kinetic energy in the bend is the dramatic increase in values of K immediately upstream of the LWD obstacle and the shift of maximum values of K toward the inner bank downstream of the obstacle. This pattern reflects the strong influence of the obstruction on the overall pattern of flow through the bend [*Daniels and Rhoads*, 2003]. The high values of K upstream of the obstacle are the result of enhancement of the outer-bank stagnation zone found in most meander bends [e.g. *Thorne and Rais*, 1984; *Markham and Thorne*, 1992] through partial damming of the flow by the woody debris. The adverse pressure gradient associated with super-elevation of the water-surface flow upstream of the obstacle causes the flow to locally stall. The adjacent streaming fluid moves around this local "mound" of water by turning abruptly toward the inner bank. High values of K are associated with a shear layer generated by the strong velocity contrast between the stagnant fluid and the streaming flow. Past work on 3-D flow in confluences has shown that lateral shear between two adjacent flows with different mean velocities dramatically increases values of K relative to flow outside the shear layer [*Sukhodolov and Rhoads*, 2001]. Deflection of the flow through the narrow opening adjacent to the LWD obstacle produces large values of K along the inner bank downstream of the obstacle. The largest values are associated with a shear layer between the accelerated flow moving through the opening and the slow-moving or re-circulating flow in the lee of the LWD obstacle.

Redistribution of turbulence energy from the outer to the inner bank is accomplished both by steering of the mean flow around the LWD obstacle and by advective redistribution of turbulence by helical motion of the mean flow. Movement of the flow around the obstacle not only directs high-velocity fluid toward the inner bank, but also locally increases curvature of the flow. The increase in curvature enhances counterclockwise helicity [*Daniels and Rhoads*, 2003], leading to inward advective transport of highly turbulent near-bed flow toward the inner bank upstream of the obstacle (see Figure 4, cross-section 7). Strong downwelling of high-velocity near-surface fluid toward the bed by helical motion probably also augments the production of turbulence kinetic energy by enhancing fluid shear near the bed.

The patterns of K observed in this study show both similarities and differences with patterns documented in past research on turbulence kinetic energy in unobstructed bends. *Anwar* [1986] found that values of K at a cross section near the apex of a bend along a small stream in the UK are greatest immediately above the bed over deepest part of the pool. The lowest values occur over the top of the point bar along the inner margin of the flow. This pattern is similar to patterns observed in this study at cross sections near the bend apex (cross-sections 4–6) upstream of the strong influence of the LWD obstruction. In contrast, a recent experimental study of turbulence in a laboratory bend with fixed vertical sidewalls and self-formed bed topography found that minimum values of K at the bend apex occur over the deepest part of the pool [*Blankaert and Graf*, 2001]. The largest values of K are located at the base and top of the inner sidewall and along the face of the outer sidewall. It is not immediately clear why patterns of K for the experimental study differ from those for *Anwar* [1986] or for this investigation. One possibility is that the presence of a vertical sidewall along the inner bank limits lateral redistribution of momentum toward the outer bank by bank-bed topographic effects, such as those associated with a fully developed point bar [*Dietrich and Smith*, 1983]. *Blankaert and Graf* [2001] did, however, document intermittent helical motion in an outer-bank stagnation zone—a feature similar to the stagnation zone along the outer bank at Madden Creek—and attributed the increase in K near the outer bank to lateral shear between this stagnation zone and fast-moving fluid in the center region of the flow—an effect similar to the lateral shear between the stagnation zone and streaming flow at Madden Creek.

The shear layer and intermittent helical motion may also account for the large degree of overlap in the contributions of different velocity components to K. Turbulence in vertically oriented shear layers generated by lateral shear is characterized by considerable overlap in domains of u' and v' [*Sukhodolov and Rhoads*, 2001], whereas helical motion, if intermittent, would involve pronounced vertical velocity fluctuations. The high degree of variability in the three components of K illustrates the strong three-dimensionality of turbulence in meander bends with LWD obstructions.

The patterns of near-bed shear stress suggest that the influence of the LWD obstruction on turbulence kinetic energy in the bend can be linked to the extant channel bed morphology and planform. The amplification of near-bed shear in the vicinity of the LWD obstruction and the shift of the zone of high bed shear stress from the outer to the inner bank are consistent both with the pattern of bed scour around the obstruction and with the widening of the channel downstream of the obstruction (Figure 1). The increase in bed shear stress from the bend apex to the area near the obstruction and decrease in bed shear stress downstream from the obstruction are conditions that should lead to sediment flux divergence, i.e. net scour, in the vicinity of the obstruction. Abrupt deflection of the flow toward the inner bank by the obstruction and redistribution of near-bed turbulence kinetic energy toward the inner bank by helical motion explains the S-shaped pattern of scour (Figure 1).

The marked shift of the zone of high bed shear stress toward the inner bank probably also accounts for the abrupt retreat of the inner bank and resultant channel widening as flow moves past the LWD obstruction (Figure 1). Redistribution of high near-bed K away from the outer bank and the shift of the zone of maximum near-bed shear stress toward the inner bank reduces the potential for erosion of the outer bank immediately downstream of the bend apex, the location where maximum rates of bank erosion typically occur in meander bends. The high degree of stability of the bend has been confirmed by examination of historical aerial photographs [*Daniels and Rhoads*, 2003] and is consistent with other studies of LWD-induced channel stabilization [*Lisle*, 1986; *Shields*, 2001]. The correlation between the observed patterns of K, near-bed shear stress and channel morphology suggest that the influence of the LWD remains persistent at higher, more effective, discharges than the flow sampled in this study. This permits the assumption that the configuration of the LWD (see Study Area description) in this bend would result in similar aggregate obstruction effects on the patterns of K and near-bed shear stress at higher stages, though actual effects may vary based on the buoyancy and permanence of the various components of the obstruction.

These results suggest that LWD obstructions along the outer banks of meander bends may provide a significant stabilizing effect on the outer bank—similar to that intended by man-made structures such as bend-way weirs. However, it is important to note that this study investigated a very stable LWD obstruction that was firmly attached to the outer bank. If the LWD obstruction were less permanent or positioned at a different location in the bend, the observed stabilizing effect may have been reduced or even absent. Furthermore, the LWD obstruction in this study appears to have initiated erosion and channel widening along the inner bank of the study bend. It is arguable that LWD structures hold great promise as bank stabilization measures, but more research is needed before they should be prescribed as remedies for bank instability.

5. CONCLUSIONS

This study has contributed to a process-based understanding of the characteristics of turbulence and shear stress in meander bends containing substantial woody vegetation obstructions. Results show that a LWD obstruction within a meander bend along Madden Creek in East Central Illinois has a pronounced influence on the structure of turbulence kinetic energy and shear stress within this bend. Major findings include the following:

(i) Depth averaged turbulence kinetic energy increases slightly towards the outer bank through the bend apex, but as flow approaches and passes the LWD obstruction, maximum values of depth averaged turbulence kinetic energy shift across the channel towards the inner bank

(ii) Patterns of turbulence kinetic energy upstream of the LWD are characterized by low values of K in the center and inner portions of the flow near the surface with values increasing towards the bed and outer bank. This pattern changes dramatically as flow encounters the LWD and a distinct shear layer develops near the outer bank along the edge of a pronounced near-bank stagnation zone. Also, as counterclockwise helical motion intensifies near the LWD, secondary circulation in the mean flow redistributes high levels of near-bed turbulence kinetic energy inward away from the outer bank.

(iii) Downstream of the LWD obstruction, a complex pattern of turbulence kinetic energy develops characterized by low values of K in the lee of the obstruction, high values of K along the inner bank where flow passes by the obstruction and maximum values of K in the turbulent wake shed from the margin of the obstruction.

(iv) Contributions of the three velocity components to turbulence kinetic energy vary widely, suggesting that turbulence in meander bends is highly three-dimensional.

(v) The influence of the LWD obstruction on K can be linked to the extant channel bed morphology and planform. This is particularly evident immediately downstream of the bend apex, where redistribution of high near-bed K away from the outer bank and the shift of the zone of maximum near-bed shear stress toward the inner bank reduces the potential for erosion of the outer bank.

Finally, it is important to recognize that the potential number of structural and spatial configurations of LWD and other types of natural biotic obstacles within diverse meander morphologies prevents generalization of the results of this particular field study beyond situations with very similar LWD scale and placement. Further research is needed to better isolate and quantify the effects of LWD obstructions on flow and morphology in meandering rivers. Both process-based field studies and hydrodynamic modeling studies of different configurations of vegetation and LWD at varying stages are needed to further quantify the range of effects that such obstructions can have on the nature of turbulence and near-bed shear stress in meander bends. In particular, longer-term studies or experimental LWD introductions are needed to address temporal issues not investigated by this research, such as the timescale required for a LWD obstruction to produce substantial hydraulic and geomorphic changes of similar magnitude to those observed in this study.

Acknowledgments. The authors are grateful to Kelly Frothingham, Marta Graves, Perry Cabot, Matt Ladewig and Kristin Jaburek for their assistance in collecting the field data. Funding for this research was provided by the U.S. Environmental Protection Agency, Water and Watersheds Program (R82-5306-010) and the University of Illinois Research Board. This manuscript benefited greatly from insightful reviews by Ellen Wohl and one anonymous reviewer.

REFERENCES

Abbe, T. B., and D. R. Montgomery, Large woody debris jams, channel hydraulics and habitat formation in large rivers, *Regulated Rivers: Research and Management*, 12, 201–221, 1996.

Anwar, H. O., Turbulent structure in a river bend, *Journal of Hydraulic Engineering*, 112 (8), 657-669, 1986.

Assani, A. A., and F. Petit, Log-jam effects on bed-load mobility from experiments conducted in a small gravel-bed forest ditch, *Catena*, 25, 117–126, 1995.

Bennett, S. J., T. Pirim, and B. D. Barkdoll, Using emergent vegetation to alter stream flow direction within a straight experimental channel, *Geomorphology*, 44, 115–126, 2002.

Blanckaert, K., and W. H. Graf, Mean flow and turbulence in open-channel bend, *Journal of Hydraulic Engineering*, 127 (10), 835–847, 2001.

Carrollo, F. G., V. Ferro, and D. Termini, Flow velocity measurements in vegetated channels, *Journal of Hydraulic Engineering*, 128 (7), 664–673, 2002.

Clifford, N. J., and J. R. French, Montoring and analysis of turbulence in geophysical boundaries: some analytical and conceptual issues, in *Turbulence: Perspectives on Flow and Sediment Transport*, edited by N. J. Cliffors, J. French, and J. Hardisty, pp. 93–120, John Wiley and Sons Ltd, Chichester, 1993.

Daniels, M. D., and B. L. Rhoads, Influence of a large woody debris obstruction on three-dimensional flow structure in a meander bend, *Geomorphology*, 51, 159–173, 2003.

Dietrich, W. E., and J. D. Smith, Influence of the point bar on flow through curved channels, *Water Resources Research*, 19 (5), 1173–1192, 1983.

Dietrich, W. E., Mechanics of flow and sediment transport in river bends, in *River Channel Environment and Process*, edited by K. Richards, pp. 179–227, Basil, Blackwell, 1987.

Ehrman, T. P., and G. A. Lamberti, Hydraulic and particulate matter retention in a 3rd-order Indiana stream, *Journal of the North American Benthological Society*, 11 (4), 341–349, 1992.

Fetherston, K. L., R. J. Naiman, and R. E. Bilbu, Large woody debris, physical processes, and riparian forest development in montane river networks of the Pacific Northwest, *Geomorphology*, 13, 133–144, 1995.

Frothingham, K. M., and B. L. Rhoads, Three-dimensional flow structure and channel change in an asymmetrical compound meander loop, Embarras River, Illinois, *Earth Surface Processes and Landforms*, in press, 2003.

Gippel, C. J., Environmental hydraulics of larger woody debris in streams and rivers, *Journal of Environmental Engineering*, 121 (5), 388–395, 1995.

Gippel, C. J., B. L. Finlayson, and I. C. O'Neill, Distribution and hydraulic significance of large woody debris in a lowland Australian river, *Hydrobiologia*, 318, 179–194, 1996.

Gregory, K. J., A. M. Gurnell, C. T. Hill, and S. Tooth, Stability of the pool-riffle sequence in changing river channels, *Regulated Rivers: Research and Management*, 9, 35–43, 1994.

Gurnell, A. M., K. J. Gregory, and G. E. Petts, The role of course woody debris in forest aquatic habitats: implications for management, *Aquatic Conservation: Marine and Freshwater Ecosystems*, 5, 143–166, 1995.

Gurnell, A. M., and R. Sweet, The distribution of large woody debris accumulations and pools in relation to woodland stream management in a small, low-gradient, stream, *Earth Surface Processes and Landforms*, 23, 1101–1121, 1998.

Keller, E. A., and F. J. Swanson, Effects of large organic material on channel form and fluvial processes, *Earth Surface Processes*, 4, 361–380, 1979.

Kim, S. C., C. T. Friedrichs, J. P. Y. Maa, and L. D. Wright, Estimating bottom stress in tidal boundary layer from acoustic doppler velocimeter data, *Journal of Hydraulic Engineering*, 126 (6), 399–406, 2000.

Lane, S. N., P. M. Biron, K. F. Bradbrook, J. B. Butler, J. H. Chandler, M. D. Crowell, S. J. McLelland, K. S. Richards, and A. G. Roy, Three-dimensional measurement of river channel flow processes using acoustic doppler velocimetry., *Earth Surface Processes and Landforms*, 23 (13), 1247–1267, 1998.

Lisle, T. E., Stabilization of a gravel channel by large stream-side obstructions and bedrock bends, Jacoby Creek, northwestern California, *Geological Society of America Bulletin*, 97 (8), 999–1011, 1986.

Lopez, F., and M. H. Garcia, Mean flow and turbulence structure of open-channel flow through non-emergent vegetation, *Journal of Hydraulic Engineering*, 127 (5), 392–402, 2001.

Markham, A. J., and C. R. Thorne, Geomorphology of gravel-bed river bends, in *Dynamics of Gravel-bed Rivers*, edited by P. Billi, R. D. Hey, C. R. Thorne, and P. Tacconi, pp. 433–457, John Wiley and Sons Ltd, New York, 1992.

Maser, C., and J. R. Sedell, *From the Forest to the Sea: The Ecology of Wood in Streams, Rivers, Estuaries, and Oceans*, 200 pp., St. Lucia Press, Delray Beach, 1994.

McKenny, R., R. B. Jacobson, and R. C. Wertheimer, Woody vegetation and channel morphogenesis in low-gradient, gravel-bed streams in the Ozark Plateaus, Missouri and Arkansas, *Geomorphology*, 13, 175–198, 1995.

Montgomery, D. R., G. E. Grant, and K. Sullivan, Watershed analysis as a framework for implementing ecosystem management, *Water Resources Bulletin*, 31 (2), 369–386, 1995.

Myers, T., and S. Swanson, Variability of pool characteristics with pool type and formative feature on small Great Basin rangeland streams, *Journal of Hydrology*, 201, 62–81, 1997.

Nezu, I., and H. Nakagawa, Three-dimensional structure of turbulence and the associated secondary currents in urban rivers, in *Environmental Hydraulics*, edited by Lee, and Cheung, pp. 379–384, Balkema, Rotterdam, 1991.

Rhoads, B. L., and A. N. Sukhodolov, Field investigation of three-dimensional flow structure at stream confluences: 1. Thermal mixing and time-averaged velocities, *Water Resources Research*, 37, 2393–2410, 2001.

Richmond, A. D., and K. D. Fausch, Characterisics and function of large woody debris in subalpine Rocky Mountain streams in northern Colorado, *Canadian Journal of Fisheries and Aquatic Sciences*, 52, 1789–1802, 1995.

Robinson, E. G., and R. L. Beschta, Course woody debris and channel morphology interactions for undisturbed streams in southeast Alaska, U.S.A., *Earth Surface Processes and Landforms*, 15, 149–156, 1990.

Shields, F. D. J., and C. J. Gippel, Prediction of effects of woody debris removal on flow resistance, *Journal of Hydraulic Engineering*, 121 (4), 341–354, 1995.

Shields, D., N. Morin, and R. A. Kuhnle, Effect of large woody debris on stream hydraulics, in *Conference on Wetland Engineering and River Restoration*, American Society of Civil Engineers, Reno, Nevada, 2001.

Sukhodolov, A., M. Thiele, and H. Bungartz, Turbulence structure in a river reach with sand bed, *Water Resources Research*, 34 (5), 1317–1334, 1998.

Sukhodolov, A. N., and B. L. Rhoads, Field investigation of three-dimensional flow structure at stream confluences, *Water Resources Research*, 37 (9), 2411–2424, 2001.

Thorne, C. R., L. W. Zevenbergen, J. C. Pitlick, S. Rais, J. B. Bradley, and P. Y. Julien, Direct measurements of secondary currents in a meandering sand-bed river, *Nature*, 315, 746-747, 1985.

Thorne, S. D., and D. J. Furbish, Influences of course bank roughness on flow within a sharply curved river bend, *Geomorphology*, 12, 241–257, 1995.

Wallace, J. B., and A. C. Benke, Quantification of wood habitat in subtropical coastal plain streams, *Canadian Journal of Fisheries and Aquatic Sciences*, 41 (11), 1643–1652, 1984.

Young, W. J., Flume study of the hydraulic effects of large woody debris in lowland rivers, *Regulated Rivers: Research and Management*, 6, 203–211, 1991.

Patterns of Wood and Sediment Storage Along Debris-flow Impacted Headwater Channels in Old-Growth and Industrial Forests of the Western Olympic Mountains, Washington

Jeremy T. Bunn and David R. Montgomery

Department of Earth and Space Sciences, University of Washington, Seattle, Washington.

We investigated the effect of hillslope forest conditions and in-channel large woody debris (LWD) on channel sediment storage and sediment transport by debris flows, using a combination of aerial-photograph interpretation and field surveys to compare the characteristics of debris-flow tracks in old-growth and industrial forests of the western Olympic Peninsula. Debris-flow initiation sites are more than four times as common in the industrial forest as in the old-growth, and debris-flow density is three times greater in the industrial forest. Along recent debris-flow tracks in both forest types over 75% of retained sediment is in contiguous deposits upstream of LWD. The volume of sediment and wood in old-growth channels is 5 to 11 times greater than in industrial-forest channels. The difference in sediment retention leads to a greater proportion of exposed bedrock in industrial-forest debris-flow tracks. In old-growth forest, large-volume sediment deposits were common in even the steepest surveyed reaches. Most of the sediment-retaining LWD in both forest types is of a diameter greater than is likely to be provided by forests that are clear-cut in short rotation. Short-rotation clearing of forest from the hillsides of headwater basins and removal of old-growth LWD from headwater channels can be expected to result in thinner hillslope soils and less sediment storage in headwater channels, leading to an industrial-forest landscape in which sediment output to higher-order channels is more tightly coupled to the rate of sediment production on hillslopes than it is under old-growth forest.

1. INTRODUCTION

In comparison to the large number of studies of sediment transport and reach-scale morphology in low-gradient alluvial channels, relatively little work has been done in steep mountain channels [*Wohl*, 2000]. Even less work has focused on debris-flow-prone headwater streams. This relatively neglected portion of channel networks is nonetheless significant; it includes most of the total channel length in a channel network and most of the drainage area in mountain drainage basins [*Sidle et al.*, 2000]. Headwater channels (and their adjacent hillslopes) are also the source of much of the sediment and woody debris that enters the fluvial system. Consequently there is growing interest in the physical and biological processes that occur in headwater channels [e.g., *May*, 2001, 2002; *Sidle et al.*, 2000; *Gomi et al.*, 2002].

The processes affecting headwater streams are linked to processes affecting adjacent hillslopes [*Church*, 2002; *Gomi et al.*, 2002]. Sediment transport and channel geometry in headwater streams are dominated by landslide-triggered debris flows rather than by fluvial transport [*Swanson and Swanston*, 1977; *Benda*, 1990; *Seidl and Dietrich*, 1992; *Gomi et al.*,

Riparian Vegetation and Fluvial Geomorphology
Water Science and Application 8
Copyright 2004 by the American Geophysical Union
10.1029/008WSA08

2002], and the frequency and magnitude of landsliding are strongly influenced by hillslope vegetation [*Sidle et al.*, 1985]. Although much of the early research on debris flows was undertaken in industrial (i.e., clear cut) forests, recent work has begun to document the ways that debris flow activity differs between old-growth and industrial forests.

Several studies have demonstrated that landslides and/or debris flows are more common or frequent in industrial forests than in old-growth forests. *Morrison* [1975, as cited in *May*, 2002] found debris flow frequency to be 8.8 times higher in clear-cuts than in forested areas. Various studies cited by *Johnson et al.* [2000] found a two to fourfold increase in landslide frequency associated with timber harvest. *Snyder* [2000] found three times more debris flow initiation sites in timber plantations than in old-growth forests. *May* [2002] found landslide density to be four times higher in second growth forest and ten times higher in clear-cuts than in old-growth forest.

Riparian and adjacent hillslope forests are the source of large woody debris (LWD), which is known to affect reach-scale morphology and hydrological processes in alluvial channels [*Fetherston et al.*, 1995; *Montgomery et al.*, 1995; *Abbe and Montgomery*, 1996, 2003]. Some studies have found that there are more pieces of LWD in old-growth streams than in

those flowing through clear-cut forests [*Murphy and Koski*, 1989; *Bilby and Ward*, 1991; *McHenry et al.*, 1998]. Others have found that there is more woody debris in logged streams [*Froehlich*, 1973; *Gomi et al.*, 2001] and one study found no significant difference in number of pieces of LWD between harvested and unharvested streams [*Ralph et al.*, 1994]. Although it has been widely reported that debris flows scour the channels through which they travel, transporting sediment and woody debris to distinct deposition zones [e.g., *Gomi et al.*, 2002], log jams are known to retain sediment in both low gradient and steep headwater channels [*Perkins*, 1989; *O'Connor*, 1994; *Montgomery et al.*, 1996], and it has been hypothesized that incorporation of LWD into the leading edge of debris flows may cause deposition in steep portions of the channel network [*Montgomery and Buffington*, 1998; *Lancaster et al.*, 2001]. *May* [1998] found that runout distance tends to be greater for debris flows that originate in or travel through clear-cut industrial forest in the Oregon Coast Range, perhaps reflecting LWD-forced deposition closer to debris-flow-initiation sites in old-growth forest.

Differences in hillslope forest condition and in-channel LWD between old-growth and industrial forests may lead to differences in landscape-scale debris flow activity and reach-

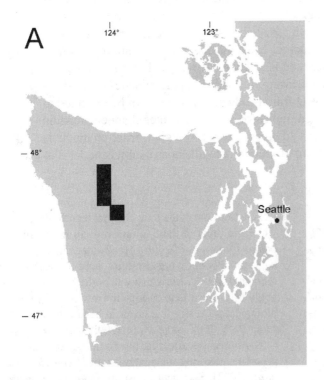
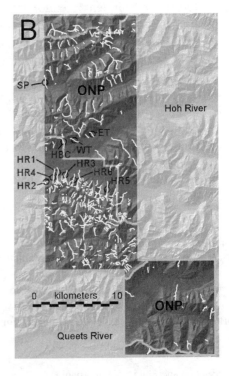

Figure 1. (A) Map showing the location of the study area on the western side of the Olympic peninsula. (B) Shaded relief map of the study area. Areas labeled ONP are within the boundaries of Olympic National Park. White lines within the study area are mapped debris flow tracks. Field survey channels are labeled with their abbreviated name (refer to Table 2).

scale morphology, sediment transport, and sediment storage. Recent studies of headwater channels in the Oregon Coast Range and British Columbia [*May*, 1998, 2002; *Johnson et al.*, 2000; *Gomi et al.*, 2001] provide some of the relatively few data sets available that address the effects of forest type and LWD on debris flow processes and the nature of headwater channels in the Pacific Northwest. The objectives of our research were a) to complement earlier efforts by extending the geographical range of existing studies, and b) to propose causal explanation(s) for observed differences in debris-flow processes between old-growth and industrial forest. Here we report the results of an aerial-photograph- and field-survey-based investigation into the effect of hillslope and riparian forests and in-channel LWD on sediment storage and transport by debris flows in steep (gradient > 10%) headwater channels of the western Olympic Mountains, Washington.

2. METHODS

2.1. Study Area

Located on the western flank of the Olympic Mountains, the study area (Figure 1) is characterized by east-west trending ridges with steep, forested hillslopes that descend to wide glacially carved valleys. Like much of the forested mountain terrain of the Pacific Northwest, these hillslopes are subject to erosion by mass wasting [*K. Schlichte*, Aerial photo interpretation of the failure history of the Huelsdonk Ridge/Hoh River area, unpublished report, Washington State Department of Natural Resources, 1991]. Study area boundaries were chosen to straddle the border between 1) old-growth forest within Olympic National Park (ONP) and 2) Washington State Department of Natural Resources (WADNR) and Forest Service lands that have been subject to timber harvest. Drainage basins on either side of the ONP boundary are similar with respect to parameters such as precipitation, channel network topology, drainage area, elevation, relief and geological substrate that are likely to affect landsliding and debris flows [*Selby*, 1993]. Annual precipitation ranges from 4 m to 5 m and falls primarily as rain between September and June [*Heusser*, 1974]. Elevation of the ridgelines ranges from 650 m to 1250 m, while main valley bottoms range from 120 m to 440 m. Local relief between ridgeline and valley bottom is typically between 650 and 700 m, but ranges from 500 m to 850 m. The valleys of the Western Olympic Mountains have experienced repeated glaciations during the Quaternary [*Heusser*, 1974] and have a pinnate drainage pattern.

The drainage areas of the headwater basins in our study area range from 0.5 to 8.3 km^2. These small basins contain first- through third-order streams that flow from ridges underlain by folded and faulted marine sandstone, siltstone, shale, and conglomerate [*Tabor and Cady*, 1978] down onto ele-

vated terraces along the mainstem rivers. During fieldwork we observed minor variation in lithology both within and between basins, but did not observe any consistent relationship between local variations in bedrock type and variations in channel form. Hillslope soils are gravelly loams, typically 0.5 to 1.5 m deep [*K. Schlichte*, op. cit.].

Old-growth forests in the region consist of western hemlock, Douglas-fir, western red cedar, Sitka spruce, and various understory species [*Edmonds*, 1998]. The industrial forests that we visited consist almost entirely of even aged stands of Douglas-fir, with a dense understory of devil's club, huckleberry and vine maple wherever the conifer canopy is thin. Digital elevation model (DEM) analysis and field observations indicate that slopes range from less than 1% on the valley bottoms to nearly vertical along the inner gorges of some creeks. Most of the primary forest has been harvested from the private, state, and tribal lands of the Olympic peninsula [*Peterson et al.*, 1997], but old-growth forest has been preserved within the boundaries of Olympic National Park. Industrial forestry in the western Olympic Mountains has thus set up a large-scale perturbation experiment, in which basins within the park serve as controls, and the similarities in lithology, slope, weather, and climate allow comparison between headwater channels in industrial and old-growth forests.

2.2. Aerial Photograph Survey

We identified slope failures and debris flow tracks on aerial photograph stereopairs and mapped them to digital orthophotos using ArcView [ESRI, 1999]. The original photographs are 1:32,000 scale, and were viewed through 3x magnification. Photographs of the entire study area at this resolution were only available from the most recent WADNR survey (OL-QT-00, flown in 2000), so we limited our survey to this one set. We mapped debris flow tracks at three defined confidence levels. "High" confidence level required that we observed a visible head scarp, open canopy along the presumed runout path, and a visible deposition lobe. "Medium-confidence" required an open canopy track with either a visible head scarp or deposition lobe. Where we observed only an open canopy track we classified the suspected debris flow track as "low-confidence." Forty percent of the mapped debris flow tracks were high-confidence, 50% were medium-confidence, and 10% were low-confidence. Over 95% of the features we mapped were ≥ 50 m long.

For the purpose of investigating landscape-scale differences in debris-flow activity, we defined the area inside ONP as "old-growth forest" and the area outside of the park as "industrial forest." Although individual drainage basins in the industrial-forest area have remnant old-growth stands and/or stream-side buffers, we consider this variation at the channel

Table 1. Landscape-Scale Parameter Definitions.

Parameter	Definition
R_P	length of the longest single debris-flow track in a basin
R_C	total length of all debris-flow tracks in a basin
D_{DF}	total length of debris-flow tracks divided by the area over which they were summed
$D_{DF}*$	ratio of D_{DF} to drainage density
I_{DF}	number of debris-flow initiation sites divided by the area over which they were summed

scale, and retain the industrial-forest classification at the landscape scale. We compared the industrial-forest and old-growth portions of the study area with respect to five parameters: Primary Runout (R_P), the length of the longest single debris-flow track; Cumulative Runout (R_C), the total length of debris-flow tracks in a basin; Debris-flow Density (D_{DF}), the total length of debris-flow tracks divided by the area over which they were summed; dimensionless Debris-flow Density ($D_{DF}*$), the ratio of D_{DF} to drainage density, and Debris-flow Initiation (I_{DF}), the number of debris-flow-initiating scarps divided by the area over which they were summed. To facilitate calculation of $D_{DF}*$ we used the FLOW module in IDRISI [*Clark Labs*, 2002] to derive a drainage network from a 10-meter-grid DEM of the study area, created by compositing individual DEMs from U.S. Geological Survey 7.5' quadrangles. We set the drainage area for channel initiation to 25,000 m² to minimize the generation of DEM-artifact channels [*Jonathan Stock*, personal communication]. To control for different proportions of upland in the old-growth and industrial-forest parts of the study area, we subtracted the mainstem valley

floors from the map area used to calculate D_{DF}, $D_{DF}*$, and I_{DF}. For reference, parameter definitions are summarized in Table 1.

2.3. Field Survey

We performed detailed field surveys of a subset of the mapped debris flows, selected on the basis of "high confidence" identification in our aerial photograph mapping, and accessibility from road or trail. We surveyed four debris flows inside the park and one outside the park that traveled through old-growth forest, four outside the park that traveled through industrial forest, and one outside the park that traversed both old-growth and industrial stands (Table 2). At the channel scale we determined forest type by the character of the riparian forest as observed in the field. Thus, a drainage basin in the landscape-scale "industrial forest" portion of the study area might contain a debris-flow track that was defined as "old-growth" for channel-scale analysis because the surveyed reaches were in a remnant old-growth patch or buffer. This scale-dependent classification is appropriate because the data collected in the field survey reflect conditions in the channel and its immediate vicinity, rather than the overall character of the landscape in which the channel is located.

We used stadia rod, 100 m fiberglass tape, Abney level, laser rangefinder, and Brunton compass to map and survey the long profile of each debris flow track. We started our surveys at the downstream end of debris-flow runout, where there was either loss of valley confinement or channel slope $\cong 10\%$, along with clear evidence of terminal debris flow deposition (the development of a distinct fan). Surveys continued upstream until we reached the initiating scarp or an impass-

Table 2. Characteristics of Surveyed Debris Flow Tracks.

Creek	Abbreviation	Basin Area (km²)	Last Debris-Flow	Surveyed Length (m)	Forest Type: Initiation	Forest Type: Runout
East Twin	ET	0.45	1997 [a]	565	Old-growth	Old-growth
Spawner	SP	1.24	2002 [b]	1480	Old-growth	Old-growth
West Twin	WT	1.07	1997 [a]	1139	Old-growth	Old-growth
Hoot	HR 5	1.23	1971-1990 [c]	907	Industrial	Old-growth
(unnamed)	HBC	1.26	unknown	612	Industrial	Old-growth
Iron Maiden	HR 1	0.58	1981-1990 [c]	1596	Industrial	Both
(unnamed)	HR 2	1.02	unknown	1193	Industrial	Industrial
Washout	HR 3	1.03	1981-1990 [c]	932	Industrial	Industrial
Dinky W.	HR 4	0.59	1971-1980 [c]	721	Industrial	Industrial
H-1070	HR 6	0.76	1981-1990 [c]	725	Industrial	Industrial

a. [*Bill Baccus*, personal communication]
b. inferred from posted date of trail washout
c. [*K. Schlichte*, 1991, op cit.]

able waterfall. At 5 to 20 m intervals (depending on morphological complexity) along the length of each debris flow track we recorded downstream bearing, elevation change, bed width, bed material, sediment depth (where it could be observed), the composition of sideslope cover within 5 m vertical distance of the channel bed, the volume and type of sideslope sediment deposits, and the diameter, length, and alignment with respect to channel centerline of functioning large woody debris (LWD_F), defined as any piece of wood that was either shielding or damming a distinct volume of sediment that formed a step ≥ 0.5 m above the adjacent bed.

We characterized bed material as bedrock or some combination of gravel (5 mm to 75 mm), cobbles (75 mm to 300 mm), boulders (> 300 mm), and/or LWD. Wood was considered to be a bed material when a noticeable proportion of the bed surface consisted of partially buried LWD such that a full bed-width discharge would flow over it. We classified sideslope cover as bedrock, consolidated regolith, loose colluvium, or vegetation. To record the volume of sideslope sediment deposits we sketched simple abstractions of their shapes and measured the corresponding dimensions with a laser rangefinder. Our method of sampling LWD_F was not exhaustive: where there were multiple pieces of LWD in a step-forming jam we measured only those that were in our judgment acting as key members, and we did not measure LWD that was retaining volumes of sediment smaller than about 2 m³, so we may have underestimated the contribution of small pieces. We are confident, however, that we adequately sampled the LWD that was directly retaining sediment in the surveyed channels.

To estimate the volume of sediment stored in the channel we interpolated the bedrock surface between exposures to approximate sediment depth at each station, assumed a parabolic cross-sectional form, and calculated the sediment volume between stations as

$$V = \frac{L}{3}\left(W_0 D_0 + \tfrac{1}{2}\Delta W D_0 + \tfrac{1}{2}\Delta D W_0 + \tfrac{1}{3}\Delta W \Delta D\right) \quad (1)$$

where V is sediment volume, L the horizontal distance between stations, W_0 is the bed width at the initial station, D_0 sediment depth at the initial station, ΔW is the change in bed width between stations, and ΔD the change in sediment depth between stations.

Volumes calculated using (1) are intermediate between those calculated using triangular and rectangular cross-sections and are more consistent with the observed parabolic cross-sectional shape of bedrock channels in the field area. Interpolation of the bedrock surface in places resulted in sediment depth ≤ 0 m for a channel segment where sediment was observed to be present. For these locations we substituted a depth of 0.4 m. This value is both consistent with sediment depths reported for Pacific Northwest channels by *Benda*

[1988] and *May* [1998] and at the low end of the range of positive sediment depths calculated for the surveyed channels. The interested reader may refer to *Bunn* [2003] for the derivation of (1) and further detail regarding profile and volume calculation.

We normalized the calculated sediment volume per unit length of channel to generate longitudinal sediment-volume profiles for each surveyed creek. To investigate reach-scale properties we divided each channel into reaches of approximately 100 m length. The actual length of defined reaches varied between 74 and 138 m, since survey stations did not always fall on 100 m increments. For each reach we calculated the overall slope (the change in elevation divided by horizontal distance along the valley centerline), the proportion of the length that was exposed bedrock, and the sediment volume normalized to unit lengths of 100 m.

We estimate that reach slopes are accurate to ±1%, since the vertical sighting error at a station almost never exceeded 10 cm and there were on average 10 stations per 100 m. The accuracy of our sediment-volume data is limited by the necessity of interpolation between bedrock exposures along the channel profile. It is possible that by imposing a constant slope underneath sediment deposits we over or underestimated sediment volumes by a factor of two in reaches under-

Table 3. Landscape-Scale Debris-Flow Track Parameters.

	Industrial Forest	Old-Growth	*p value*
Landscape Characteristics			
Upland Area (km²)	144	244	
Drainage Density (km/km²)	3.02	2.79	
Debris Flow Density			
Primary D_{DF} (km/km²)	0.55	0.30	
Cumulative D_{DF} (km/km²)	1.37	0.44	
D_{DF}* (dimensionless)	0.45	0.16	
High Confidence D_{DF}*	0.18	0.06	
Debris Flow Initiation			
I_{DF} (no./km²)	3.89	0.82	
High Confidence I_{DF}	1.06	0.24	
Primary Runout Distance			
Median R_P (km)	0.99	0.87	
Mean R_P (km)	1.11	1.02	*0.49*
Standard Deviation (km)	0.65	0.79	*0.11*
N	68	68	
Cumulative Runout Distance			
Median R_C (km)	2.16	1.08	
Mean R_C (km)	2.80	1.48	*0.0001*
Standard Deviation (km)	2.49	1.25	*< 0.0001*
N	68	68	

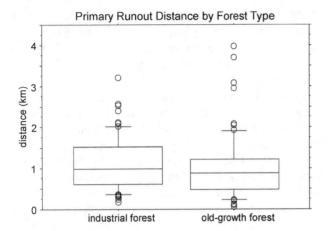

Figure 2. Primary runout distance distributions. The median primary runout is 0.99 km for industrial forest (N = 68) and 0.87 km for old-growth (N = 68). The distributions are positively skewed; the mean primary runout is 1.11 km for industrial forest and 1.02 km for old-growth.

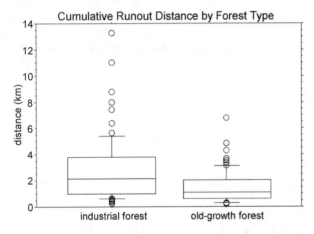

Figure 3. Cumulative runout distance distributions. The median cumulative runout is 2.16 km for industrial forest (N = 68) and 1.08 km for old-growth (N = 68). The distributions are positively skewed; the mean cumulative runout is 2.80 km for industrial forest and 1.48 km for old-growth.

lain by bedrock steps [see *Bunn* 2003 for details]. Any errors are systematic across all sampled creeks, however, so relative comparison should be unaffected by them.

2.4. Statistical Methods

We used the two-sided t-test for differences in means and two-sided F-test for differences in variances. Variables were log-transformed where necessary to obtain approximately normal distributions for hypothesis testing. To guard against pseudo-replication we compared the distribution of reach-scale parameters within each creek to the distribution across all creeks in its forest type. We found no significant channel effect and so considered each reach an independent sample for hypothesis testing.

3. RESULTS

3.1. Landscape Scale

In our study area, debris flows were more common in the industrial forests outside of ONP than they were in the old-growth forests inside its borders (see Figure 1). This is represented by differences in I_{DF} and D_{DF}, which are recorded in Table 3. Industrial forest I_{DF} is > 4 times greater than old-growth I_{DF}, whether calculated using all source areas or only those of high-confidence debris flow tracks. Primary D_{DF} is 1.8 times greater in the industrial forest than in the old-growth. D_{DF} calculated on the basis of cumulative runout length is 3.1 times higher in the industrial forest. Whether calculated on the basis of all mapped debris flows or only high confidence debris flows, D_{DF}^* is approximately three times higher in the industrial forest than in the old growth.

Most creeks identified as debris flow tracks on the aerial photographs were not field checked, but of nine creeks in old-growth forest that were identified as debris flow tracks in the aerial photograph analysis and visited in the field, only four were found to have the characteristics of catastrophic debris flow disturbance (e.g., vegetation scoured from the channel and sideslopes, fresh deposits of poorly-sorted angular clasts jumbled up with woody debris, a terminal debris fan). The other five creeks lacked evidence of recent debris flow activity, and were misinterpreted as debris flow tracks in the aerial-photograph survey due to their relatively open canopies and the presence of local side-slope failures along their lengths. All of the industrial forest channels that were visited in the field were proper debris flow tracks, and based on ridge-top visual reconnaissance we estimate that most (if not all) of the industrial forest channels identified as debris flow tracks on air photos have indeed experienced debris flow disturbance. Hence, we may have underestimated the differences between forest classes with respect to D_{DF} and I_{DF}, but we certainly have not overestimated it.

Although debris flow tracks are more common in the industrial forest than in the old-growth, individual debris flows do not appear to travel farther along channels in the industrial forest. The distribution of R_P is similar for the industrial forest and old-growth areas (Figure 2), and although the distribution is shifted towards lower values for old-growth forest, mean R_P in the old growth is not significantly lower than that in the industrial forest (Table 3). In contrast, the difference in R_C distribution between the old-growth and industrial forests

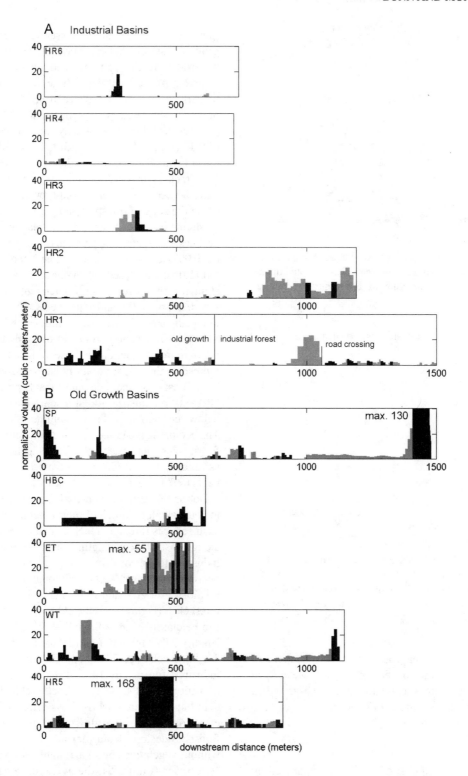

Figure 4. Sediment volume profiles. Black areas represent mixed sediment and LWD deposits, grey areas represent sediment only. Sediment volumes are normalized by length, so area in the figure is proportional to volume in the field. (A) Industrial forest channels and HR1, which travels through old-growth forest in its upper reaches and industrial forest downstream of 650 m. (B) Old-growth channels.

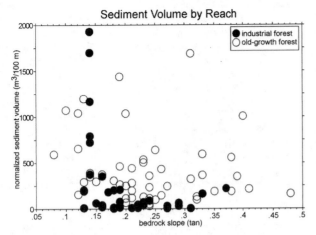

Figure 5. Sediment volume versus reach average (bedrock) gradient. The diameter of the markers represents the estimated uncertainty in slope, see text for discussion of uncertainty in estimated sediment volume.

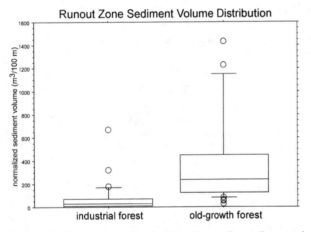

Figure 6. Sediment volume distributions. The median sediment volume is 28 m³/100 m for industrial forest (N = 35) and 239 m³/100 m for old-growth (N = 43). The distributions are positively skewed; the mean sediment volume is 75 m³/100 m for industrial forest and 606 m³/100 m for old-growth.

(Figure 3) reflects the greater proportion of the channel network affected by debris flows outside of the park. Mean R_C in the industrial forest is almost twice that of the old-growth, a difference that is highly significant (Table 3).

3.2. Channel Scale

There are discrete deposits of sediment separated by stretches of bedrock channel along the surveyed debris flow runout paths in both forest types, as shown by the sediment volume profiles (Figure 4). There are fewer sediment deposits in industrial forest debris flow tracks, however, and they tend

to be of shorter length and lesser volume than those in old-growth channels. Where wood is present as a bed-forming material there can be large accumulations of sediment well upstream of the final deposition zone, a pattern that is particularly evident in the profiles of SP, HBC, WT, and HR5, all of which occur within old-growth stands. In both forest types sediment volume varies widely for reaches below a gradient of 0.15 (Figure 5). Several of the low-gradient reaches plotted in Figure 5 are located at the downstream end of the debris flow tracks, where terminal deposition begins, but the differences between industrial and old-growth forests are most apparent in the zone of transport and partial deposition above 0.15 gradient. In the industrial forest reaches sediment volume declines with increasing gradient (log-log regression R = -0.40, p = 0.007) and is consistently below 250 m³/100 m at gradients > 0.15. In contrast, sediment volume in the old growth does not significantly correlate with reach slope (log-log regression R = -0.26, p = 0.06) and at gradients > 0.15 is higher than that of equally steep reaches in the industrial forest.

Debris-flow runout zones in old-growth forest generally retain more sediment than those in industrial forest. Figure 6 represents the distribution of channel sediment volumes for each forest class, excluding terminal deposition reaches. Over 90% of the industrial forest reaches have sediment volumes below the median sediment volume of old-growth reaches. The geometric mean sediment volume for reaches in industrial forest is less than one tenth that of reaches in old-growth forest, and the total sediment volume per unit length for all channels is 7.7 times higher in old-growth forest (Table 4).

Debris flows in the industrial forest leave channels scoured to bedrock along much of their length, but in the old growth such scour is rare, and reaches tend to retain sediment cover along most of their length. Lower sediment volumes in the industrial forest correspond to greater proportions of exposed bedrock in steeper reaches, although there is no simple correlation between reach slope and bedrock exposure in either forest class (Figure 7). In the old-growth forest 60% of reaches can be classified as "alluvial" (< 25% bedrock exposure by length), 35% "transitional" (25% = bedrock < 75%) and only 5% "bedrock" (≥ 75% bedrock exposure), whereas in the industrial forest only 14% of reaches are alluvial, 40% are transitional, and 46% are bedrock (Table 4).

In our study area debris flows differ only slightly in their effect on channel margins in old-growth versus industrial forests. Sideslope scour (defined as the length of sideslopes without vegetative cover) is roughly equivalent between forest classes (Figure 8); the average proportion of scoured sideslopes for old-growth reaches is 41%, while for industrial forest reaches it is 53%. Approximately 20% of reaches in both forest classes have scoured sideslopes along 75% or more of their length. There is a higher proportion of reaches

Table 4. Channel- and Reach-Scale Runout-Zone Parameters

	Industrial Forest	Old-Growth	p value
Reach Bed Character			
Mean Bedrock Proportion	63%	25%	< 0.0001
N	35	43	
Alluvial Reaches	14%	60%	
Transitional Reaches	40%	35%	
Bedrock Reaches	46%	5%	
Sideslope Character			
Mean Scour Proportion	53%	41%	0.07
N	35	43	
< 25% Scour	17%	37%	
25-75% Scour	63%	42%	
≥ 75% Scour	20%	5%	
Reach Sediment Volume (m³/m):			
Total	0.78	6.02 (3.64)[a]	
Reach Geometric Mean	0.23	2.71 (2.44)[a]	< 0.0001
Standard Deviation	1.27	14.53 (3.50)[a]	< 0.0001
N	35	43 (35)	
Average Deposit Volume (m³/m):			
Directly Associated w/ LWD	0.51	6.54 (2.79)[a]	
Contiguous Upstream of LWD	0.34	1.56 (1.98)[a]	
Not Associated with LWD	0.22	0.13 (0.15)[a]	
Functional Large Woody Debris			
LWD_F / 100 m	1.88	3.45	
Length Geometric Mean (m)	4.6	7.1	0.002
Standard Deviation (m)	4	9.4	< 0.0001
N	48	127	
Diameter Geometric Mean (cm)	55	60	> 0.3
Standard Deviation (cm)	34	45	0.004
N	79	179	

a.   Values in parentheses exclude HR5

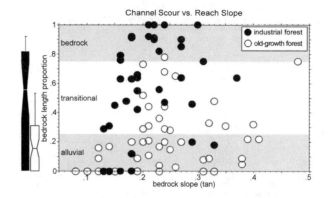

Figure 7. Channel bedrock proportion versus reach average (bedrock) gradient. The diameter of the markers approximately represents the estimated uncertainty in slope and bedrock proportion. In the box plots to the left of the figure the middle line represents the median, the notch represents the 95% confidence interval for the median, and the box spans the 25th to 75th percentiles.

Most of the sediment retained along debris flow tracks in both forest types is stored in contiguous deposits directly upstream of concentrations of LWD (Figure 4). The huge wood and sediment deposit upstream of the 500 m mark in HR5 more than doubles the volume of sediment in the surveyed old-growth runout-zone reaches. When data from HR5 are excluded from the analysis, 57% of old-growth sediment volume is directly associated with wood, 40% is in contiguous deposits upstream of concentrations of LWD, and only 3% is not associated with LWD at all. In industrial-forest debris-flow tracks 47% of old-growth sediment volume is directly associated with wood, 32% is in contiguous deposits upstream of concentrations of LWD, and 21% is not associated

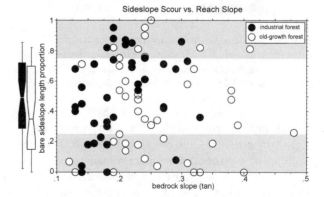

Figure 8. Scoured sideslope proportion versus reach average (bedrock) gradient. The diameter of the markers approximately represents the estimated uncertainty in slope. In the box plots to the left of the figure the middle line represents the median, the notch represents the 95% confidence interval for the median, and the box spans the 25th to 75th percentiles.

with < 25% sideslope scour in the old-growth, and a correspondingly lower proportion with between 25% and 75% scour, but the difference in means is only marginally significant (p = 0.07). Sideslope scour is weakly correlated with channel bedrock exposure (linear regression R = 0.42, p = 0.0001).

Diameter of Functioning LWD by Forest Type

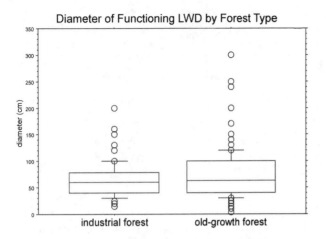

Figure 9. LWD_F diameter distributions. The median LWD_F diameter is 0.60 m for both industrial (N = 79) and old-growth forest (N = 179). The distributions are slightly positively skewed; the mean LWD_F diameter is 0.63 m for industrial forest and 0.72 m for old-growth.

Length of Functioning LWD by Forest Type

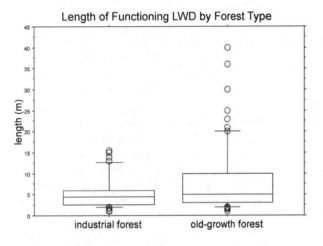

Figure 10. LWD_F length distributions. The median length of LWD_F is 5.0 m for industrial (N = 48) and 6.3 m for old-growth forest (N = 127). The distributions are positively skewed; mean LWD_F length is 5.7 m for industrial forest and 10.4 m for old-growth.

with LWD at all. Deposits in old-growth are typically larger than those in industrial-forest channels. Although LWD-free deposits are on average 1.5 times more voluminous in industrial-forest debris-flow tracks, mixed sediment-and-wood deposits are between 5 and 13 times larger in old-growth channels (depending on whether or not HR5 is included in the analysis) and on average 5 to 6 times as much sediment is stored in deposits upstream of LWD concentrations (Table 4).

There is more LWD in old-growth debris-flow tracks, and it is directly involved in the retention of substantial volumes of sediment. The difference in LWD load is reflected in the

greater volume of mixed sediment-and-wood deposits in the old-growth reaches. It is also reflected in our recording 1.8 times as many pieces of LWD_F per 100 m in old-growth channels as in industrial-forest channels. This difference is marginally significant (p = 0.087), but underestimates the true difference in LWD load because in many locations in the old-growth there were multiple LWD_F pieces, and we only measured the key members.

In addition to the difference in amount of LWD_F, there are minor differences in its characteristics between forest types. The distributions of LWD_F diameter (Figure 9) have the same lower bounds in both forest classes, but there are larger diameter pieces in the old growth. There is, however, no significant difference in the geometric mean diameter of LWD_F between forest classes (Table 4). Most of the LWD that is functioning to retain sediment in both old-growth and industrial forests is of a diameter typical of old-growth forests and greater than can be expected to be recruited from industrial forests managed in short rotation [Montgomery et al., 2003]; in both forest classes 50% of LWD_F is more than 0.60 m in diameter, 75% over 0.40 m, and 90% more than 0.30 m in diameter. LWD_F pieces are significantly longer on average in the old-growth forest (Figure 10, Table 4). In addition, the maximum length of industrial-forest LWD_F is 15.5 m, while in the old-growth it is 40 m. There is no notable difference in the orientation of LWD_F with respect to the channel; the numbers of parallel and perpendicular pieces are about equal in both forest types.

4. DISCUSSION

4.1. Landscape Scale

Our observation that mean primary runout length (R_P) is equivalent between forest types contrasts with the results of *May* [1998], who found that runout distance tends to be greater for debris flows that originate in or travel through clear-cut industrial forest. Like *May* [1998], however, we did find greater average cumulative runout (R_C) resulting from the greater number of source areas per basin in industrial forests. The difference in findings stems, at least in part, from the difference in drainage network structure. In a dendritic network such as that of *May's* [1998] Oregon Coast Range study area there are many opportunities for debris flows to stop at channel intersections, and indeed *Benda and Cundy* [1990] found that they could predict terminal deposition based on a threshold intersection angle of 70°. In pinnate drainage networks such as that of the western Olympic Mountains such high-angle tributary junctions are rare and primary runout lengths are largely determined by the distances between failure-prone headwater hollows and the intersection of low-order channels with unconfined, low-gradient mainstem valleys. This distance is a function of drainage basin geometry, and in our study area does not substantially differ between the industrial and old-growth basins.

While we did not attempt to inventory all landslides, we did find that debris flow inducing landslides are more common in the industrial forests in our study area, as represented by over four times higher I_{DF} and approximately three times higher cumulative D_{DF}. Although landslides under old-growth canopy may be undercounted in some aerial-photograph-based inventories [*Pyles and Froehlich*, 1987; and see *Robison et al.*, 1999 for an extensive discussion], we found that we over-counted debris flows in the old-growth due to our method of inferring the presence of debris-flow tracks from canopy openings along headwater channels. Because we surveyed at only one point in time, we cannot directly address recurrence interval, but the fact that debris flow scars are more common in the industrial forests suggests that debris-flow-inducing landslides are more frequent.

Many authors have observed that the rate of landsliding increases after clear-cutting [e.g., *Johnson et al.*, 2000; *Snyder*, 2000; *May*, 2002]. Slope-stability analysis can be used to explain this increase as a consequence of decreased soil cohesion following timber harvest, and to predict the depth of soil loss that can be expected to result from this effect. One of the most widely used slope stability models is the infinite-slope model [*Selby*, 1993], in which

$$FS = \frac{[C + (\rho_S - m\rho_W)gD\cos\theta\tan\phi]}{(\rho_S gD \sin\theta)} \quad (2)$$

where FS is the factor of safety, C is cohesion, ρ_S is the density of soil, ρ_W the density of water, m is the saturated fraction of the depth to the failure plane, g is the acceleration due to gravity, D is the slope-normal depth to the failure plane, θ is the slope of the surface (and failure plane), and ϕ is the internal friction angle of the material. Rearranging (2) to solve for soil depth at $FS = 1$, results in

$$D = \frac{C}{\rho_S g (\sin\theta - \cos\theta\tan\phi) + m\rho_W g \cos\theta\tan\phi}. \quad (3)$$

Using (3) it is possible to predict the maximum depth of stable soil for a given combination of slope, soil density, cohesion, friction angle, and saturation. For our calculations we used $\rho_S = 2000$ kg m^{-3}, $\rho_W = 1000$ kg m^{-3}, $\phi = 35°$ and $m = 0.5$, representing partial saturation of the soil mantle. In Figure 11 we plot the maximum depth of stable soil against slope for cohesion of 10 and 2 kPa to span the expected range between minimum cohesion for old-growth and recently clear-cut forests respectively [e.g., *Schaub*, 1999; *Montgomery et al.*, 2000; *Schmidt et al.*, 2001]. The expected change in the max-

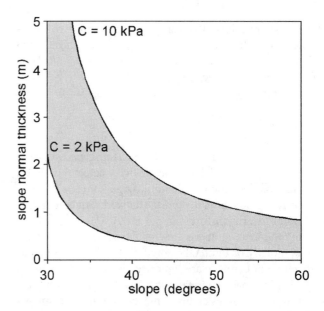

Figure 11. Predicted maximum stable soil thickness vs. slope. Plotted curves are representative of hillslopes in old-growth ($C = 10$ kPa) and clear-cut ($C = 2$ kPa) forests that fail at partial saturation ($m = 0.5$). The shaded area between the curves represents the change in equilibrium soil thickness expected to result from clear-cut timber harvest.

imum depth of stable soil between old-growth and clear-cut hillslopes, assuming sufficient time to achieve equilibrium, is between 1.5 and 0.7 m for slopes between 40° and 60°. Landsliding should be more common in episodically or periodically clear-cut forests, because soil mantle thickness is out of equilibrium during post-harvest periods of low cohesion.

Landsliding in the clear-cut portion of our study area occurred both in hillslope hollows and on planar hillsides [e.g., *Logan et al.*, 1991]. This is also to be expected, because soil production declines with soil depth and becomes negligible at about 1 m [*Heimsath et al.*, 1997, 1999], which is shallower than the maximum stable depth for slopes < 48° under old-growth forest (see Figure 11). Hence, under an old growth forest with high cohesion, the soil on planar and divergent slopes will not tend to become deep enough to fail, while on convergent slopes (i.e., hollows) soils will tend to increase in depth due to creep until they exceed the stable depth and eventually fail [*Dietrich and Dunne*, 1978; *Dietrich et al.*, 1995]. Only once the cohesion of the soil has been reduced after timber harvest would one expect to see widespread failure on planar and divergent slopes [*Montgomery et al.*, 1998]. Although it is generally thought that protecting headwater hollows from disturbance is sufficient to reduce the potential for slope failure [K. Schlichte, op cit.], this result suggests that steep planar slopes are particularly vulnerable to failure after clear-cut timber harvesting. It also suggests that hillslopes that are subject to clear-cut timber harvest will contribute more sediment to the channel system than those left in old-growth forest, as reported by many authors [e.g., *Roberts and Church*, 1986; *Logan et al.*, 1991; *Montgomery et al.*, 2000; *Guthrie*, 2002].

4.2. Channel scale

Based in part on the results of their landscape evolution model, *Lancaster et al.* [2001] argue that where debris flows are common and wood is plentiful, large volumes of sediment can be stored high in drainage networks. In both industrial and old-growth forest in our study area most of the sediment volume in debris-flow impacted headwater channels is directly or indirectly stored by LWD. Mixed wood-and-sediment deposits are typically five to thirteen times larger in old-growth debris flow tracks and there are on average at least 1.8 times as many pieces/100 m of sediment-retaining LWD. Not surprisingly, there is five (averaged by channel) to eleven (averaged by reach) times as much sediment in old-growth as there is in industrial-forest debris flow runout zones. We found sediment volumes on the order of those predicted by *Lancaster et al.* [2001] in four out of five (HR5, HR1, HBC, SP) of the old-growth channels we surveyed. Because they have low surface gradients and small drainage areas, these deposits are unlikely to be rapidly dispersed by fluvial processes [*Perkins*, 1989; *O'Connor*, 1994].

There were multiple deposits in each channel where sediment was retained upstream of LWD concentrations. It is unclear whether these concentrations of LWD and sediment reflect pre-existing log jams that stopped part of a debris flow, or dynamic snout deposition during the debris flow itself, as described by *Parsons et al.* [2001] for experimental debris flows and inferred by *Whipple* [1994] for debris flows lacking LWD. In any case, we found that most reaches along debris-flow tracks in wood-rich old-growth forest channels retain sediment, and that bare bedrock is relatively rare. This observation contrasts with the results of *May* [2001] for the Oregon Coast Range and the widespread assertion that debris flows scour headwater streams to bedrock, but we did find extensive bedrock exposure in the relatively wood-depleted industrial forest channels.

4.3. Synthesis

We suggest that the differences in both debris flow frequency and channel sediment storage between old-growth and industrial forests represent a fundamental shift in the sediment production and transport system (Figure 12), akin to that suggested by *Gabet and Dunne* [2002] for California grasslands. Under old-growth forest, short-term excess sediment delivered to the channel network can be stored in LWD-rich headwater channels, so sediment supply to higher order streams is buffered from the stochastic inputs of mass wasting. Overall sediment production is low (due to the thickness of the soil mantle), sediment storage in the channels is high, and sediment output from headwater streams is low. When forests are clear-cut and the effective cohesion of the soil declines, the equilibrium hillslope soil thickness decreases and planar and divergent slopes (as well as hollows) become prone to instability, resulting in an accelerated rate of landsliding. The sediment delivered to the streams by mass-wasting episodes is not retained in steep headwater reaches due to a lack of LWD-mediated storage capacity, and is instead delivered *en masse* to lower gradient reaches by debris flows. This state of low sediment production and high sediment output is necessarily transient, and can only last until the hillslopes have shed their excess sediment, at which point sediment production will be higher due to thinner soils, channels will have little storage capacity (assuming a continued dearth of old-growth class LWD), and sediment flux from the hillslopes will be rapidly transmitted through headwater streams to the rest of the drainage network.

5. CONCLUSIONS

Industrial forestry has a profound influence on sediment transport by mass movement in the western Olympics. Forest clearing both increases the frequency and changes the effect of debris flows, resulting in fundamentally different equilibrium hillslope soil depth, channel sediment storage, and reach morphology in industrial versus old-growth forests. Sustained short-rotation harvest of forest from the hillsides of headwater basins and removal of old-growth LWD from headwater channels can be expected to lead eventually to a stable state in which soil depth on hillslopes and sediment storage in headwater channels is relatively low, and sediment output to higher order reaches is dependent on and tightly coupled to the rate of sed-

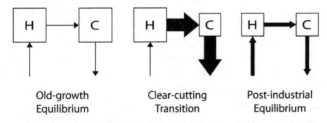

Figure 12. Conceptual model. The boxes represent storage on hillslopes (H) and in headwater channels (C). The up arrows represent weathering flux, the right arrows mass-wasting, and the down arrows debris flow and fluvial transport out of the headwater channels. The relative size of the elements represents the relative magnitudes of stores and fluxes.

iment production on the hillslopes. During the transition from old-growth to industrial forest, sediment storage on hillslopes is diminishing, sediment storage in channels is low, and sediment output from headwater channels is likely to be high.

Acknowledgments. This research was supported by USDA Forest Service Cooperative Agreement # PNW 99-3032-2-CA and the University of Washington Department of Earth and Space Sciences. Bill Baccus facilitated our work in Olympic National Park. Harvey M. Greenberg provided GIS data and assistance. Special thanks go to Byron Amerson, Suzanne Osborne, Dave Trippett, Peter Wald, Chris Brummer, Simon DeSzoeke, Suzaynn Schick, and Oliver Deschler for providing invaluable assistance during sometimes difficult and hazardous fieldwork. The comments of Jeffrey D. Parsons, Joanne Bourgeois, and reviewers Melinda D. Daniels and Thomas E. Lisle greatly improved the quality and readability of this paper.

REFERENCES

Abbe, T. B., and D. R. Montgomery, Large woody debris jams, channel hydraulics, and habitat formation in large rivers, *Regulated Rivers: Research and Management,* 12, 201–221, 1996.

Abbe, T. B., and D. R. Montgomery, Patterns and processes of wood debris accumulation in the Queets River basin, Washington, *Geomorphology,* 51, 81–107, 2003.

Benda, L. E., *Debris flows in the Tyee formation of the Oregon Coast Range,* unpublished, M.S. thesis, Univ. of Washington, Seattle, 1988.

Benda, L. E., The influence of debris flows on channels and valley floors in the Oregon Coast Range, U.S.A., *Earth Surf. Proc. Land.,* 15, 457–466, 1990.

Benda, L. E., and T. W. Cundy, Predicting deposition of debris flows in mountain channels, *Can. Geotech. J.,* 27, 409–417, 1990.

Bilby, R. E., and J. W. Ward, Characteristics and function of large woody debris in streams draining old-growth, clear-cut, and second-growth forests in southwestern Washington, *Can. J. Fish. Aq. Sci.,* 48, 2499–2508, 1991.

Bunn, J. T., *Patterns of Wood and Sediment Storage Along Debris-flow Impacted Headwater Channels in Old-Growth and Industrial Forests,* unpublished, M.S. thesis, Univ. of Washington, Seattle, 2003.

Church, M., Geomorphic thresholds in riverine landscapes, *Freshwater Biol.,* 47, 541–557, 2002.

Dietrich, W. E., and T. Dunne, Sediment budget for a small catchment in mountainous terrain, *Zeitschrift für Geomorphologie, Suppl.* 29, 191–206, 1978.

Dietrich, W. E., R. Reiss, M. L. Hsu, and D. R. Montgomery, A process-based model for colluvial soil depth and shallow landsliding using digital elevation data, *Hydrol. Proc.,* 9, 383–400, 1995.

Edmonds, R. L., *Vegetation patterns, hydrology, and water chemistry in small watersheds in the Hoh River Valley, Olympic National Park,* U.S. Dept. of the Interior, National Park Service, Denver, 1998.

Fetherston, K. L., R. J. Naiman, and R. E. Bilby, Large woody debris, physical process, and riparian forest development in montane river networks of the Pacific Northwest, *Geomorphology,* 13, 133–144, 1995.

Froehlich, H. A., *Natural and man-caused slash in headwater streams,* Oregon Logging Handbook 33, Pacific Logging Congress, 1973.

Gabet, E. J., and T. Dunne, Landslides on coastal sage-scrub and grassland hillslopes in a severe El Niño winter: The effects of vegetation conversion on sediment delivery, *Geol. Soc. Amer. Bull.,* 114(8), 983–990, 2002.

Gomi, T., R. C. Sidle, and J. S. Richardson, Understanding processes and downstream linkages of headwater systems, *Bioscience,* 52(10), 905–916, 2002.

Gomi, T., R. C. Sidle, M. D. Bryant, and R. D. Woodsmith, The characteristics of woody debris and sediment distribution in headwater streams, southeastern Alaska, *Can. J. For. Res.,* 31, 1386–1399, 2001.

Guthrie, R. H., The effects of logging on frequency and distribution of landslides in three watersheds on Vancouver Island, British Columbia, *Geomorphology,* 43, 273–292, 2002.

Heimsath, A. M., W. E. Dietrich, K. Nishiizumi, and R. C. Finkel, The soil production function and landscape equilibrium, *Nature,* 388, 358–361, 1997.

Heimsath, A. M., W. E. Dietrich, K. Nishiizumi, and R. C. Finkel, Cosmogenic nuclides, topography, and the spatial variation of soil depth, *Geomorphology,* 27, 151–172, 1999.

Heusser, C. J., Quaternary vegetation, climate, and glaciation of the Hoh River Valley, Washington, *Geol. Soc. Amer. Bull.,* 85, 1547–1560, 1974.

Johnson, A. C., D. N. Swanston, and K. E. McGee, Landslide initiation, runout, and deposition within clearcuts and old-growth forests of Alaska, *J. Amer. Water Resour. Assoc.,* 36, 1097–1113, 2000.

Lancaster S. J., S. K. Hayes and G. E. Grant, Modeling sediment and wood storage and dynamics in small mountainous watersheds, in *Geomorphic Processes and Riverine Habitat,* edited by J. M. Dorava, D. R. Montgomery, B. B. Palcsak, and F. A. Fitzpatrick, pp. 85–102, American Geophysical Union, Washington, DC, 2001.

Logan, R. L., K. L. Kaler, and P. K. Bigelow, *Prediction of sediment yield from tributary basins along Huelsdonk Ridge, Hoh River, Washington,* Washington Division of Geology and Earth Resources Open File Report 91-7, Washington State Department of Natural Resources, Olympia, 1991.

May, C. L., *Debris Flow Characteristics Associated with Forest Practices in the Central Oregon Coast Range,* unpublished, M.S. thesis, Oregon State Univ., Corvallis, 1998.

May, C. L., *Spatial and Temporal Dynamics of Sediment and Wood in Headwater Streams in the Central Oregon Coast Range,* unpublished, Ph.D. thesis, Oregon State Univ., Corvallis, 2001.

May, C. L., Debris flows through different forest age classes in the central Oregon Coast Range, *J. Amer. Water Resour. Assoc.,* 38, 1097–1113, 2002.

McHenry, M. L., E. Shott, R. H. Conrad, and G. B. Grette, Changes in the quantity and characteristics of large woody debris in streams of the Olympic Peninsula, Washington, U.S.A. (1982–1993), *Can. J. Fish. Aq. Sci.,* 55, 1395–1407, 1998.

Montgomery, D. R., and J. M. Buffington, Channel Processes, Classification, and Response, in *River Ecology and Management: Lessons from the Pacific Coastal Ecoregion,* edited by R. J. Naiman and R. E. Bilby, pp. 13–42, Springer-Verlag, New York, 1998.

Montgomery, D. R., J. M. Buffington, R. D. Smith, K. M. Schmidt, and G. Pess, Pool spacing in forest channels, *Water Resour. Res.,* 33, 1097–1105, 1995.

Montgomery, D. R., K. Sullivan, and H. M. Greenberg, Regional test of a model for shallow landsliding, *Hydrol. Proc.,* 12, 943–955, 1998.

Montgomery, D. R., K. M. Schmidt, H. M. Greenberg, and W. E. Dietrich, Forest clearing and regional landsliding, *Geology,* 28, 311–314, 2000.

Montgomery, D. R., T. B. Abbe, J. M. Buffington, N. P. Peterson, K. M. Schmidt, and J. D. Stock, Distribution of bedrock and alluvial channels in forested mountain drainage basins, *Nature, 318,* 587–589, 1996.

Montgomery, D. R., T. M. Massong, and S. C. S. Hawley, Influence of debris flows and log jams on the location of pools and alluvial channel reaches, Oregon Coast Range, *Geol. Soc. Amer. Bull.,* 115, 78–88, 2003.

Morrison, P. H., *Ecology and Geomorphological Consequences of Mass Movements in the Alder Creek Watershed and Implications for Forest Land Management,* unpublished, B.A. thesis, Univ. of Oregon, Eugene, 1975.

Murphy, M. L., and K. V. Koski, Input and depletion of woody debris in Alaska streams and implications for streamside management, *N. Amer. J. Fish. Man.,* 9, 423–436, 1989.

O'Connor, M. D., *Sediment Transport in Steep Tributary Streams and the Influence of Large Organic Debris,* unpublished, Ph.D. Thesis, Univ. of Washington, 1994.

Parsons, J. D., K. X. Whipple, and A. Simoni, Experimental study of the grain-flow, fluid-mud transition in debris flows, *J. Geol.,* 109, 427–447, 2001.

Perkins, S. J., Landslide deposits in low-order streams – their erosion rates and effects on channel morphology, in *Proceedings of the Symposium on Headwaters Hydrology,* edited by W. W. Woessner and D. F. Potts, pp. 173–182, American Water Resources Association, Bethesda, 1989.

Peterson, D. L., E. G. Schreiner, and N. M. Buckingham, Gradients, vegetation and climate: spatial and temporal dynamics in the Olympic Mountains, U.S.A., *Global Ecol. Biogeog. Lett.,* 6, 7–17, 1997.

Pyles, M. R., and H. A. Froehlich, Rates of Landsliding as Impacted by Timber Management Activities in Northwestern California, *Bull. Assoc. Eng. Geol.,* 24, 425–431, 1987.

Ralph, S. C., G. C. Poole, L. L. Conquest, and R. J. Naiman, Stream channel morphology and woody debris in logging and unlogging basins of western Washington, *Can. J. Fish. Aq. Sci.,* 51, 37–51, 1994.

Roberts, R. G., and M. Church, The sediment budget in severely disturbed watersheds, Queen Charlotte Ranges, British Columbia, *Can. J. For. Res.,* 16, 1092–1106, 1986.

Robison, E. G., K. Mills, J. Paul, L. Dent, and A. Skaugset, *Storm impacts and landslides of 1996: final report,* Forest Practices Technical Report No. 4, Oregon Department of Forestry, 1999.

Schaub, T. S., *Incorporating Root Strength Estimates into A Landscape-Scale Slope Stability Model Through Forest Stand Age Inversion From Remotely Sensed Data,* unpublished, M.S. thesis, Univ. of Washington, Seattle, 1999.

Schmidt, K. M., J. J. Roering, J. D. Stock, W. E. Dietrich, D. R. Montgomery, and T. S. Schaub, Root cohesion variability and shallow landslide susceptibility in the Oregon Coast Range. *Can. Geotech. J.,* 38, 995–1024, 2001.

Seidl, M. A., and W. E. Dietrich, The problem of channel erosion into bedrock, in *Functional Geomorphology,* Catena Supplement 23, 101–124, 1992.

Selby, M. J., *Hillslope Materials and Processes,* Oxford University Press, Oxford, 1993.

Sidle, R. C., A. J. Pearce, C. L. O'Loughlin, *Hillslope Stability and Land Use,* Water Resources Monograph 11, American Geophysical Union, Washington, D.C., 1985.

Sidle, R. C., Y. Tsuboyama, S. Noguchi, I. Hosoda, M. Fujieda, and T. Shimizu, Streamflow generation in steep headwaters: A linked hydro-geomorphic paradigm, *Hydrol. Proc.,* 14, 369–385, 2000.

Snyder, K. U., *Debris Flows and Flood Disturbance in Small Mountain Watersheds,* unpublished, M.S. thesis, Oregon State Univ., Corvallis, 2000.

Swanson, F. J., and D. N. Swanston, Complex mass-movement terrains in the western Cascade Range, Oregon, *Rev. Eng. Geol. 3; Landslides,* 113–124, 1977.

Tabor, R. W., and Cady, W. M., *Geologic map of the Olympic Peninsula,* U.S. Geological Survey Miscellaneous Investigations Series Map, I-993, 1978.

Whipple, K. X., *Debris flow fans: process and form,* unpublished, Ph.D. thesis, Univ. of Washington, Seattle, 1994.

Wohl, E., *Mountain Rivers,* Water Resources Monograph 14, American Geophysical Union, Washington D.C., 2000.

Jeremy T. Bunn and David R. Montgomery, Department of Earth and Space Sciences, Box 351310, 63 Johnson Hall, University of Washington, Seattle, WA 98195.

Root-Soil Mechanics and Interactions

Donald H. Gray

Department of Civil and Environmental Engineering, University of Michigan, Ann Arbor, Michigan.

David Barker

*Division of Civil Engineering, Environmental Health and Safety Management,
Nottingham Trent University, Nottingham, UK.*

Plant roots play an important role in stabilizing slopes and stream banks. Specific hydro-mechanical processes can be identified through which vegetation affects stability in both beneficial and detrimental ways. Woody vegetation improves shallow mass stability mainly by increasing the shear strength of the soil via root reinforcement and by a buttressing effect from well anchored stems. The most effective restraint is provided where roots penetrate across the soil mantle into fractures or fissures in the underlying bedrock or where roots penetrate into a residual soil or transition zone whose density and shear strength increase with depth. Techniques and procedures are summarized herein for determining root architecture and distribution in soils. The mechanical or reinforcing effect of plant roots on the stability of slopes can be described and accounted for in a systematic manner. Root fibers reinforce a soil by transfer of shear stress in the soil matrix to tensile resistance in the fiber inclusions. Simple force equilibrium models are useful for identifying parameters that affect root reinforcement and predicting the amount of strength increase from the presence of fibers in a soil. Plant roots tend to respond to unfavorable stress conditions in a self-correcting manner through a bio-adaptive process termed edaphoecotropism. This process allows roots to escape or avoid unfavorable site conditions and may enhance their slope stabilization role.

1. INTRODUCTION

Vegetation affects both the surficial and mass stability of slopes in significant and important ways. Various hydro-mechanical influences of vegetation including methods for predicting and quantifying their magnitude and importance on the stability of upland slopes are described elsewhere [*Gray and Sotir*, 1996; *Coppin and Richards*, 1990; *Greenway*,

1987]. For the most part vegetation has a beneficial influence on the stability of slopes; however, it can occasionally affect stability adversely or have other undesirable impacts, e.g., obstruct views, hinder slope inspection, or interfere with flood fighting operations on levees. A number of strategies and techniques can be invoked to maximize benefits and minimize liabilities of vegetation of slopes [*Gray and Sotir*, 1996; *Gray*, 2001]. These include such procedures as the proper selection and placement of vegetation in addition to management techniques such as pruning, coppicing, and landform grading.

The main purpose of this chapter is to discuss specifically the mechanical role of woody plant roots in reinforcing soil and affecting its constitutive behavior. To the extent that root

Riparian Vegetation and Fluvial Geomorphology
Water Science and Application 8
10.1029/008WSA09

fibers do indeed affect the constitutive behavior and strength of soils, it is also important to know how these roots occur and are distributed in the ground. Accordingly, some information is also presented on root architecture and distribution in slopes and streambanks. Finally, information will be presented on edaphoecotropism in woody plant roots. Edaphoecotropism [*Vanicek*, 1973] refers to the ability of plant roots to adapt to or avoid stress, vis-à-vis mechanical obstacles, lack of moisture, light, and deep burial. This property of roots, and of live woody tissue in general, is especially significant with regard to biotechnical slope protection or the combined use of plants and structures for protecting stream banks and levees.

2. ROOT ARCHITECTURE AND DISTRIBUTION

2.1. Depth and Distribution of Root Systems

Deeply penetrating vertical taproots and sinker roots provide the main contribution to the stability of slopes, vis-à-vis resistance to shallow sliding. Mechanical restraint against sliding only extends as far as the depth of root penetration. In addition, the roots must penetrate across the failure surface to have a significant effect. The influence of root reinforcement and restraint for different slope stratigraphies and conditions has been summarized by *Tsukamoto and Kusuba* [1984]. The most effective restraint is provided where roots penetrate across the soil mantle into fractures or fissures in the underlying bedrock or where roots penetrate into a residual soil or transition zone whose density and shear strength increase with depth. The stabilizing effect of roots is lowest when there is little or no penetration across the shear interface. However, even in these cases lateral roots can play an important role by maintaining the continuity of a root-permeated soil mantle on a slope.

Root morphology studies require careful excavation and can be difficult and expensive to undertake, particularly in the case of large mature trees. Because of oxygen requirements, the roots of most trees tend to be concentrated near the surface. As a rough rule of thumb the mechanical reinforcing or restraining influence of roots on a slope is probably limited to a zone about 1.5 m from the surface. Studies by *Patric et al.* [1965] in a loblolly pine plantation showed that 80 to 90% of the roots in their test plots were concentrated in the upper 0.9 m. The bulk of the near-surface roots were laterals; in contrast, roots below 0.9 m were generally oriented vertically.

Root area ratios (RAR) were measured by *Shields and Gray* [1993] as a function of depth in a sandy levee along the Sacramento River in California for a variety of woody plant species. Root area ratio refers to the fraction of the total cross-sectional area of a soil that is occupied by roots. RAR versus depth curves were presented for two mutually perpendicular transects adja-

cent to a group of elderberry bushes. Root area ratios tended to decrease exponentially with depth and few roots were encountered below a depth of 1.2 m in either transect.

2.2. Root Spread

Tree roots can spread out for considerable distances; in one reported instance [*Kozlowski*, 1971] roots of poplars growing in a sandy soil extended out 6.4 m. The extent of root spread is normally reported in relative multiples of the tree height or crown radius. Root spreads reported in the technical literature typically range from 1 to 3 times the crown radius. A useful rule of thumb is that a root system will spread out a distance at least equal to the 1.5 times the radius of the crown. The hydraulic influence of a tree, i.e., significant soil moisture reductions caused by evapo-transpiration, can be active to distance of at least 1 times the tree height [*Biddle*, 1983]. These findings have implications with regard to both slope stability and safe placement of structures adjacent to trees growing on compressible soils.

2.3. Factors Affecting Root Development

Root development and structure are affected initially by genetic disposition but ultimately are governed more by environmental and edaphic conditions [*Sutton*, 1969]. *Henderson et al.* [1983] have noted that root systems tend to grow wide

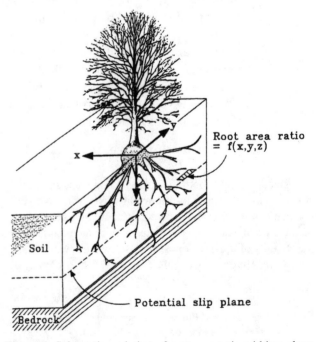

Figure 1. Schematic variation of root area ratio within a slope [Greenway, 1987].

and deep in well drained soils as opposed to developing a flat, plate-like structure in a surface soil underlain by a more dense or rocky substratum.

The degree to which roots are able to penetrate underlying bedrock depends to a large extent on the nature and extent of discontinuities (e.g., joints and fractures) in the bedrock. Trees growing in shallow, coarse-textured soils developed on granitic bedrock, for example, can develop sinker and taproots that penetrate into fissures and fractures in the underlying bedrock.

2.4. Root Structure and Distribution: Experimental Methods

Various methods for determining root structure and distribution are described in detail by *Bohm* [1979]. Root area ratios or root biomass concentration as a function of depth are required in order to estimate rooting contributions to soil shear strength. This ratio varies spatially in three dimensions, as shown schematically in Figure 1. The root area ratio of greatest interest coincides with the critical sliding surface. Normally this surface is oriented parallel to the slope or bedrock-soil interface.

One approach to estimate RAR is simply to recover large samples of root-permeated soil from various depths and measure the root biomass per unit volume at each depth by sieving the soil, recovering, and weighing the roots. Root biomass per unit volume can be converted to an equivalent root area ratio if the unit weight or density of the roots is known.

Root area ratios and root distribution can be measured directly in an excavated trench using the "profile wall" method [*Bohm*, 1979]. In this method the roots exposed in the vertical side of a trench are carefully mapped by means of a gridded, acetate overlay, as shown in Figure 2.

Figure 2. Profile wall method used to determine root distribution and area ratio as function of depth. Exposed roots are mapped on gridded acetate overlay.

Table 1. Maximum and minimum root area ratios RAR measured in trench excavations in a sandy channel levee.

Site Number	Site Description	RAR (%) Perpendicular Trench	Parallel Trench
2	Control	0.010 – 0.58	0.010 – 2.02
3	Dead Oak	0.001 – 0.24	0.001 – 0.40
4	Live Valley Oak	0.060 – 0.83	0.008 – 0.13
5	Willow	0.004 – 0.34	0.001 – 0.32
7	Elderberry	0.070 – 1.10	0.006 – 0.16
8	Black Locust	0.001 – 0.85	0.001 – 0.12

The range in root area ratios measured using the profile wall method along two transects oriented perpendicular and parallel to the crest of a sandy levee along the Sacramento River in Northern California is listed in Table 1. The transects or trenches passed adjacent to groves of different tree species. Note that the maximum root area ratio was about 2%. This probably represents the upper limit for area ratios of root-permeated soils in field and laboratory tests. These root area ratios and their variation with depth can be used to obtain an estimate of the rooting contribution to shear strength with depth.

Perhaps the best way to ascertain root architecture and distribution is to exhume or uncover the roots in-situ. This is accomplished by cutting and removing the trunk and then exposing the roots by removing the surrounding soil. The soil can be removed by hydraulic washing or alternatively by using an air jet device known as an "Airspade." Soil particles surrounding roots are dislodged and removed by means of high pressure air jets at the tip of the airspade. Virtually the entire root system of a tree can be exhumed more or less intact and studied in this manner as shown in Figure 3.

3. ROOT STRENGTH

3.1. Factors Affecting Strength

Wide variations in tensile strength of roots have been reported in the technical literature depending on species and such site factors as growing environment, season, root diameter, and orientation. *Greenway* [1987] compiled an excellent review of root strength and factors affecting it. With regard to the influence of seasonal effects, *Hathaway and Penny* [1975] reported that variations in specific gravity and lignin/cellulose ratio within poplar and willow roots produced seasonal fluctuations in tensile strength. *Schiechtl* [1980] observed that roots growing in the uphill direction were stronger than those extending downhill in response to gravitational effects.

3.2. Ranges in Root Tensile Strength and Tensile Modulus

Root tensile strengths have been measured by a number of different investigators. Nominal tensile strengths reported in the technical literature are summarized in Table 2 for selected shrub and tree species. Tensile strengths vary significantly with diameter and method of testing (e.g., in a moist or air dry state). Accordingly, the values listed in Table 2 should be considered only as rough or approximate averages. Nevertheless, some interesting trends can be observed in the tabulated strength values. Tensile strengths can approach 70 MPa but appear to lie in the range of 10 to 40 MPa for most species. The conifers as a group tend to have lower root strengths than deciduous trees. Shrubs appear to have root tensile strengths at least comparable to that of trees. This is an important finding because equivalent reinforcement can be supplied by shrubs at shallow depths without the concomitant liabilities of trees resulting from their greater weight, rigidity, and tendency for wind throwing. This could be an important consideration, for example, in streambank or levee slope stabilization. Willow species, which are frequently used in soil bioengineering stabilization work, have root tensile strengths ranging from approximately 14 to 35 MPa.

It is important to recognize that root tensile strength is affected as much by differences in size (diameter) as by species. Several investigators [*Turnanina*, 1965; *Burroughs and Thomas*, 1977; *Nilaweera*, 1994] have reported a decrease in root tensile strength with increasing size (diameter). Roots are no different in this regard than fibers of other materials, which exhibit a similar trend. The variation in root tensile strength with root diameter for several tropical hardwood species is shown in Figure 4. Root tensile strengths vary from approximately 8 to 80 MPa for root diameters ranging from 2 to 15 mm. A decrease in root diameter from 5 to 2 mm can result in a doubling or even tripling of tensile strength.

Figure 3. Root architecture and distribution of white oak exhumed in-situ using an "Airspade."

Table 2. Nominal tensile strength T_R of selected tree and shrub species [adapted from *Schiechtl*, 1980].

Species	Common Name	T_R (MPa)
Tree Species		
Abies concolor	Colorado white fir	11
Acacia confusa	Acacia	11
Alnus firma var. *multinervis*	Alder	52
Alnus incana	Alder	32
Alnus japonica	Japanese alder	42
Betula pendula	European white birch	38
Nothofagus fusca	Red beech	32
Picea sitchensis	Sitka spruce	16
Picea abies	European spruce	28
Pinus densiflora	Japanese red pine	33
Pinus lambertiana	Sugar pine	10
Pinus radiata	Monterey pine	18
Populus deltoides	Poplar	37
Populus euramericana 1488	American poplar	33
Pseudotsuga mensieii	Douglas fir (Pacific Coast)	55
Pseudotsuga mensieii	Douglas fir (Rocky Mountains)	19
Quercus robur	Oak	20
Sambucus callicarpa	Pacific red elder	19
Salix fragilis	Crack willow	18
Salix helvetica	Willow	14
Salix matsudana	Willow	36
Salix purpurea (Booth)	Purple willow	37
Tilia cordata	Linden	26
Tsuga heterophylla	Western hemlock	20
Shrub Species		
Castanopsis chrysophylla	Golden chinkapin	18
Ceanothus velutinus	Ceanothus	21
Cytisus scoparius	Scotch broom	33
Lespedeza bicolor	Scrub lespedeza	71
Vaccinium spp.	Huckleberry	16

The tensile modulus of roots is also of some interest because in many cases the full tensile strength of the roots is not mobilized. Instead, the amount of mobilized tensile resistance will be a function of the modulus and amount of tensile strain or elongation in the root. Only limited data on tensile modulus of roots are available. *Hathaway and Penny* [1975] presented typical stress-strain curves for several riparian species of poplar and willow. They tested root specimens, without bark, that had been air dried and then rewetted by soaking prior to testing. The ultimate breaking strains,

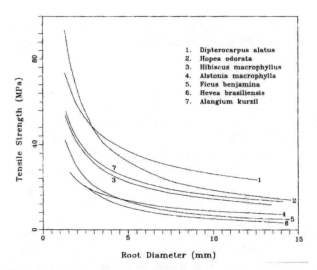

Figure 4. Relationship between root tensile strength and root diameter for several tropical hardwood species [Nilaweera, 1994].

Young's moduli, and tensile strengths measured in these tests are presented in Table 3.

4. ROOT/FIBER SOIL REINFORCEMENT

4.1. Force-Equilibrium Models

Important investigations have been carried out during the past two decades that have greatly improved our understanding of root reinforcement of soils and the contribution of roots to slope stability. These studies include modeling of root-fiber soil interactions, laboratory testing of fiber/soil composites, and in-situ shear tests of root-permeated soils. Relatively simple and straightforward force equilibrium models [*Waldron*, 1977; *Waldron and Dakessian*, 1981; *Wu et al.*, 1979] provide useful insights into the nature of root-fiber soil interactions and the contribution of root fibers to soil shear strength.

Table 3. Tensile strength and stress-strain behavior of some poplar (*Populus*) and willow (*Salix*) roots [*Hathaway and Penny*, 1975].

Species	Tensile Strength (MPa)	Young's Modulus (MPa)	Ultimate Strength (%)
Populus I-78	45.6	16.4	17.1
Populus I-488	32.3	8.4	16.8
Populus yunnamensis	38.4	12.1	18.7
Populus deltoides	36.3	9.0	12.4
Salix matsundana	36.4	10.8	16.9
Salix Booth	35.9	15.8	17.3

More sophisticated models based on the deformational characteristics of fiber reinforced composites [*Shewbridge and Sitar*, 1989, 1990] and statistical models that take into account the random distribution and branching characteristics of root systems have also been developed [*Wu et al.*, 1988a, b].

4.1.1 Strength contribution. Root fibers increase the shear strength of soil primarily by transferring shear stresses that develop in the soil matrix into tensile resistance in the fiber inclusions via interface friction along the length of imbedded fibers. This process is shown schematically in Figure 5 for an imbedded fiber oriented initially perpendicularly to the shear surface. When shear occurs the fiber is deformed as shown. This deformation causes the fiber to elongate, provided there is sufficient interface friction and confining stress to lock the fiber in place and prevent slip or pullout. As the fiber elongates it mobilizes tensile resistance in the fiber. The component of this tension tangential to the shear zone directly resists shear, while the normal component increases the confining stress on the shear plane.

The assumption of initial fiber orientation perpendicular to the shear surface requires further discussion. Root fibers have many orientations and are unlikely to be oriented perpendicular to the shear failure surface. Furthermore, both theoretical analyses and laboratory studies [*Gray and Ohashi*, 1983] have shown that a perpendicular orientation is not the optimal orientation. Fibers oriented initially at an acute angle (< 90°) to the failure surface or in the direction of maximum principal tensile strain result in the highest increase in shear strength. This orientation corresponds to the angle of obliquity (45° + $\phi/2$, where ϕ the angle of internal friction of the soil), or approximately 60° in most sands. Conversely, an oblique orientation with the shear surface (> 90°) can actually result in a shear

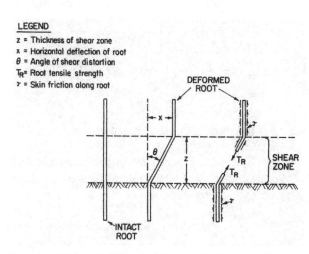

Figure 5. Schematic diagram of perpendicular root-fiber reinforcement model [Gray and Ohashi, 1983].

strength decrease because the fibers initially go into compression rather than tension, which results in a loss of normal stress on the failure surface. The simple, perpendicular model is actually a very useful simulation because it yields an average estimate of all possible orientations. This finding is supported by both laboratory studies on sand/fiber mixtures [*Gray and Ohashi*, 1983] and by statistical studies of sands with randomly distributed fibers [*Maher and Gray*, 1990].

Based on this perpendicular model the increase in shear strength of the fiber/soil composite will be given by the following expression:

$$\Delta S = t_R(\sin \theta + \cos \theta \tan \phi) \qquad (1)$$

where ΔS is the shear strength increase, θ is the angle of shear distortion in the shear zone, ϕ is the angle of internal friction of the soil, and t_R is the mobilized tensile stress of root fibers per unit area of soil.

The mobilized tensile stress of root fibers t_R will depend upon the amount of fiber elongation and fixity of the fibers in the soil matrix. Full mobilization can occur only if the fibers elongate sufficiently and if imbedded root fibers are prevented from slipping or pulling out. The latter requires that the fibers be sufficiently long and frictional, constrained at their ends, and/or subjected to high enough confining stresses to increase interface friction. Accordingly, three different response scenarios are possible during shearing of a fiber-reinforced soil composite, namely fibers break, stretch, or slip.

4.1.2 Fiber break mode. Shear strength increase from full mobilization of root-fiber tensile strength requires calculation of the average tensile strength of the root-fibers T_R and the fraction of soil cross-section occupied by roots A_R/A. The mobilized tensile stress per unit area of soil in this case is given by:

$$t_R = T_R(A_R/A) \qquad (2)$$

The angle of shear distortion (Figure 5) is given by:

$$\theta = \tan^{-1}(x/z) \qquad (3)$$

where x is the shear displacement and z is the shear zone thickness.

The fraction of soil cross-section occupied by roots, namely the "root area ratio," can be determined by counting roots by size class within a given soil as:

$$A_R/A = (\Sigma n_i a_i)/A \qquad (4)$$

where n_i is the number of roots in size class i and a_i is the mean cross-sectional area of roots in size class i.

Accounting for the variation in root-fiber tensile strength with root diameter, mean tensile strength of roots T_R can be determined by:

$$T_R = (\Sigma T_i n_i a_i)/(\Sigma n_i a_i) \qquad (5)$$

where T_i is the strength of roots in size class i.

By substituting Equation (2) into (1) the predicted shear strength increase from full mobilization of root tensile strength will be given by:

$$\Delta S = T_R(A_R/A)[\sin \theta + \cos \theta \tan \phi] \qquad (6)$$

The value of the bracketed term [$\sin \theta + \cos \theta \tan \phi$] in Equation (6) is relatively insensitive to normal variations in θ and ϕ so *Wu et al.* [1979] proposed an average value of 1.2 for this term. Equation (6) can then be simplified to:

$$\Delta S = 1.2 T_R(A_R/A) \qquad (7)$$

Thus, the predicted shear strength increase depends entirely on the mean tensile strength of the roots and the root area ratio. This model assumes that the roots are well anchored and do not pull out under tension. The root fibers must be long enough and/or subjected to sufficient interface friction for this assumption to be satisfied. If a simple uniform distribution of bond or interface friction stress between soil and root is assumed, the minimum root length L_{min} required to prevent slippage and pullout is given [*Gray and Ohashi*, 1983] by:

$$L_{min} = T_R D/4\tau_b \qquad (8)$$

where D is the root diameter and τ_b is the limiting bond or interface friction stress between root and soil.

The bond stress between root-fibers and soil can be estimated from the confining stress acting on the fibers and the coefficient of friction. For vertical fibers this bond stress varies with depth and can be calculated [*Gray and Ohashi*, 1983] as follows:

$$\tau_b = h\gamma(1 - \sin \phi)f \tan \phi \qquad (9)$$

where h is the depth below the ground surface, γ is the soil density, and f is the coefficient of friction between the root fiber and soil. The coefficient of friction between soil and wood ranges from 0.7 to 0.9. The rough texture and kinky shape of roots means that their friction coefficients will likely lie closer to the high end.

Roots will generally exceed the length criterion given in Equation (8) except close to the ground surface where the confining stress and hence the bond stresses will be low. This claim is supported by field observations where a preponderance of broken roots, compared to roots that have pulled out, can be seen in landslide scars or failure surfaces.

4.1.3 Fiber stretch mode. Lack of sufficient fiber elongation coupled with strain compatibility requirements may prevent mobilization of root-fiber tensile or breaking strength. In this case the calculation of the mobilized tensile strength t_R will be governed by the amount of fiber elongation and the fiber tensile modulus E_R. A force-equilibrium analysis yields the following expression for the mobilized tensile stress per unit area of soil:

$$t_R = k\beta(A_R/A) \tag{10}$$

$$k = (4z\tau_b E_R/D)^{1/2}; \beta = (\sec\theta - 1)^{1/2} \tag{11}$$

where z is the thickness of the shear zone.

Equation (1) assumes a linear tensile stress distribution in the fiber, zero at the ends to a maximum value at the shear plane. A parabolic stress distribution would yield a slightly higher value [*Waldron*, 1977]. By substituting Equation (10) and (11) into Equation (1) the predicted shear strength increase from mobilization of root tensile resistance from stretching will be given by:

$$\Delta S = k\beta(A_R/A)[\sin\theta + \cos\theta\tan\phi] \tag{12}$$

This expression reveals that shear strength increases vary inversely with the square root of the fiber diameter. Accordingly, at equal root area ratios, numerous small diameter fibers will be more effective than a few large fibers.

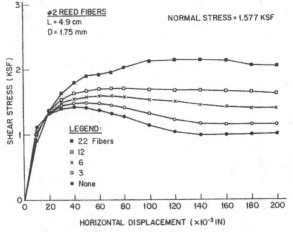

Figure 6. Effect of increasing amounts of fiber reinforcement on the stress-strain behavior of a dry sand in direct shear. Fibers are oriented perpendicular to the failure plane [Gray and Ohashi, 1983].

4.1.4 Fiber slip mode. If the fibers are very short, unconstrained, and subject to low confining stresses, they will tend to slip or pull out when the soil/fiber composite is sheared. They will nevertheless continue to contribute a reinforcing increment. At incipient slippage, the maximum tension in a root-fiber T_N is given by:

$$T_N = 2\tau_b L/D \tag{13}$$

where L is the root length in which the maximum stress occurs at the center.

The shear strength increase or reinforcement from (n) slipping roots of one size class is given by:

$$\Delta S = (\pi\tau_b nLD/2A)[\sin\theta + \cos\theta\tan\phi] \tag{14}$$

If there are (j) slipping root size classes with (n_i) roots in each size class, then the shear strength increase is given by:

$$\Delta S = (\pi\tau_b/2A)[\sin\theta + \cos\theta\tan\phi]\Sigma n_i L_i D_i \tag{15}$$

Under field conditions roots occur in different sizes and lengths, and can have different tensile strengths and degrees of fixity. Accordingly, all three mechanisms may occur simultaneously. *Waldron and Dakessian* [1981] present procedures for systematically accounting for each. These models are idealizations of actual conditions, but they show what parameters are important and how they affect shear strength. Furthermore, the trends and relationships predicted by these simple force-equilibrium models have been validated by laboratory studies.

4.2. Laboratory and In-Situ Tests

The presence of fibers (roots) in a soil increases the shear strength of the soil in ways predicted by the force-equilibrium models described in the previous section. Fiber reinforcement in a sandy, cohesionless soil is manifested in both the stress-strain behavior of the soil/fiber composite and in the failure envelopes as shown in Figures 6 and 7, respectively.

Fiber reinforcement tends to increase the peak stress and reduce the amount of post peak stress loss in dense soils. The failure envelopes in fiber reinforced sands tend to be bilinear as depicted in Figure 7. The initial part of the envelope is steep and then bends over and becomes parallel to the unreinforced envelope. The break point coincides with the critical confining stress. Below this stress the fibers tend to slip, while above this stress the fibers lock in and stretch. The breakpoint in the envelope shifts to the left, i.e., the critical confining stress is reduced, as the fiber length increases and the fibers lock in place more easily.

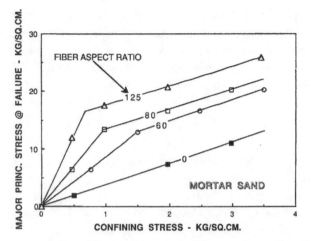

Figure 7. Effect of fiber additions on the failure envelopes of a well graded, angular, dry sand reinforced with randomly distributed fibers with different length/diameter (aspect) ratios. Fiber/soil composites tested in triaxial compression [Maher and Gray, 1990].

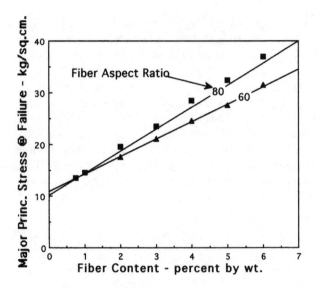

Figure 9. Shear strength increases vs. fiber content for a dune sand tested in triaxial compression with randomly distributed fibers with different length/diameter (aspect) ratios [Maher and Gray, 1990]

A bilinear or curvilinear failure envelope is a trademark of all fiber reinforced soils regardless of the type of test or reinforcement. Bilinear failure envelopes with a sharp, well defined break in the envelope are particularly well manifested in angular, well graded sands as shown in Figure 7. Extrapolation of the second part of the envelope to the ordinate results in a cohesion intercept. In the case of dry, cohesionless sands reinforced with root fibers this intercept defines a shear strength increase sometimes referred to as the "root cohesion."

Laboratory tests show that the shear strength increase or root cohesion is proportional to the amount of fiber or root area

ratio as shown in Figure 8 for both dry dune sand with oriented fibers tested in direct shear and in Figure 9 for the same sand with randomly oriented fibers tested in triaxial compression. This observation is consistent with predictions from the force equilibrium models in the previous section and appears to hold at root area ratios up to 2% or weight concentrations up to 5% the range of practical interest for most root permeated soils. Similar relationships have been noted in field or in-situ tests on root permeated soils. *Endo and Tsuruta* [1969] determined the reinforcing effect of tree roots on soil shear strength by running large scale, in-situ direct shear tests on soil pedestals

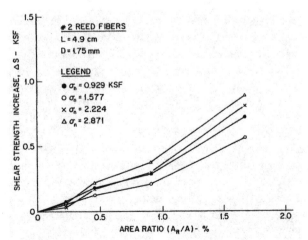

Figure 8. Shear strength increases vs. fiber content for a dune sand with oriented fibers tested in direct shear at different confining stresses [Gray and Ohashi, 1983].

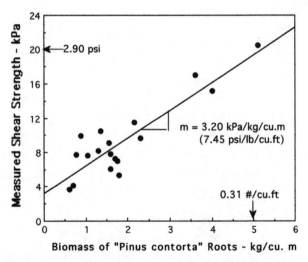

Figure 10. Results for in-situ direct shear tests on sand permeated with pine roots [Ziemer, 1981].

Figure 11. Photographs showing ability of live woody tissue to flow around and engulf (a) wire mesh fencing and (b) metal pole.

of a clay loam containing live roots of young European alder trees (*Alnus glutinosa*). *Ziemer* [1981] conducted in-situ direct shear tests on sands permeated with live roots of pine trees (*Pinus contorta*). The largest roots exposed in this shear cross section were under 17 mm. He also observed an approximately linear increase in shear strength with increasing root biomass, as shown in Figure 10.

5. ENVIRONMENTAL ADAPTATION IN WOODY ROOTS

Plant roots tend to respond to unfavorable stress conditions, vis-à-vis light, lack of moisture, deep burial, and mechanical obstacles (rocks, hard layers, etc.), in a self-correcting manner through a remarkable bio-adaptive process termed edaphoecotropism [*Vanicek*, 1973]. This process allows roots to escape or to adapt to unfavorable site conditions. Deep burial and a need for oxygen, can trigger the growth of secondary roots or adventitious rooting at a higher elevation from the buried trunk. This phenomenon has been observed [*Gray and Leiser*, 1982] in beech trees growing in coastal dunes along Lake Michigan whose trunks are buried by accreting sand and like-

wise by redwood trees growing on alluvial flats along Redwood Creek in California that become buried by sediment.

Edaphoecotropic reorientation of main root branches in trees and shrubs is important both in the safe anchoring of woody plants in the soil and conversely as a way of increasing the stability of the soil or slope anchorage itself. An example of the latter effect is the tendency of buttress or supporting roots on the downhill side of trees to develop larger diameters. Live woody tissue in general has the ability to literally flow around and engulf obstacles (stones, wire mesh, metal posts, etc.) without any disruption or dislocation of the obstacle itself as shown in Figure 11.

Concerns have occasionally been voiced about the danger of incorporating plants into porous protective streambank structures or bank armor, such as riprap and gabion mattresses, lest the structure be damaged or disrupted by stems and roots of woody plants. This concern has been greatly exaggerated; instead of disrupting the structure, woody root tissue tends to flow around and engulf structural components [*Gray and Sotir*, 1996] thereby binding individual armor units or components into a monolithic, unitary mass. In the case of vegetated bank armor the plant roots also

Figure 12. Photograph showing compatibility between porous gabion structure and live willows growing in the structure.

extend through the armor layer into the underlying soil thus increasing the lift off resistance to hydraulic drag forces [*Shields*, 1991]. Accordingly, live willow poles and cuttings can be inserted into and around porous bank armor and they will grow there in a mutually beneficial and compatible manner as shown in Figure 12.

6. CONCLUSIONS

One of the primary mechanisms by which woody vegetation improves mass stability against shallow slope failures is via root reinforcement. The most effective restraint is provided where roots penetrate across the soil mantle into fractures or fissures in the underlying bedrock or where roots penetrate into a residual soil or transition zone whose density and shear strength increase with depth. Both exhumation and profiling techniques that are described herein can be used to determine root architecture and distribution in soils.

Root fibers reinforce a soil by transfer of shear stress in the soil matrix to tensile resistance in the fiber inclusions. Simple force equilibrium models are useful for identifying parameters that affect root reinforcement and predicting the amount of strength increase from the presence of fibers in a soil. These models take into account three possible forms of failure, namely the root fibers slip, stretch, or break.

Plant roots tend to respond to unfavorable stress conditions in a self-correcting manner through a bio-adaptive process termed edaphoecotropism. This process allows roots to escape or to avoid unfavorable site conditions and to enhance their slope stabilization role.

REFERENCES

Biddle, P. G., Patterns of soil drying and moisture deficit in the vicinity of trees on clay soils, *Geotechnique*, 33, 107–126, 1983.

Bohm, W., Methods of studying root systems, *Ecological Services No. 33*, Berlin, Springer-Verlag, 1979.

Burroughs, E. R., and B. R. Thomas, Declining root strength in Douglas fir after felling as a factor in slope stability, *Research Paper INT-190*, Intermountain Forest and Range Experiment Station, 27 pp., US Forest Service, Ogden, UT, 1977.

Coppin, N. J., and I. Richards, *Use of Vegetation in Civil Engineering*, Butterworths, Kent, 1990.

Endo, T., and T. Tsuruta, The effect of tree roots upon the shearing strength of soil, *Annual Rept. of Hokkaido Branch, Tokyo Forest Experiment Stn.*, Vol. 18, 168–179, 1969.

Gray, D. H., How to maximize the benefits of slope plantings, *Landscape Architect and Specifier News*, 17, 32–37, 2001.

Gray, D. H., and R. Sotir, *Biotechnical and Soil Bioengineering Slope Stabilization*, Wiley, New York, 1996.

Gray, D. H., and A. Lieser, *Biotechnical Slope Protection and Erosion Control*, Van Nostrand Reinhold, New York, 1982.

Gray, D. H., and H. Ohashi, Mechanics of fiber reinforcement in sands, *Journal of Geotechnical Engineering*, 109, 335–353, 1983.

Greenway, D. R., Vegetation and slope stability, in *Slope Stability*, edited by M. F. Anderson and K. S. Richards, Wiley, New York, 1987.

Hathaway, R. L., and D. Penny, Root strength in some *Populus* and *Salix* clones, *New Zealand Journal of Botany*, 13, 333–344, 1975.

Henderson, R., E. D. Ford, J. D. Deans, and E. Renshaw, Morphology of the structural root system of Sitka spruce 1: Analysis and quantitative description, *Forestry*, 56, 122–135, 1983.

Kozlowski, T. T., *Growth and Development of Trees*, 520 pp., Academic Press, New York, 1971.

Maher, M., and D. H. Gray, Static response of sands reinforced with randomly distributed fibers, *Journal of Geotechnical Engineering*, 116, 1661–1677, 1990.

Nilaweera, N. S., *Effects of tree roots on slope stability*, unpublished thesis for degree of Doctor of Technical Science, Asian Institute of Technology, Bangkok, Thailand, 1994.

Patric, J. H., J. E. Douglass, and J. D. Hewlett, Soil water absorption by mountain and piedmont forests, *Soil Science Society of America Proceedings*, 29, 303-308, 1965.

Schiechtl, H. M., *Bioengineering for Land Reclamation and Conservation*, 404 pp., University of Alberta Press, Edmonton, Canada, 1980.

Shewbridge, S. E., and N. Sitar, Deformation characteristics of reinforced sand in direct shear, *Journal of Geotechnical Engineering*, 115, 1134–1147, 1989.

Shewbridge, S. E., and N. Sitar, Deformation based model for reinforced sand in direct shear, *Journal of Geotechnical Engineering*, 116, 1153–1157, 1990.

Shields, F. D., Woody vegetation and riprap stability along the Sacramento River mile 84.5 to 119, *Water Resources Bulletin*, 27, 527–536, 1991.

Shields, F. D., and D. H. Gray, Effects of woody vegetation on the structural integrity of sandy levees, *Water Resources Bulletin*, 28, 917–931, 1993.

Sutton, R. F., Form and development of conifer root systems, *Technical Communication No. 7*, Commonwealth Agricultural Bureau, England, 1969.

Tsukamoto, Y., and O. Kusuba, Vegetative influences on debris slide occurrences on steep slopes in Japan, in *Proceedings, Symposium on Effects of Forest Land Use on Erosion and Slope Stability*, Environment Policy Institute, Honolulu, Hawaii, 1984.

Turmanina, V. I., On the strength of tree roots, *Bulletin Moscow Society Naturalists*, 70, 36–45, 1965.

Vanicek, V., The soil protective role of specially shaped plant roots, *Biological Conservation*, 5, 175–180, 1973.

Waldron, L. J., The shear resistance of root-permeated homogeneous and stratified soil, *Soil Science Society of America Proceedings*, 41, 843–849, 1977.

Waldron, L. J., and S. Dakessian, Soil reinforcement by roots calculation of increased soil shear resistance from root properties, *Soil Science*, 132, 427–35, 1981.

Wu, T. H., W. P. McKinell, and D. N. Swanston, Strength of tree roots and landslides on Prince of Wales Island, Alaska, *Canadian Geotechnical Journal*, 16, 19–33, 1979.

Wu, T. H., R. M. Macomber, R. T. Erb, and P. E. Beal, Study of soil-root interactions, *Journal of Geotechnical Engineering*, 114, 1351–1375, 1988a.

Wu, T. H., P. E. Beal, and C. Lan, In-situ shear test of soil-root systems, *Journal of Geotechnical Engineering*, 114, 1376–1394, 1988b.

Ziemer, R., Roots and shallow stability of forested slopes, *International Association of Hydrological Sciences*, 132, 343–361, 1981.

D. Barker, Division of Civil Engineering, Environmental Health and Safety Management, Nottingham Trent University, Nottingham, NG1 4BU, United Kingdom.

D. H. Gray, Department of Civil and Environmental Engineering, The University of Michigan, Ann Arbor, MI, 48104.

Advances in Assessing the Mechanical and Hydrologic Effects of Riparian Vegetation on Streambank Stability

Natasha Pollen

Department of Geography, King's College London, UK and Channel and Watershed Processes Research Unit, USDA-ARS National Sedimentation Laboratory, Oxford, Mississippi.

Andrew Simon

Channel and Watershed Processes Research Unit, USDA-ARS National Sedimentation Laboratory, Oxford, Mississippi.

Andrew Collison

Philip Williams & Associates, San Francisco, California.

Streambank instability poses a number of economic and ecological problems. As sediment has been reported to be one of the principal contaminants of rivers in many areas, a high priority of river managers is to stabilize streambanks to prevent additional sediment being added to the channels. Riparian vegetation plays a number of roles in the protection of streambanks from erosion by the processes of particle entrainment and mass wasting, and its use in stabilization has a number of possible benefits, but is often neglected in favor of hard stabilization measures, such as concrete and riprap, that are more easily quantifiable and remain constant over time. This study seeks to investigate the importance of the assumptions previously made in calculations of soil reinforcement by roots, and aims to study how root networks, and the contribution to soil strength made over time varies both mechanically and hydrologically. Results show that previous methods used to estimate root reinforcement may have overestimated values by up to 91%. A new Fiber-Bundle model (RipRoot) is proposed here to reduce reinforcement overestimations. Hydrologic reinforcement by evapotranspiration may also be important. Results show that soil cohesion values were increased by 1.0 to 3.1 kPa due to reductions in matric suction by trees of just two-years old. The net effects of mechanical and hydrologic reinforcement have also been investigated, with results showing that even during the wettest time of the year, when evapotranspiration effects are negligible, the mechanical reinforcement from the root networks maintains some degree of stability.

Riparian Vegetation and Fluvial Geomorphology
Water Science and Application 8
Copyright 2004 by the American Geophysical Union
10.1029/008WSA10

1. INTRODUCTION

Unstable streambanks affect a wide range of stream users due to the additional sediment supplied to channels by mass-wasting processes and fluvial entrainment of bank material, and the resulting retreat of banks. Streambank instability poses a number of economic and ecological problems including loss of land, and destabilizing of structures such as bridges, whilst the addition of additional sediment to the channels causes downstream aggradation and impairs water quality. The US Environmental Protection Agency (EPA) recently reported that sediment is one of the principal pollutants of rivers in the USA, both in terms of sediment quantity and sediment quality due to adsorbed contaminants [*EPA*, 2002]. EPA estimate that in Mississippi alone, there are 72,256 km of rivers that are impaired by sediment [*EPA*, 2002], with impairment being defined as a condition where erosion and sediment transport rates and amounts are so great that biologic communities, or other designated stream uses are adversely affected. *Simon and Thorne* [1996] showed that in the loess area of the Midwest United States, streambank material contributed as much as 80% of the total sediment load eroded from incised channels. Streambank stabilization of sediment-impaired channels is thus a high priority for river managers, and with growing environmental concern over the 'hard-stabilization', measures such as steel and stone, that are commonly used by engineers, these same managers are becoming increasingly interested in the possible use of vegetation in providing stabilization.

Riparian vegetation plays a number of roles in protecting streambanks from erosion by particle entrainment and mass wasting. For example, the above ground parts of vegetation protect banks from shear stresses exerted by flowing water [*Hickin*, 1984], whilst below ground parts of the vegetation play an important role in anchoring vegetation, and increasing soil strength through the production of a soil-root matrix [*Thorne*, 1990]. In addition, riparian areas are ecologically significant as they provide paths for migration of species, and represent zones of relatively high species diversity due to their position at the land-water interface [*Malanson*, 1993]. As such these areas should be protected and restored where prudent, rather than replaced with concrete and other hard engineering alternatives. Vegetation as a means of stabilizing streambanks, is also much more cost effective than hard engineering alternatives, with the main cost involved being related to labor rather than materials costs.

However, riparian vegetation can also have adverse effects on streambank stability, and the net balance of the benefits and disadvantages are hard to quantify [*Simon and Collison*, 2002]. The main problems encountered are related to the dynamic nature of vegetation, its inherent natural variability, and issues concerning the investigation of root networks, which often requires destructive and invasive studies. Therefore, whilst vegetation has been widely used in stream restoration projects, its use in streambank stabilization schemes has often been limited. The problems involved in estimating the stabilizing effects of vegetation on streambank stability must be over-

come before its use in stabilizing schemes will become more extensive. The overall objective of this study is to estimate the mechanical root reinforcement of vegetation of different types and ages, and the hydrologic effects riparian species have on streambank stability.

2. BACKGROUND

Riparian vegetation has both mechanical and hydrologic impacts on streambank stability, some of which are positive and some of which are negative. The mechanical effects are for the most part beneficial. Roots anchor themselves into the soil to support the above ground parts of the vegetation, and in doing so, produce a reinforced soil matrix in which stress is transferred from the soil to the roots, increasing the overall shear strength of the matrix [*Greenway*, 1987]. As soil is strong in compression but weak in tension, and conversely, plant roots are weak in compression but strong in tension, when the two are combined they produce a matrix of reinforced earth, which is stronger than the soil or the roots separately [*Thorne*, 1990]. The roots are hence able to increase the confining stress of the soil and provide reinforcement by transferring shear stress in the soil to tensile resistance in the roots [*Gray and Sotir*, 1996].

The disadvantageous mechanical impacts of vegetation on soil stability are associated with the forces exerted from the above ground parts of the vegetation. The weight of the vegetation, in particular mature trees, produces a surcharge on the slope or streambank, increasing the driving forces acting in the downslope direction, and reducing the soil stability.

The beneficial hydrological impacts of vegetation on soil stability also include processes that occur above and below ground. During rainfall, the vegetative canopy intercepts rainfall, thereby decreasing the amount of water available for infiltration. Vegetation also removes water from the soil due to extraction of water from the root zone for use in the processes occurring in the above ground biomass [*Dingman*, 2001]. Pore-water pressures in the soil hence remain lower, and the likelihood of mass failure is reduced [*Selby*, 1993]. Reduction of pore-water pressures and the associated increase in matric suction within streambanks have been found to provide significant amounts of additional strength, in some cases greater than the strength provided by the soil-root matrix [*Simon et al.*, 1999; *Simon and Collison*, 2002].

The hydrologic disadvantages of vegetation on soil stability are related to the way in which soil infiltration characteristics are altered both at the soil surface and deeper within the soil profile. At the surface, canopy interception and stem flow tend to concentrate rainfall locally around the stems of plants, creating higher local pore-water pressures [*Durocher*, 1990]. The presence of stems and roots at the soil surface can also act to disturb the imbrication of the soil, hence increasing infiltration capacity. The infiltration capacity of the soil may also be increased as a result of increased desiccation cracking in certain soils, particularly those containing a high clay content [*Lambe and Whitman*, 1969]. An increase in the soil infiltration capacity creates higher pore water pressures inside the streambank, hence reducing its stability.

Once the water has entered the soil, the roots present may also channel water to greater depths more quickly [*Collison and Anderson*, 1996; *Simon and Collison*, 2002]. For example, *De Roo* [1968] notes that roots often contract as they become drier, and that this combined with shrinkage of drying soil, creates a gap between the roots and the soil. This gap serves to protect the plant's roots from rapid water loss as the resistance of water movement from the roots to the soil is increased, but may also act as a macropore, channeling water at higher speeds and to greater depths in the soil profile [*Collison and Anderson*, 1996; *Martinez-Meza and Whitford*, 1996]. In most cases this allows water to build up quicker and deeper in the soil, nearer to a potential shear zone, causing loss of matric suction and increasing pore-water pressures. To some extent, the degree to which the preferential flow of water affects soil stability depends on the soil materials and stratigraphy present.

Although studies have been carried out to assess the overall impacts of vegetation on hillslope stability, fewer studies have attempted to quantify the effects of vegetation on streambanks. As there are problems with the direct transfer of results from hillslope studies to streambanks, little is known about the mechanical and hydrological impacts of riparian vegetation on streambank stability. In contrast to hillslopes, streambanks tend to be steeper and shorter with a more varied profile [*Abernethy and Rutherfurd*, 2000]. In addition, most of the research on the effects of vegetation on hillslope stability has been carried out in upland areas, and mainly for use in the forestry industry in areas such as NW USA [for example, *Ziemer and Swanston*, 1977]. Therefore most of the data available does not relate to riparian species and often only the contribution made by mature trees is considered. Exceptions to this general trend do exist however, and include the studies that have been carried out in Mississippi by *Simon and Collison* [2002], *Pollen*, [2001], *Easson and Yarbrough* [2002] and in Australia by *Abernethy and Rutherfurd* [1998; 2000].

2.1. Quantification of Mechanical Reinforcement by Root Networks

The theory for quantifying root reinforcement stems from literature developed to calculate the increased strength added to other composite materials used in the construction industry, such as reinforced concrete. *Greenway* [1987] notes that

the magnitude of root reinforcement depends on root growth and density, root tensile strengths, root tensile modulus values, root tortuosity, soil-root bond strengths, and the orientation of roots to the principal direction of strain.

In order to quantify the effects of root reinforcement on soil strength, two methods have commonly been used. In the first method, the values collected from in-situ shear tests of root-permeated soils have been used to replace the value of the soil strength alone [e.g. *Wu et al.*, 1988]. However, in-situ shear tests present a number of problems. For example, isolating a block of root permeated soil to shear is not an easy task, and the soil and the anchoring of the roots may be disturbed before shearing is undertaken. The second method involves the development of physically-based force-equilibrium models. One such model is the simple perpendicular root model developed by *Waldron* [1977]. This root reinforcement model is based on the Coulomb equation in which soil shearing resistance is calculated from cohesive and frictional forces:

$$S = c + \sigma_N \tan\phi \qquad (1)$$

where σ_N is the normal stress on the shear plane, ϕ is angle of internal friction, and c is total cohesion

To extend Equation 1 for root-permeated soils, *Waldron* [1977] assumed that all roots extended vertically across a horizontal shearing zone, and the roots acted like laterally loaded piles, so tension was transferred to them as the soil sheared. The modified Coulomb equation therefore became:

$$S = c + \Delta S + \sigma_N \tan\phi \qquad (2)$$

where ΔS is increased shear strength due to roots

In this simple root model, the tension developed in the root as the soil is sheared is resolved with a tangential component resisting shear and a normal component increasing the confining pressure on the shear plane and ΔS is represented by:

$$\Delta S = T_r (\sin\theta + \cos\theta \tan\phi) \qquad (3)$$

where T_r is average tensile strength of roots (kPa), θ is the angle of shear distortion.

Gray [1974] reported that results of several studies on root-permeated soil showed that the angle of internal friction of the soil appeared to be affected little by the presence of roots. Sensitivity analyses carried out by *Wu et al.* [1979] showed that the value of the bracketed term in (3) is fairly insensitive to normal variations in θ and ϕ (40 to 90° and 25 to 40°, respectively) with values ranging from 1.0 to 1.3. A value of 1.2 was therefore selected by *Wu et al.* [1979] to replace the bracketed term leaving:

$$\Delta S = T_r \, (A_R \, /A) \times 1.2 \qquad (4)$$

where $A_R \, / A$ is root area ratio (dimensionless), A is the area of the soil (m^2), and A_R is the area of roots (m^2)

2.1.1 Assumptions and limitations of Wu et al's equation [1979]. Four important assumptions are made in the simple root-reinforcement model of *Wu et al.* [1979]. First, the model assumes that the roots are perpendicular to the slip plane. However, the angles of the roots in relation to the direction of the force applied to the soil are important, as this dictates the distribution of stresses within the root volume and the maximum tensile strength reached before failure of the root occurs [*Niklas*, 1992]. Extended models allowing for inclined roots have been developed, but *Gray and Ohashi* [1983] have shown from laboratory tests that perpendicular orientations of reinforcing fibers provide comparable reinforcement to randomly oriented fibers. This lends support to the use of the simple perpendicular root model, where it may be assumed that the roots are randomly orientated in the soil.

Second, the model assumes that the full tensile strength of all the roots is mobilized when the soil shears. However, laboratory and field strength testing of streambank materials and riparian roots show that root strength is typically mobilized at much larger displacements than soil strength [*Pollen et al.*, 2002]. In a soil, peak strength is typically mobilized in the first few millimeters of strain, and then decreases to a residual value reflecting particle realignment that minimizes shear resistance. In roots, some displacement is required to straighten the tortuosity before strain is taken up and they then stretch first elastically and then non-elastically [*Collison et al.*, 2001]. This suggests that peak root strengths may not be fully mobilized at the time of maximum soil instability, and that the banks may fail before the full theoretical contribution from roots is achieved. Over prediction of increased shear strength may occur [*Waldron and Dakessian* 1981], but analysis of the stress-displacement characteristics of roots has been limited to date.

Third, the model assumes that the roots are well anchored and do not pull out when tensioned. Laboratory and field shear tests have shown that, as with the failure of other composite materials, root failure occurs by two mechanisms: pull-out (slipping due to bond failure) or rupture (tension failure) [*Coppin and Richards*, 1990]. The predominant failure mechanism is a function of the variations in material properties and the geometries of the fibers and the matrix [*Beaudoin*, 1990]. If pull-out failure is an important mechanism in the mass failure of root-reinforced streambanks, then it is necessary to know the typical forces required for this to take place.

Fourth, the model assumes that the soil fails by shearing. However, in the case of a streambank cantilever failure, it is the tensile strength of the soil that is more important than the shear strength of the soil. When a streambank fails by cantilever, the mechanics are similar to those of a fiber-reinforced composite that is placed in tension. In such a case, the stress-strain curve can be separated into three distinct regions in which first the matrix and the fibers take up the load applied together, second the matrix cracks, and third all of the load is transferred to the fibers until the fibers fail either in tension, or by pull-out [*Beaudoin*, 1990; *Kutzing and Konig*, 1999)]. In this case, the simple addition of the maximum tensile strengths of the roots, to the maximum tensile strength of the soil would again provide an overestimation of the strength provided by the roots.

2.1.2 Research questions and the scope of this study. The first part of this study examines the material properties of roots and soil in field and laboratory situations, so that the assumptions made by force-equilibrium models such as *Wu et al.* [1979] can be tested. In addition, root architecture studies show how the density of roots of different sizes varies through the soil profile and with the growth of riparian trees. The second part of the study examines hydrologic interactions between riparian vegetation and streambank stability. This was investigated by collecting pore-water pressure data under different vegetative plots at the Goodwin Creek research site in northern Mississippi. An additional experiment conducted at the USDA-ARS National Sedimentation Laboratory addresses the effects of different riparian species on soil pore-water pressures by growing specimens of each species in isolated soil monoliths. The final part of the study was to examine the relative and total effects of mechanical and hydrologic effects of vegetation, on streambank stability.

2.1.3 Modeling streambank stability. Almost all streambank stability models are limit equilibrium models where driving forces and resisting forces are calculated to determine a Factor of Safety [*Duncan*, 1992]. The model used in this study was an updated version of the static 2-dimensional limit equilibrium wedge-failure model developed by *Simon et al.* [1999], which was itself a refinement of models previously developed by *Osman and Thorne* [1988] and *Simon et al.* [1991], so that the failure plane was no longer constrained to pass through the bank toe, and the confining pressure of the flow was taken into account. The model considers the balance between the forces resisting movement of the bank (RF), and those forces driving movement of the bank (DF) for up to five soil layers, to produce a factor of safety (F_s) value, where a value of less than 1 means the bank is unstable (RF<DF) and a value of greater than 1 means the bank is stable (RF>DF). Erosion of the bank toe and different vegetative covers are also included in the model.

This particular Bank Stability model was developed for use with cohesive, multi-layered banks. The model accounts for the geotechnical resisting forces by using the failure criterion of the Mohr-Coulomb equation [Simon et al., 1999] for the saturated part of the failure surface (Eq. 2). For the unsaturated part of the failure surface the criterion modified by Fredlund et al. [1978] is used:

$$S_r = c' + (\sigma - \mu_a) \tan \phi' + (\mu_a - \mu_w) \tan \phi^b \qquad (5)$$

where S_r is shear strength (kPa), c' is effective cohesion (kPa), σ is normal stress (kPa), μ_a is pore air pressure (kPa), and μ_w is pore water pressure (kPa). The angle ϕ^b describes the rate of increase in shear strength with increasing matric suction. The geotechnical driving force is given by the term:

$$F = W \sin\beta \qquad (6)$$

where, F is driving force acting on bank material (N), W is weight of failure block (N), and β is angle of the failure plane (degrees).

3. EXPERIMENTAL METHODS

3.1. Measuring Root Tensile Strengths and Stress-Displacement Characteristics

Root tensile strength and stress-displacement were measured in the field using a device based on a design by Abernethy [1999], comprising a metal frame with a winch attached to a load cell and displacement transducer. Different size roots were attached to the load cell and displacement transducer and placed in tension until the root broke, and the diameter of each root was recorded. The species tested were Eastern Sycamore (Plantanus occidentalis), River Birch (Betula nigra), Sweetgum (Liquidamber stryaciflua), Black Willow (Salix nigra), Sandbar Willow (Salix exigua), Longleaf Pine (Pinus palustris Miller), Switch grass (Panicum virgatum), Gamma grass (Tripsacum dactyloides), Western Cottonwood (Populus fremontii), Douglas Spirea (Spirea douglasii), Himalayan Blackberry (Rubus discolor), and Oregon Ash (Fraxinus latifolia). All of these species were selected because of their common occurrence in riparian areas of the USA.

The maximum load applied to each root before breaking and its diameter were used to calculate the tensile strength of each root. The stress-displacement characteristics of each root were also studied using the displacement transducer

data, and these were compared to the stress-displacement characteristics of direct shear-box tests run on root-permeated and non-root-permeated soil samples taken at Goodwin Creek from the upper layer of an outside meander bend consisting of moderately cohesive, clayey-silt. The comparison between soil and root loading rates were used to test the assumption made by Wu et al. [1979] that all of the roots break simultaneously, and at the same time that the soil reaches its peak strength.

Root-soil friction was also measured using the pulling device in order to assess the relative forces required for roots to be pulled out of the soil and for roots to be broken. The load cell measured the force required to either pull the root out of the soil, or for it to break. Root-soil friction experiments were carried out at a site similar to Goodwin Creek at different times of the year (April and July 2002) to test under varying soil moisture conditions.

3.2. Recording Root Size and Frequency Distributions

The root system of woody or herbaceous riparian vegetation was examined and recorded using a method similar to the wall-profile method of Bohm [1979]. The roots were exposed by digging a trench measuring approximately 1.5 m across by 1 m depth, at a distance of approximately 0.5 m from the main stem or stems. The roots were then cut back to the face of the trench wall and a grid was placed against the wall of the trench so that the root diameters and depths could be recorded. A range of tree ages (3 to 10 years) was selected for each species using dendrochronology. The same site was used for all ages of each species to minimize the number of variables changing between specimens.

Initial estimates of ΔS provided by the roots of each specimen studied were calculated using the equation of Wu et al. [1979]. These values were then averaged over 1 m depth of soil to produce the average reinforcement value added to this depth by each tree/shrub/clump of grass to be included in the Bank Stability Model. Such averaging also provides a good way of comparing the effects of different specimens and species.

3.3. Hydrologic Monitoring of Riparian Species

Soil monoliths were prepared at the USDA-ARS National Sedimentation Laboratory, Oxford, MS, using soil from the same source to ensure the same growth conditions were replicated in each monolith. Each monolith was placed outdoors on a wooden base to allow for free drainage. Riparian trees (Eastern Sycamore, Black Willow and River Birch) were planted in the summer of 2000. The two healthiest trees of each species, along with two control monoliths (bare soil), were instrumented with tensiometers in Febru-

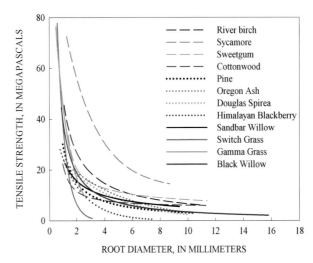

Plate 1. Root tensile strength versus root diameter for twelve common riparian species of the USA.

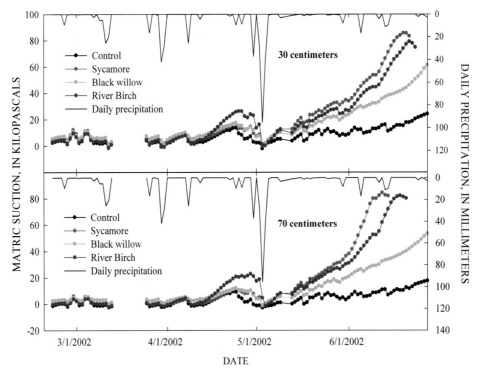

Plate 2. Matric suction values for the period February 18 to June 27, 2002 in soil monoliths containing bare soil (control) and three riparian woody species.

ary 2002. A rain gauge and tensiometers at depths of 0.3 m and 0.7 m below the soil surface were installed in each monolith.

4. RESULTS AND DISCUSSION

4.1. Root Tensile Strengths

Root tensile strengths decreased non-linearly with increasing root diameter for all species tested. Apart from E. Sycamore the tensile strengths of the roots of different riparian woody and herbaceous species tested in this study to date do not differ significantly (p = 0.95) (Plate 1). It is surprising that the tensile strengths of the roots of different riparian species lie within the same range given the different root textures and the range of ages of the roots tested. The breaking mechanism of the roots tested also varied; sometimes it was the epidermis, or 'skin' of the root that broke and then slipped off the cortex of the root, and sometimes the whole root broke. Whilst both breaking mechanisms were observed, the root structure of some species made them more prone to root 'skinning'.

Of 45 Sandbar Willow roots tested, only 6 of them failed by skinning (13%) but for Western Cottonwood, 18 out of 60 (30%) roots tested broke by skinning. However, when the data was analyzed, the force taken to 'skin' the roots was not significantly different to that required to break them. The structural differences of roots of different species and ages appears to have little effect on the breaking stresses.

4.2. Root Architecture Investigations

The root studies have shown a variety of root architectures within the young riparian trees studied. The growth and forms of those systems, although determined to some extent by species type, are largely controlled by variations in site conditions, in particular the drainage of the bank material and the position of the vegetation in relation to the water table. For example, many of the young trees excavated on sandbars deposited in times of high flow exhibited a deep taproot extending to the water table, with many small roots growing at the taproot's apex. In contrast, most of the saplings excavated in the bank toe region had a flat, plateroot system as the devel-

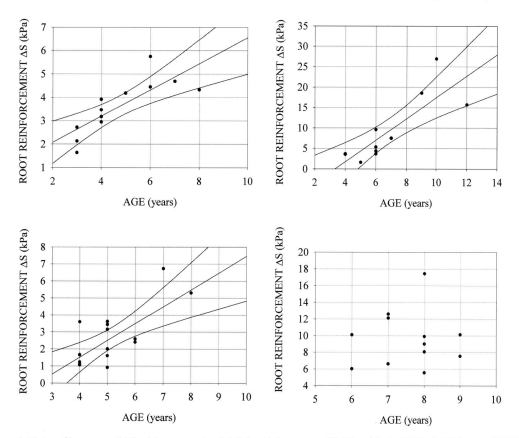

Figure 1. Rates of increase of ΔS with tree age (p=0.95) for A) Sweetgum, B) River birch, C) Black Willow and D) Eastern Sycamore. The horizontal line on D) represents the mean value for the trees studied: no significant trend with age existed for this species.

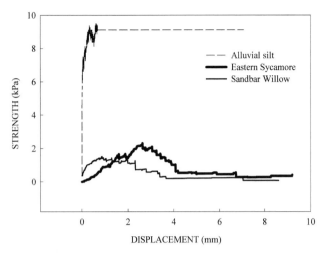

Figure 2. Cumulative stress-strain curves for 30 roots of each species, compared to the stress-strain curve of a streambank soil sample.

opment of a taproot was not necessary since the water table was close to the soil surface.

The more established riparian trees positioned on bank tops, or higher terraces, tended to show a heartroot system. Results of the root mapping carried out for Eastern Sycamore, River Birch, Sweetgum, Black Willow, Sandbar Willow, Longleaf Pine, Switch grass, Gamma grass, Western Cottonwood, Douglas Spirea, Himalayan Blackberry, and Oregon Ash, all show that the total number of roots decreases exponentially with increasing depth through the soil profile. However, when the root data was divided into different root diameter size classes, it was noted that the variations in root numbers with soil depth were different for roots of the different size classes. As the tensile strength of the roots is dependent on their diameter, knowledge of these varying distributions of different size roots is important if we are to calculate the variation of ΔS through the soil profile. In the case

of the smaller root size classes (diameters <1 mm, 1 to 2 mm, 2 to 3 mm, 3 to 5 mm) the trend was the same as that of the overall number of roots: there was an exponential decline in numbers of small roots with increasing soil depth, and the number of these smaller roots increases in older specimens. The distribution of the larger diameter roots (5 to 10 mm, >10 mm) differed from that of the smaller roots: the larger roots tended to be concentrated within particular layers of the soil, often at greater depth than the smaller roots which were concentrated towards the top of the soil profile. This is likely due to the presence of zones of high nutrient or moisture content [*pers. comm., Balch,* 2001]. As with the smaller diameter roots, the number of larger diameter roots increased with the age of the specimen, but there were much fewer larger roots, which explains why the total number of roots within the soil profile still shows an exponential decline with increased soil depth.

4.3. Calculations of ΔS Using the Root Reinforcement Model of Wu et al. (1979)

Values of ΔS for each tree sampled were averaged over 1m and plotted against tree age (Figure 1). These graphs show the results for four of the tree species studied. For River Birch the rate of increase in ΔS was 1 kPa/yr for the first five years and 2.6 kPa/yr subsequently, for Black Willow ΔS was 0.5 kPa/yr, and for Sweetgum the rate was 0.8 kPa/yr. After 10 years the average ΔS values were: River Birch 18 ± 5 kPa, Black Willow 7 ± 2.5 kPa, and Sweetgum 7 ± 1.5 kPa. The relationship between age and reinforcement was not significant for Sycamore, but the mean strength for the data set was 9 kPa. Although significant linear relationships exist between age and ΔS for Sweetgum, River Birch and Black Willow at 95% confidence level, there is considerable scatter within the datasets.

Table 1.

Vegetation	Scenario	ΔS (kPa)	% Full Mobilization	% Overestimation	F_s
No Vegetation	No mechanical root reinforcement	0	0	0	0.91
		3	100	-	1.07
Sandbar Willow	Sum of the ultimate tensile strength of all the roots				
	1. ΔS at peak of cumulative root strength curve	1.5	50	50	1
	2. ΔS at peak soil strength	1.36	45	55	0.99
		5.6	100	-	1.31
Eastern Sycamore	Sum of the ultimate tensile strength of all the roots				
	1. ΔS at peak of cumulative root strength curve	2.3	41	59	1.03
	2. ΔS at peak soil strength	0.52	9	91	0.94

Table 1. Different Factor of Safety values for a 2 m high streambank composed entirely of silt, under different vegetative treatments and assuming different root and soil mobilization scenarios.

Although linear trend lines have been fitted to the data, with more available data, more complex root growth patterns and changes in ΔS over time may be seen. For example, root growth may follow the sigmoidal shape growth curve, which is characteristic of so many growth functions in biological organisms. It should be noted that age is not necessarily the best predictor of tree development and root growth. Not all trees of same age are the same size, due to variations in growing conditions between sites and even along the same streambank. Comparing increased soil strength due to roots with tree diameter breast height should provide better relationships.

4.4. Investigation of the Assumptions Made by the Simple Root Reinforcement Model of Wu et al. (1979)

4.4.1 Material properties of roots and soil. Initial data analysis of the root stress-displacement curves of Sandbar Willow and Eastern Sycamore showed that the roots of the two species took up strain at different rates. The Eastern Sycamore roots held an average load of 15.15 kg before they snapped, but were typically displaced a larger distance before this occurred (mean displacement = 35.7 mm). In contrast, the Sandbar Willow roots withstood smaller loads before snapping (mean maximum load = 11.6 kg), and the overall displacement of these roots was less (mean displacement = 18.9 mm). The rates of initial uptake of stress were also different, with the Sandbar Willow roots taking up more stress in the first few millimeters of displacement. The cumulative strength provided by 30 roots of each species was calculated by summing the strengths of each individual root as displacement increased. The cumulative plots of root strength for Eastern Sycamore and Sandbar Willow (Figure 2) show that the strength provided

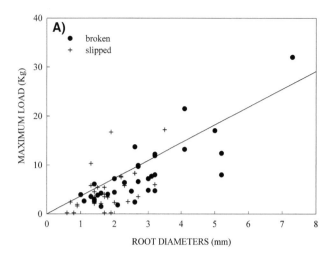

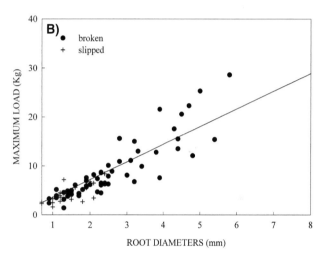

Figure 4. Soil-root friction results: A) April and B) July.

increased to a peak value, although some of the roots snap before this point, shown by sudden drops in the cumulative strength (Figure 2). The results of the first shear-box sample run for bank material at Goodwin Creek, showed that peak soil strength is reached at 6.8 mm displacement (Figure 2), which suggests that differences in the uptake of strain in roots may be crucial in predicting accurately the increased ΔS provided by roots.

The cumulative stress-displacement curves in Figure 2, show that the roots do not all fail simultaneously, and the peak in cumulative root strength may not occur at the same displacement as peak soil strength. The initial ΔS values estimated using *Wu et al.* [1979] in this study are therefore overestimations.

Two situations other than full mobilization are actually likely to occur. First, peak soil strength is reached within the first centimeter of soil displacement, but the roots continue to

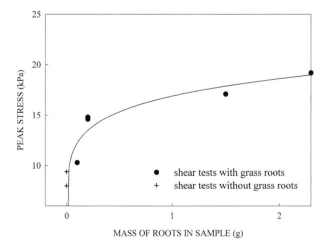

Figure 3. Peak stresses reached in root-permeated and non-root-permeated direct shear-box tests.

hold the soil together to a larger displacement. The strength of the matrix is the residual strength of the soil added to the peak in cumulative strength provided by the roots. Second, the streambank fails after peak soil strength is reached despite the presence of the roots. This occurs because the displacement of the soil increases the driving forces acting on the bank to a point where the roots cannot hold the soil in place, in which case some roots break and others are simply pulled out of the soil as the block fails. The strength of the matrix is the peak soil strength added to the cumulative strength provided by the roots at the displacement where the peak soil strength is reached.

The ΔS values calculated for these two different scenarios show markedly different results from the estimations of ΔS used to date (Table 1). Original estimates of ΔS were 3 kPa for Sandbar Willow and 5.6 kPa for Eastern Sycamore. If scenario 1 holds true, then these values become 1.5 kPa for Sandbar Willow and 2.3 kPa for Eastern Sycamore, showing overestimation of 50% and 59% respectively. When these values are included in the Bank Stability model [*Simon et al.*, 1999], a simulation of a 2m high silt streambank shows that the F_s values are reduced from the original predictions of 1.07 for Sandbar Willow to 1.00 and from 1.31 for Eastern Sycamore to 1.03. If scenario 2 holds true, then the overestimation in the ΔS values from the original values is even greater. The estimated ΔS values were 1.36 kPa for Sandbar Willow and 0.52 kPa for Eastern Sycamore, showing overestimation of 55% and 91% respectively. These overestimations of ΔS were more pronounced for Eastern Sycamore due to a difference in the rate of uptake of stress with displacement. Although Eastern Sycamore roots reach a higher ultimate tensile strength than the Sandbar Willow roots (2.3 kPa and 1.5 kPa respectively), the

Sandbar Willow roots had taken up 1.36 kPa by the time the soil reaches peak strength, whereas the Eastern Sycamore roots had only taken up 0.52 kPa (Figure 2).

Preliminary results from direct shear-box tests carried out on root-permeated and non-root-permeated soil samples confirm that although the presence of roots does increase the values of ΔS (Figure 3), the magnitudes of these values of ΔS are overestimated by the *Wu et al.* [1979] equation. In the case of the first of the root-permeated soil samples tested, the value of ΔS measured by the shear box was just 34 % of that estimated using the *Wu et al.* [1979] equation for the roots in that soil sample (5 kPa versus 20 kPa).

4.4.2 Soil-root friction properties. Results of tests to determine root-soil friction forces are shown in Figure 4. It can be seen that above a certain root diameter, all of the roots broke. Below this threshold some roots broke and some were pulled out of the soil whole. This suggests that there is a threshold diameter above which the friction between the soil and roots exceeds the root tensile strength and so the root breaks. Below the threshold, the roots were pulled out of the soil if the force required to break the root-soil friction bond was less than the force required to break the root. It is possible that in the instances where roots below the threshold diameter broke, the roots may have been branched. The presence of such branches would have increased the surface area in contact with the soil, hence increasing the root-soil friction bond to an extent where the force required to break the root was smaller than that to pull the root out of the soil.

The root-soil friction tests conducted in April and July were carried out under soil moisture contents of 21.1% and 11.3% respectively. The gradients of the trend lines are almost the same (3.64 for the April tests and 3.61 for the July tests), even under the different soil moisture conditions. This is because the strengths of the roots, are independent of soil moisture. However, the change in soil moisture did affect the threshold diameter above which all roots broke. During the April tests the threshold value was approximately 3.5 mm whereas in the July tests the threshold value was approximately 2.4 mm.

The threshold diameter separating those roots that break from those that can be pulled out of the soil, changed under different soil moisture conditions suggesting that as the soil dries out the frictional bonds between the roots and the soil become stronger (Figure 4). More roots break when a drier soil is sheared rather than being pulled out, as the frictional force required to pull the roots out exceeds the breaking force required at a smaller root diameter. This supports the results of *Ennos* [1990] who found that the leek radicals tested were more easily extracted whole from the soil under wetter soil conditions.

The forces required to pull the roots out of the soil did not differ significantly from those required to break the roots

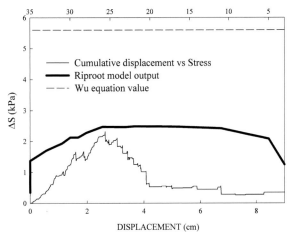

NUMBER OF ROOTS STILL INTACT THROUGH MODEL RUN OF RIPROOT

— Cumulative displacement vs Stress
— Riproot model output
– – – Wu equation value

Figure 5. Comparison of ΔS values for Eastern Sycamore, from the cumulative displacement-stress curve, the RipRoot Model and the Wu equation.

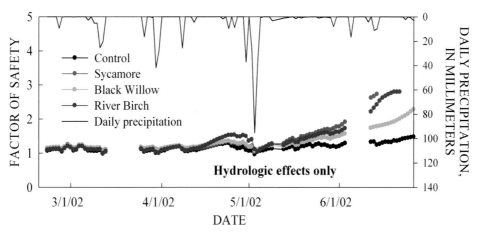

Plate 3. Factor of safety values for a 4.7 m high bank using pore-water pressure data from the soil monoliths and geotechnical and bank geometry data from the Goodwin Creek research bendway.

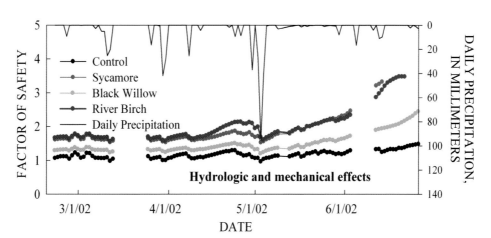

Plate 4. Factor of safety values for a 4.7 m high bank using pore-water pressure data from the soil monoliths and geotechnical and bank geometry data from the Goodwin Creek, Mississippi research bendway.

Table 2.

Soil monolith	0.3 m (kPa)	0.7 m (kPa)	Average increase in matric suction (kPa)	Associated increase in apparent cohesion (kPa)
Control	9.4	5.2	-	-
Black Willow	14.7	11.4	5.8	1.0
River Birch	26.4	23.7	17.8	3.1
Sycamore	21.1	17.1	11.8	2.1

Table 2. Average values of matric suction (in kPa) for soil monoliths during spring 2002 showing greater matric suction and associated increases in apparent cohesion for monoliths with woody vegetation. Shear strength values based on the assumption that fb = 10.0° (Simon et al. 2000).

(t-test carried out at 95% confidence level). This result occurs under both soil moisture conditions tested, and suggests that whilst root-pull out during soil failure does occur, the forces required for this to happen are so close to the values required for root breaking to occur, that in calculations of the reinforcement provided to the soil by roots it is possible to simply assume that all roots break as the soil shears. Thus, the assumption made in the root reinforcement model of *Wu et al.* [1979], that the only root failure mechanism of importance is breaking, is an acceptable assumption, in the case of the streambank material tested in this study.

4.5. Development of an Alternative to the Wu Equation (1979): The RipRoot Model

With the information obtained from the previous sections it became clear that, the assumption that all of the roots break at once and at the same displacement was questionable. In response to this a root model (RipRoot) based on the fiber bundle models (FBM's) used in the field of materials science was developed. The first of these models was developed by *Daniels* [1945] and subsequent models have built on this to incorporate the most important aspects of composite material damage during loading [*Hidalgo et al.*, 2001]. The basic principle of an FBM is that the maximum load withstood by a bundle of fibers, is less than the sum of each of their strengths. This is because when a load is applied to the bundle the fibers will not all break at the same time.

FBM's take this into account by following simple rules: an initial load is added to the bundle, containing a number (n) of parallel fibers. Although at first the load is distributed equally between the n fibers, once the load is increased sufficiently for a fiber to break, the load that was carried by the fiber is redistributed to the remaining (n-1) intact roots, each of which then bears a larger share of the load. If this redistribution of load causes any further roots to break, further redistribution of load occurs (in this type of model this is known as an avalanche effect), and so on until no more breakages occur. Another increment of load is then added to the system, and the process is repeated until either all of the fibers have been broken, or the matrix containing the fibers fails.

RipRoot employs a global load sharing method to determine redistribution of stresses after breakages. That simply means that the load from the broken roots is applied equally to the remaining intact roots. More spatially complex models also exist in which local load sharing is applied. In this type of FBM the load is redistributed to those fibers closest to the point of fiber rupture. RipRoot makes one other main assumption, which is that the elastic properties of all of the roots are the same. The model uses root diameters input by the user to calculate root tensile strengths from the curves in Plate 1. These values are then used by the model to calculate cumulative load supported by the roots, and the breakages and redistributions of stress as load is added to the soil matrix.

An example of the initial results from the RipRoot model is shown in Figure 5. When root diameter data for the 30 Eastern Sycamore roots from Figure 2 were input to RipRoot, the peak in ΔS was 2.48 kPa, compared with a peak on the cumulative displacement curve of 2.31 kPa. Both of these methods take into account progressive breaking of roots. The value given by the *Wu et al.* [1979] equation for the same 30 roots was 5.6 kPa, and represents the cohesion value if all of the roots were to break at once The preliminary findings from the RipRoot model provide lower estimates of ΔS, dramatically reducing the overestimation of the Wu equation, and producing values much closer to those estimated by the cumulative stress-displacement curves in Figure 2.

4.6. Hydrologic Effects of Different Riparian Species: Soil Monoliths

Pore-water pressure data from depths of 0.3 and 0.7 m within individual soil monoliths containing Black Willow, River Birch, Eastern Sycamore, and bare soil are shown for the period February, through June, 2002 (Plate 2). This period was selected because it represents the wettest and most critical period for streambank stability. With rainfall, all of the tensiometers showed decreases in matric suction (negative pore-water pressure) or increases in positive pore-water pressure reflecting the addition of water. However, both the magnitude of the changes and the absolute values attained within the soil monoliths differed by treatment. At both 0.3 m and 0.7 m depths, the soil in the control monoliths became the wettest during and after rainfall indicating the role of the woody species in maintaining matric suction and enhancing shear strength (see equation 5). During late February and early March before leaves appeared on stems and branches, there appeared to be little difference in matric suction values between individual species. This lack of significant difference between the vegetated monoliths and the controls is in part, a function of the young age of the specimens and the lack of a carryover of high values of matric suction from the previous summer that can be typical in more mature stands of trees [*Simon and Collison*, 2002]. New growth appeared in early April with a theoretical corresponding increase in evapotranspiration. This change is seen as a steepening of the drying trend and a further departure from the control monoliths between precipitation events (i.e. River Birch during mid-April; Plate 2). The increasing effect of evapotranspiration is seen as the difference in matric suction drying-values and trends at both depths between the control and individual species monoliths during May and June. Sycamore and River Birch create the largest matric suction values during this period (80 kPa).

Rapid decreases in matric suction are typical of vegetated, dry soils [*Simon and Collison*, 2002] and are observed for the River Birch monoliths during the large precipitation event in early May (Plate 2). This is due to macropores developed around roots and minor contributions from stemflow. Still the resulting matric suction values in the vegetated monoliths are greater than those in the control monolith.

Over the period of monitoring, River Birch showed the greatest overall effect on matric suction values at both 0.3 and 0.7 m, Black Willow the least (Table 2). Averaging the difference in matric suction values at the two depths and multiplying this difference by tan ϕ^b provides an indication of the average increase in apparent cohesion that each of the woody species would provide to a hypothetical streambank (Table 2). River Birch would provide a 310% increase in apparent cohesion due to matric suction, followed by Sycamore (200%) and Black Willow (100%). These values are significant in light of typical values of effective cohesion in many silt-clay systems.

The findings of these experiments have important implications for streambank stability, especially during the winter and spring because this is the wettest time of the year. It is at this critical time of year that any additional strength provided to the streambank by matric suction or other means is particularly crucial. To test the hydrologic roles of each of the riparian species in enhancing bank strength and streambank stability, tensiometric data from the soil monoliths was input into the dynamic version of the bank-stability model developed by

Simon et al. [1999]. This is the same model used in earlier parts of this chapter with the exception that the time-series pore-water pressure data from the soil monoliths are used as inputs to evaluate the effect of hydrologic changes for each of the woody species on streambank stability. The geotechnical and bank-geometry data used for the model are those from the research bendway site on Goodwin Creek [*Simon et al.*, 1999, 2000; *Simon and Collison*, 2002]. The modeled bank in the bendway is 4.7 m high with a bank angle of 80°, and is composed of 5 layers with effective cohesions (*c'*) from top to bottom of 1.4, 2.7, 2.7, 6.3 and 6.3 kPa, respectively. Hydrologic data from the 0.3 m tensiometers are placed in the uppermost layer while data from the 0.7 m tensiometers are placed in the next layer down. The modeled failure-plane angle is 60°.

To evaluate the hydrologic effects of each species relative to the control soil monoliths. The model was initially run for the period February 18 to June 27, 2002 using the geotechnical data from the bendway site and the hydrologic data from the soil monoliths (Plate 3). Much like the comparisons of the matric suction data, little differentiation is possible in early spring, although minimum F_s values of the control monoliths in response to precipitation are consistently lower than the vegetated monoliths. Again this lack of significant differences is due at least in part to either (1) the absence of higher matric suction values under trees that carryover from the previous summer or (2) these higher values of matric suction from the previous summer were lost during the winter, before the soil monoliths were instrumented. It can be seen that River Birch and Sycamore provide the greatest hydrologic benefits in terms of enhancing streambank stability (Plate 3). By the end of June F_s values for Sycamore and River Birch are near 3.0, and for Black Willow, near 2.5, much greater than the F_s in the control monoliths. These results provide clear justification for the beneficial role of woody riparian vegetation to enhance streambank stability via increases in matric suction even for trees as young as two years old.

4.7. Net Effects of Different Riparian Species on Streambank Stability: Soil Monoliths

By combining the bank-stability modeling results using hydrologic data from the soil monoliths (Plate 3) with ΔS values due to root reinforcement, we are able to evaluate the net effects of individual species expressed in terms of the factor of safety. For demonstration purposes we again use the 4.7 m high streambank on Goodwin Creek. Plate 4 shows why even during the wettest period of the year when matric suction values for the vegetated scenarios are only marginally better than those of the control, vegetated streambanks are more likely to remain stable; the mechanical reinforcement provided by the root networks maintains some degree of stability even during the wettest periods. The minimum F_s values for the modeled control situation falls below 1.0 in early May indicating a failure in the simulated bank. In contrast, all of the simulations for vegetated banks remain stable (F_s values > 1.0) for the entire period. Owing to their greater root reinforcement effects, Sycamore and River Birch stand out as providing the most beneficial effects compared to Black Willow throughout the simulation period.

5. CONCLUSIONS

Fieldwork and laboratory experiments carried out have shown that streambank stabilization by vegetation roots involves a complex interaction between the root and soil systems. Mechanically, investigations of the forces required to pull roots out of the soil rather than break them have shown that the values are statistically inseparable. The assumption made by *Wu et al.* [1979] that all the roots fail by the breaking mechanism is a valid assumption for the soil type tested in this study. However, the assumption that all of the roots break simultaneously and at the same displacement that peak soil strength is reached is invalid based on the experimental data herein, producing large overestimations of ΔS. The magnitude of this overestimation varies according to the uptake of load by the roots of different species, compared to the soil type in question. Simple add-on factors representing the increased soil strength contribution provided by roots may overestimate the resulting increase in streambank stability. Fiber bundle models such as the one developed in this study, may provide more reliable estimates of root reinforcement.

Hydrological reinforcement is provided from the above ground process of interception by the vegetative canopy, and below ground by the removal of soil moisture via evapotranspiration. Young trees can remove enough water from the upper meter of the soil to reduce pore-water pressures significantly, providing and additional 1 to 3 kPa of apparent cohesion to the soil which, during the critical wet periods of the year may prove sufficient to resist mass failure. These effects will only increase as the trees get older. By combining the hydrologic effects of individual species with those due to root reinforcement (ΔS), factor of safety values for the simulated streambank on Goodwin Creek indicate streambank stability throughout the spring of 2002 due the significant increases in shear strength provided by woody vegetation. River Birch and Sycamore provide the greatest benefits compared to Black Willow.

This study goes some way in advancing knowledge of root reinforcement and hydrologic modifications of streambanks by riparian vegetation, but in order to be able to produce more accurate quantification of ΔS, a complete picture of the way in which roots and soil interact during shearing must be gained.

Future studies aim to study in greater detail the interactions between soil and roots during the shearing process, and to further model the combined mechanical and hydrologic effects on streambank stability.

Acknowledgments. Funding for this study was provided by USDA-ARS Discretionary Research Funds, US Army Corps of Engineers, Vicksburg District, and a studentship from the School of Social Science and Public Policy, King's College London. The authors would like to thank Paul Comper, Joe Dickerson, Jo Simpson, and Geoff Waite for their help with field data collection, and Brian Bell and Mark Griffith for their help with the construction and maintenance of the soil monoliths. Many thanks also go to the reviewers of this manuscript, Angela Gurnell and Massimo Rinaldi, who provided many useful comments and suggestions.

REFERENCES

Abernethy, B., *On the Role of Woody Vegetation in Riverbank Stabiliy,* Unpublished Ph.D. thesis, Monash University, Melbourne, 1999.

Abernethy, B., and Rutherfurd, I. D., Where along a river's length will vegetation most effectively stabilize streambanks? *Geomorphology*, 23, 55–75, 1998.

Abernethy, B., and I. D. Rutherfurd, The effect of riparian tree roots on the mass-stability of riverbanks, 270 pp., *Earth Surface Processes and Landforms*, 25, 921–937, 2000.

Beaudoin, J. J., *Handbook of Fiber-Reinforced Concrete*, 332 pp., Noyes Publications, New Jersey, 1990.

Bohm, W., *Methods of Studying Root Systems*, 270 pp., Springer-Verlag, Berlin, 1979.

Collison, A. J. C., and M. G. Anderson, Using a combined slope hydrology/stability model to identify suitable conditions for landslide prevention by vegetation in the humid tropics, *Earth Surface Processes and Landforms*, 21, 737–747, 1996.

Collison, A. J. C., N. Pollen, and A. Simon, Mechanical reinforcement and enhanced cohesion of streambanks using common riparian species, *Eos. Trans. AGU*, 82(47), Fall Meet. Suppl., Abstract H31A-023, 2001.

Coppin, N. J., and I. G. Richards, *Use of Vegetation in Civil Engineerin,* 292 pp., Butterworths, London, 1990.

Daniels, H.E., The statistical theory of the strength of bundles of threads I., *Proc. R. Soc. London, Ser. A.* 183, 405–435, 1945.

De Roo, H. C., Tillage and Root Growth, in *Root Growth*, edited by W. J. Whittington, pp.339-357, Butterworths, London, 1968.

Dingman, S. L., *Physical Hydrology (2nd Edition)*, 646 pp., Prentice Hall, N.J., 2001.

Duncan, J. M., State of the art: static stability and deformation analysis, in *Proceedings of the ASCE Specialty Conference on the stability and performance of Slopes and Embankments II, Berkeley,* edited by R. B. Seed and R. W. Boulanger, pp. 227–266, ASCE Special Publication 31, 1992.

Durocher, M. G., Monitoring spatial variability in forest interception, *Hydrological Processes,* 4, 215–229, 1990.

Easson, G., and L. D. Yarbrough, The effects of riparian vegetation on bank stability, *Environmental and Engineering Geoscience,* 8(4), 247–260, 2002.

Ennos, A. R., The anchorage of leek seedlings: The effect of root length and soil strength, *Annals of Botany,* 65, 409–416, 1990.

Fredlund, D. G., Morgenstern, N. R., and Widger, R. A., The shear strength of unsaturated soils, *Canadian Geotechnical Journal,* 15, 313–321, 1978.

Gray, D. H., Reinforcement and stabilization of soil by vegetation, *Journal of Geotechnical Engineering,* 100(GT6), 695–699, 1974.

Gray, D. H. and H. Ohashi, Mechanics of fiber reinforcement in sand, *Journal of Geotechnical Engineering,* 109, 335–353, 1983.

Gray, D. H., and R. B. Sotir, *Biotechnical and Soil Bioengineering Slope Stabilization: A practical Guide for Erosion Control,* 378 pp., John Wiley and Sons Inc, New York, 1996.

Greenway, D. R., Vegetation and Slope Stability, in *Slope Stability,* edited by M. G. Anderson, and K. S. Richards, pp. 187–230, John Wiley and Sons Ltd, New York, 1987.

Hickin, E. J., Vegetation and river channel dynamics, *Canadian Geographer,* 28(2), 111–126, 1984.

Hidalgo, R. C., F. Kun, and H. J. Herrmann, Bursts in a fiber bundle model with continuous damage, *Physical Review E,* 64, 2001.

Kutzing, L., and G. Konig, Design principals for steel fiber reinforced concrete—A fracture mechanics approach, *LACER,* 4, 175-184, 1999.

Lambe, T. W., and R. V. Whitman, *Soil Mechanics,* 553 pp., John Wiley and Sons, New York, 1969.

Malanson, G. P., *Riparian Landscapes,* 296 pp., Cambridge University Press, Cambridge, 1993.

Martinez-Meza, E., and W. G. Whitford, Stemflow, throughfall and channelization of stemflow by roots in three Chihuahuan desert shrubs, *Journal of Arid Environments,* 32, 271–287, 1996.

Niklas, K. J., *Plant Biomechanics: An Engineering Approach to Plant Form and Function,* 607 pp., The University of Chicago Press, London, 1992.

Osman, A. M., and C. R. Thorne, Riverbank stability analysis I: Theory, *Journal of Hydraulic Engineering,* 114(2), 134–150, 1988.

Pollen, N., *The Mechanical Reinforcement of Riverbanks by the roots of four Northern Mississippi Riparian Species,* Unpublished undergraduate dissertation, 120 pp., King's College London, 2001.

Pollen, N., A. J. C. Collison, and A. Simon, Advances in assessing the mechanical contribution of riparian vegetation to streambank stability, in *Proceedings of the Earth and Water Resources Institute 2002 Conference (EWRI),* Virginia, USA, 2002.

Selby, M. J., *Hillslope materials and processes,* 450 pp., Oxford University Press, Oxford, 1993.

Simon, A., and C. R. Thorne, Channel adjustment of an unstable coarse-grained alluvial stream: Opposing trends of boundary and critical shear stress, and the applicability of extremal hypothesis, *Earth Surface Processes and Landforms,* 21, 155–180, 1996.

Simon, A., A. Curini, S. Darby, and E. Langendoen, Streambank mechanics and the role of bank and near-bank processes in incised channels, in *Incised River Channels,* edited by S. Darby and A. Simon, pp. 123–152, John Wiley and Sons, New York, 1999.

Simon, A., A. Curini, S. Darby, and E. Langendoen, Bank and near-bank processes in an incised channel, *Geomorphology,* 35, 193–217, 2000.

Simon, A., and A. J. C. Collison, Quantifying the mechanical and hydrologic effects of riparian vegetation on streambank stability, *Earth Surface Processes and Landforms,* 27, 527–546, 2002.

Thorne, C. R., Effects of vegetation on riverbank erosion and stability, in *Vegetation and Erosion,* edited by J. B. Thornes, pp.125–143, John Wiley and Sons Inc, Chichester, 1990.

USEPA, http://www.epa.gov/OWOW/tmdl/sed_miles.html, 2002.

Waldron, L. J., The shear resistance of root-permeated homogeneous and stratified soil, *J. Soil Science Soc. Amer.,* 41, 843-849, 1977.

Waldron, L. J., and S. Dakessian, Soil reinforcement by roots: Calculation of increased soil shear resistance from root properties, *Soil Science.* 132, 427–435, 1981.

Wu, T. H., W. P. McKinnell III, and D. N. Swanston, Strength of tree roots and landslides on Prince of Wales Island, Alaska, *Canadian Geotechnical Journal,* 16, 19–33, 1979.

Wu, T. H., P. E. Beal, and C. Lan, In-Situ shear test of soil-root systems, *Journal of Geotechnical Engineering,* 114 (GT12), 1376–1394, 1988.

Ziemer, R. R. and D. N. Swanston, *Root strength changes after logging in SE Alaska,* Pacific Northwest Forest Experimental Station: USDA Forest Service Research Note, Pacific Northwest, 306, 1977.

A. J. C. Collison, Philip Williams and Associates, 720 California St., Suite 600, San Francisco, CA 94108.

N. Pollen, and A. Simon, Channel and Watershed Processes Research Unit, USDA-ARS National Sedimentation Laboratory, PO Box 1157, Oxford, MS 38655.

The Influence of Trees on Stream Bank Erosion: Evidence from Root-Plate Abutments

Ian D. Rutherfurd and James R. Grove

School of Anthropology, Geography, and Environmental Studies, and Cooperative Research Center for Catchment Hydrology, University of Melbourne, Australia

Stream bank erosion often isolates the root plate of a riparian tree on a pedestal of sediment jutting out from the stream bank. To our knowledge, these root-plate abutments have not been formally described. Apart from being a landform in their own right, abutments are of interest because their morphology integrates the complex effects of trees on bank erosion processes. From measuring seven abutments formed along the Acheron River in southeastern Australia, we conclude the following: (1) Roots from a single tree increase the resistance of impinging banks in a semi-circle centered on the trunk. The abutment has a radius that is always smaller than (usually less than half) the canopy radius. This relationship holds for four dominant riparian tree species along the Acheron River, situated on gravel and sandy-loam banks that are from 1 m to 4 m high. (2) All abutments are deeply undercut, with most of the abutment formed of a 0.5 m to 1 m thick overhanging plate of finer sediments reinforced by roots. The deviation of the curve of the concave stream bank, at the bank toe below trees, indicates that trees provide some bank toe strengthening, even when the bank is nearly 4 m high. However, the strengthening from single trees is not enough to materially alter the migration rate of a meander bend. (3) The bed is deepened at the tip of the abutment by up to 30% of the bank height. Thus the abutments have a secondary effect on channel morphology.

1. INTRODUCTION

Trees potentially affect all stream bank erosion processes [see a review in *Simon and Collison*, 2002]. As a result it has proven remarkably difficult to isolate (and quantify) the effect of trees on bank erosion rates. One laborious method is to measure erosion rates on vegetated and unvegetated banks. This has led to the conclusion that vegetated banks

Riparian Vegetation and Fluvial Geomorphology
Water Science and Application 8
10.1029/008WSA11

migrate more slowly than unvegetated banks [*Pizzuto*, 1984; *Beeson and Doyle*, 1996; *Micheli and Kirchner*, 2002b], and that vegetated streams are narrower [*Hey and Thorne*, 1986]. Another approach is to develop numerical, geo-mechanical models that represent the many variables involved [*Abernethy and Rutherfurd*, 2000; *Abernethy and Rutherfurd*, 2001; *Micheli and Kirchner*, 2002a; *Simon and Collison*, 2002]. A third approach is to use the shape and size of eroded tree abutments as a relative measure of the overall effect of trees on erosion rates.

Consider a concave (outer) bank, of uniform resistance, in a meandering stream. The bend migrates toward a large, isolated tree on the top of the stream bank (Figure 1). The median migration rate for a large global dataset of meandering rivers is 1.6% of channel width per year [*Walker and Rutherfurd*, 1999]. If the root plate of the tree decreases the erosion rate, it will perturb the smooth arc of the bend, isolating a pedestal of sediment that we call an abutment (Figures 1 and 2). Surprisingly, we have found no direct reference to these features in the literature, although *Davis and Gregory* [1994] describe erosion caused by flow deflection around a root-plate.

We theorize that the form of abutments, relative to the circular root-plate of a tree, will integrate the many effects of a tree on erosion processes (Figure 2). Although closely spaced trees would protect the full bank length (Figure 3), we will restrict our discussion here to abutments formed around the root plate of large, isolated trees. In this paper, we investigate the morphology of seven abutments along the Acheron River in southeastern Australia. We use the abutments to test three hypotheses about the effects of tree roots on erosion.

2. HYPOTHESES

2.1. Hypothesis 1

The radius of the tree canopy will be the same size as the radius of the abutment. Mapping of root plates has led to the suggestion that the below-ground parts of a tree cover about the same area as the above–ground parts, i.e. the canopy

Figure 2. An abutment formed around a single willow tree on the Kiewa River, northeastern Victoria. Note that the abutment perturbs the otherwise smooth arc of the eroding bend, as does a second tree in the background (flow direction is into the photo).

[*Carbon et al.*, 1980; *Abernethy and Rutherfurd*, 2001]. The canopy diameter is defined as the 'drip line,' or the greatest width of the tree canopy projected onto the ground. It is reasonable to assume that the various effects of trees on erosion rates are restricted to the drip line of the tree. For example, *Abernethy and Rutherfurd* [2001] found that the apparent cohesion introduced by tree roots had fallen from 120 kPa at the trunk to only 25 kPa at the drip line. Thus, if Hypothesis 1 is supported, it will mean that the stream bank will begin to be perturbed by the root plate of a tree when the eroding bank reaches the drip line.

2.2. Hypothesis 2

The face of the abutment will be more concave (more undercut) than adjacent banks up or downstream. Root density below trees falls rapidly with depth, with reviews suggesting that most roots are found in the top 0.5 m to 1.0 m of the soil profile [*Jackson et al.*, 1996; *Tufekcioglu et al.*, 1999] and in the top 0.2 m to 0.5 m below large Eucalyptus trees [*Gray and Sotir*, 1996; *Abernethy and Rutherfurd*, 2001; *Laclau et al.*, 2001]. As a result, apparent cohesion provided by roots declines rapidly in the first meter of depth [*Shields and Gray*, 1992]. The shape of the bank-face of the abutment, compared to the shape of the banks up and downstream of the abutment, will indicate the relative effect of trees on the resistance of different parts of the bank face. More specifically, we hypothesize that the face of the abutment will be more concave than the banks up or downstream. Long-term erosion rates are controlled by erosion at

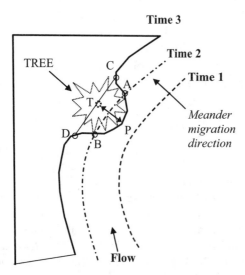

Figure 1. Schematic representation of the development of an abutment around a root plate following bend migration in uniform sediments. At Time 2, the abutment forms an arc (in plan) subtended by a secant formed by the curve of the bend, forming a chord (A – B) through the root plate of the tree. The length of that chord reaches a maximum (C – D as does the radius of the abutment, T – P) as the bend cuts toward the tree trunk. Point P marks the apex of the abutment.

the bank toe [*Thorne*, 1982]. If roots do not affect the toe, then it is unlikely they will affect long-term erosion rates.

2.3. Hypothesis 3

A relatively deeper pool will form at the tip of an abutment. Abutments will affect patterns of bed and bank erosion on a bend. Abutments resemble engineered groynes [*Przedwojski*, 1995; *Sukhodolov et al.*, 2002] and spur dikes [*Kuhnle et al.*, 1999], and we would expect them to have a similar effect on patterns of scour and deposition in the stream. Following the model studies of *Kuhnle et.al.* [1999] on spur-dikes, we would expect a scour hole to form at the tip of the abutment. In the *Kuhnle et al.* experiments, the maximum scour at the tip of the spur dike more than doubles the channel depth.

2.4. Scope of the Study

Note that abutments are distinct from the mechanism of buttressing, in which a downslope tree retards a rotational failure [*Thorne*, 1990]. In addition, we are considering the case of trees growing on the bank top, not on the bank face. The root plate of trees on the bank face grow parallel with the bank, confusing the erosion effects. Finally, abutments can tell us a great deal about relative erosion rates between the bank affected by the tree, and the banks up and downstream. Abutments can tell us little about absolute erosion rates.

3. METHODS

The dimensions of abutments were measured along a typical meandering stream in southeastern Australia. We selected the Acheron River because it is unregulated for most of its length, the riparian vegetation has not been cleared, and the stream is not so large that any effects of trees would be overwhelmed by other factors. The river is described by *Gordon* [1996]. The headwaters of the Acheron are located 70 km NE of Melbourne (37°30′S, 145°45′E) at an elevation of 1220 m and it has a catchment area of 740 km^2. The river flows 60 km north to join the Goulburn River at an elevation of 190 m. The channel is meandering, ranging from gravel banks in the headwaters to silty-clay banks on the plains. Maximum bankfull dimensions are 30 m wide and 5 m deep. Mean annual rainfall mid-catchment, at Marysville, is 1376 mm, with a strong winter peak, that produces 528 mm of runoff. The average mean daily flow in the lower catchment at Taggerty is 10.3 m^3/s. The annual coefficient of variation of discharge is 0.44 [between 1946 and 1987], which is typical for streams of this region.

The river length was divided into four zones based on physiography and landuse change. Zone 1 is a steep, narrow

Figure 3. Multiple abutments formed by erosion of the outer bank of the Goulburn River, northern Victoria (near McCoys Bridge, flow is from left to right).

mountain stream with native, dense sclerophyllous forest vegetation, whilst Zone 4 is a shallow sloped, wide lowland stream with sparse native riparian tree cover and intense pastoral landuse. Zones 2 and 3 are intermediate areas between these two extremes, mainly delimited by access points. Within each zone a reach was randomly selected. Bends were inspected in each reach in order to find prominent examples of isolated abutments. We also had to reject sites where some other factor (such as flow diversion from large woody debris) could perturb the smooth profile of the stream bank. It is important to emphasize here that we were not randomly sampling bends to isolate the general influence of trees on erosion. Instead, we were selecting the most prominent examples of abutments that we could find in order to identify the maximum effect of individual trees on erosion processes.

The character of an abutment will be a random function of when we visit it. We may find a small abutment because the stream has only recently migrated into the edge of the root plate. There might be no abutment because it has recently eroded away or because one never existed. Similarly, roots that are freshly exposed will tell you most about the processes of erosion. Exposed roots eventually die and break off. Thus, we restricted ourselves to the most prominent abutments.

After we had found an abutment, we made the following measurements. The bank profile and bank face were surveyed using the method of *Hudson* [1982]. The root density at each measurement point was visually recorded on a scale of 1 to 5 using a percentage cover estimation chart [*Gordon et al.*, 1992]. This method does not distinguish between fine and coarse roots. The complex network of roots at the abut-

Figure 4. Undercut abutment at Site 1. Note the roots beneath the overhanging block and that the back of the undercut coincides with the gravels at the base of the bank.

ment face was difficult to sample (Figure 4) and it was considered that trying to measure all root diameters, or subsample, for each soil-root interface on the bank, would add little extra information. The bank material was described using either, field texture for fine bank material, or particle sizing using digital photographs for mean gravels size following the method of *Adams* [1979]. We also measured the dimensions of the adjacent stream channel (width and mean-depth), the positions and dimensions of all trees on the bend canopy diameter at the drip line, Diameter at Breast-Height (DBH), and maximum height. The abutment radius was measured by identifying the radius of the circle that fits the arc of the abutment. This is the same method used to define the radius of curvature of a meander bend [e.g. *Hickin and Nanson*, 1984].

4. RESULTS

Prominent abutments are not common on the Acheron River even though riparian trees are common. Despite walking many kilometers of channel (as well as boating the lower reaches), we were only able to find seven abutments that satisfied the selection criteria: five along the Acheron River and two on its major tributary, the Steavenson River (Table 1). As with many lowland streams in southeast Australia, the dominant riparian trees were exotic willows (*Salix fragilis*). These blanketed the bank with fine roots and did not form abutments. There are numerous riparian trees with canopies intersecting the bank-line, but the vast majority of these have not developed abutments.

Abutments 1 through 4 are illustrated as examples (Figures 5 to 8).

4.1. Description of Surveyed Sites

4.1.1 Site 1. This site was located in the confined upland reach of the Acheron River, 10 km from the divide. The bed and lower banks consisted of gravels (Figure 4).

The smooth arc of the bend was clearly perturbed for a bank length of 3 m around the root-plate of a lone hazel pomaderris (*Pomaderris aspera*; Tree 3 in Figure 5), producing an abutment of 1.5 m radius. The bank material is the same for the full length of the bend. Therefore, the deviation in the bank profile is a product of the increased resistance in the root-plate zone and not a result of other variations in force or resistance. The abutment is deeply undercut, by up to 1 m, with a dense mat of fine roots hanging down from the undercut block (Figures 5B and 5C). Note that the undercut is formed entirely in the gravel layer, with the overhanging block consisting of sandy-silt. The exposed roots varied in diameter from 1 mm to 10 mm (Figure 4). The position of

Table 1. Description of abutment sites

Site Number	Location and *Distance Downstream from Headwater (km)*	Max Bank Height (m)	Banktop Channel Width (m)	Bank material UB = Upper Bank; LB = Lower Bank
1	Acheron River *10.1*	1.65	8	UB: Fine sandy-loam LB: coarse gravel layer, 20 cm d_{50}.
2	Steavenson River *17.8*	1.55	10	UB: Sandy loam (2% clay; fine- medium sand) LB: mean gravel size -5.36ϕ
3	Steavenson River *17.6*	1.2	10	UB: Sandy loam (2% clay; fine- medium sand) LB: mean gravel size -1.27ϕ
4	Acheron River *29.4*	2.1	16	UB: Silty sand (clay 5%; silt 45%; fine sand 50%) LB: mean gravel size -1.05ϕ
5	Acheron River *29.5*	2.1	16	UB: Silty sand (clay 5%; silt 45%; fine sand 50%) LB: mean gravel size -1.05ϕ
6	Acheron River *53.2*	4.1	18	Sandy loam
7	Acheron River *54.5*	3.7	20	Loamy sand (8-10% clay)

maximum undercutting deviated from the bend curvature, suggesting that tree roots still affect erosion resistance at the bank toe, even in gravels.

4.1.2 Site 2. An abutment of 3 m radius has formed around a large Eucalypt (Tree 1 in Figure 6A). The abutment is undercut by up to 2 m, with both the undercut depth and the abutment size being greater upstream of the tree trunk. As for Site 1, the undercut has been cut into gravels below the root-impregnated sandy-silt unit. Note that the line of silver wattles (*Acacia dealbata*, Tree 2 in Figure 7A) is 5 m from the river bank, but produce no deflection of the bank profile.

4.1.3 Site 3. The prominent abutment (3.5 m radius) at Site 3 has formed around the root plate of a dead silver wattle. The site is 200 m upstream of Site 2. The landholder believes that

the tree has been dead for several years, yet the root plate is still able to resist erosion of the abutment. Again, the stream has removed the gravels under the root plate, so that undercutting extends as far back as the tree trunk (Figure 7A).

4.1.4 Sites 4 and 5. Figure 8B shows an abutment formed around the root plate of a large River Peppermint (*Eucalyptus elata*) at Site 4. The abutment at Site 5 is in the background. Site 4 is some 30 km from the headwaters. As with Sites 1 to 3, abutment 4 is undercut more than 2 m into gravels. The undercut is so deep that the toe of the bank follows the arc of the bend. This means that the abutment is restricted to the top of the bank (Figure 8A).

4.1.5 Sites 6 and 7. Sites 6 and 7 are over 51 km from the headwaters, and River red gum (*E. camaldulensis*) is the

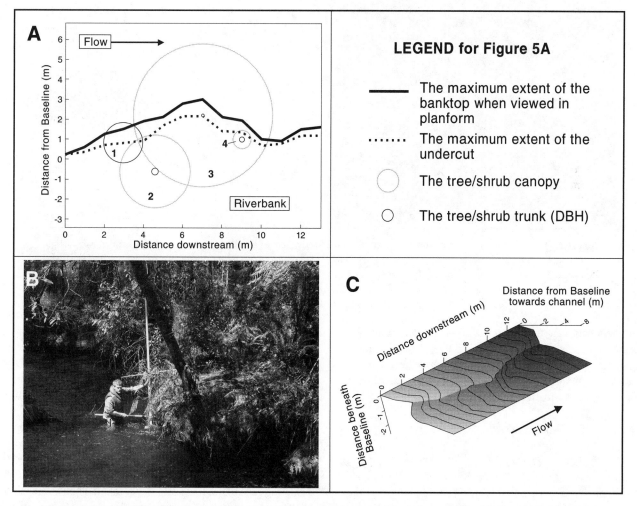

Figure 5. Details of the abutment at Site 1, 10 km from the headwaters of the Acheron River. (A) Planform view of the abutment, the numbers indicate: (1) Tree fern Cyathea Sp., (2) Silver Wattle Acacia dealbata, (3) Hazel pomaderris Pomaderris aspera, and (4) Multistemmed Hazel pomaderris. (B) Photograph of site looking upstream. (C) 3–dimensional surface showing undercutting of the abutment.

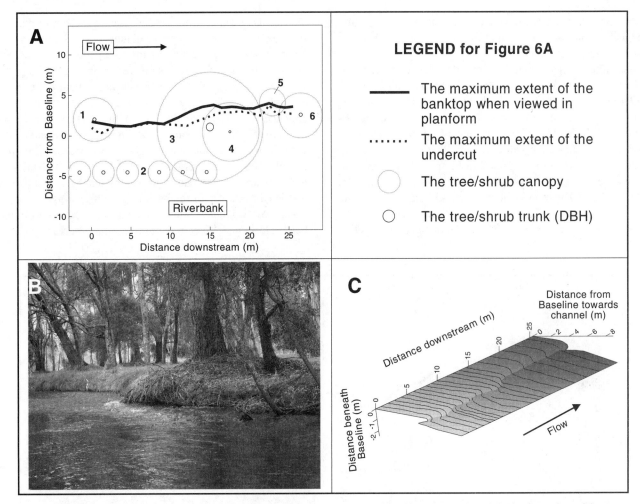

Figure 6. Details of the abutment at Site 2, 17.8 km from the headwaters of the Steavenson River. (A) Planform view of the abutment, the numbers indicate: (1,4,5,6) Acacia (Acacia Sp.), (2) Silver Wattle (Acacia dealbata), and (3) Eucalyptus (Eucalyptus Sp.). (B) Photograph of site looking upstream. (C) 3–dimensional surface showing undercutting of the abutment.

dominant riparian tree in these lowland sites. The bank height at these sites has increased to over 3 m, and there is no longer gravel at the bank toe. Despite the absence of gravel, the abutments at Sites 6 and 7 are undercut up to 1.5 m into the sandy-loam material.

4.2. Hypothesis Testing: Results

4.2.1 Hypothesis 1: The radius of the tree canopy will be the same size as the radius of the abutment. In general, larger canopies were associated with larger abutments (Figure 9), although the sample size is small. However, all of the abutments were considerably smaller than the canopy, with the radius of the abutments being half that of the canopy on average (Figure 9). A similar allometric relationship exists

for the radius of the abutment and the diameter of the tree trunk, with the bank being perturbed at about five times the trunk diameter. Despite walking many kilometers of river bank, we found no clear deviations of the bank top associated with trees that were larger than the radius of the canopy.

Another explanation for the dimensions of the root plate may be that the eroding bank reached the drip line when the tree was smaller (say half its present diameter). As the root plate deformed the bank line, the tree continued to grow, increasing the canopy size while the bank cut around the existing root plate. It is more likely that the bank reached the root plate when the canopy was close to its present size. Using methods described in *Walker and Rutherfurd* [1999], the median migration rate for the Acheron River should be tens of centimeters per year. Thus, the maximum time it

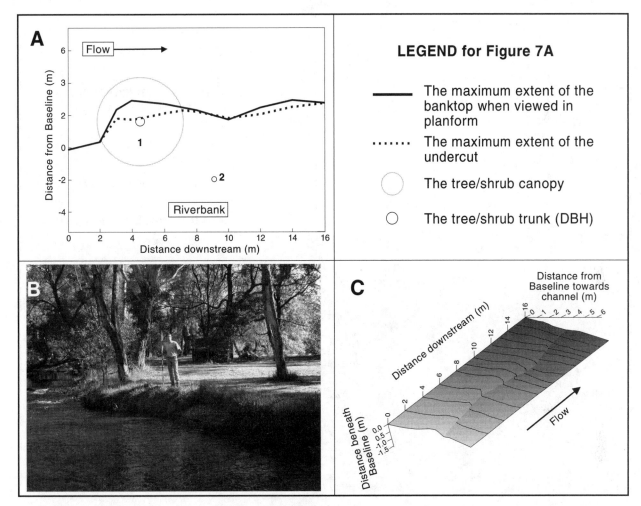

Figure 7. Details of the abutment at Site 3, 17.6 km from the headwaters of the Steavenson River. (A) Planform view of the abutment, the numbers indicate: (1) Dead Silver Wattle (Acacia dealbata), (2) Trunk of an unidentified tree species. (B) Photograph of site looking upstream. (C) 3–dimensional surface showing undercutting of the abutment. (Note that the canopy diameter of the dead silver wattle at Site 3 is an estimate based on measurements from nearby trees.)

would take to erode the abutments surveyed here would be twenty years. This is not sufficient time for the canopies to double in diameter in slow growing red gums. In addition, the root plate tends to be larger than the canopy rather than smaller. Therefore, we conclude that the trees sampled here do not measurably perturb the migrating bank until they have progressed almost half-way through the root plate.

4.2.2 Hypothesis 2: The face of the abutment will be more concave (more undercut) than adjacent banks. This hypothesis was supported at all sites. All of the abutments were undercut more than the banks up or downstream. At Site 1, for example, the banks at either end of the abutment are vertical, with no undercutting (Figure 10). In five of the seven sites, there was some undercutting up and downstream of the abutment, but considerably less than at the abutments. Our quali-

tative classification supports the proposition that root density falls rapidly with depth (Figure 11). Although the undercuts at Sites 1 to 5 were clearly associated with the contact between the gravel and the overlying sandy-loam (Figures 5 to 8), the undercuts at sites 6 and 7 were in uniform fine material. In four of the abutments, the length of the overhang was similar to the height of the bank. Note that abundant roots are exposed in the undercuts (Figure 4), implying that the sediment removed was well impregnated with roots. Therefore, the main explanation for the presence of the abutments is that roots support the overhanging block as well as increasing the resistance of the block to scour. Clearly, the roots also strengthen the toe of the bank because in all but one of the abutments the toe of the bank deviated from the bank profile, implying some increased resistance at the toe.

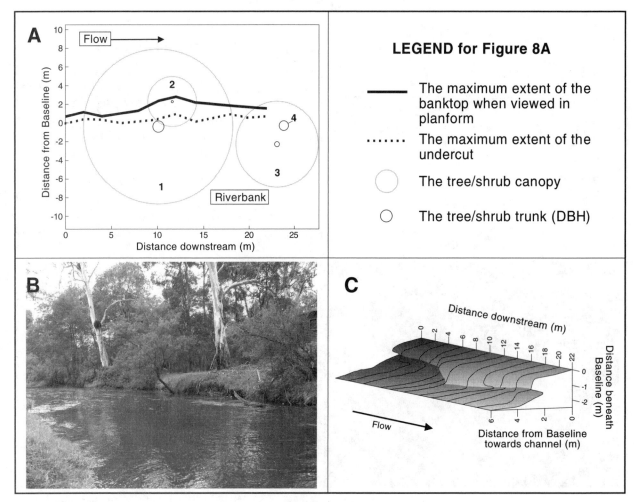

Figure 8. Details of the abutment at Site 4, 29.4 km from the headwaters of the Acheron River. (A) Planform view of the abutment, the numbers indicate: (1) River Peppermint (Eucalyptus elata), (2,3) Silver Wattles (Acacia dealbata), and (4) Tree stump of an unidentified species. (B) Photograph of site looking upstream. (C) 3-dimensional surface showing undercutting of the abutment. (Note that the canopy diameter of the dead silver wattle at Site 3 is an estimate of what it would have been, based on measurements from nearby trees.)

Roots exposed in the undercuts below the abutments indicate the erosion mechanism. In all cases, the presence of a fine mesh of roots suggests that erosion was by removal of individual particles or peds, rather than by mass failure of a cohesive block of sediment.

4.2.3 Hypothesis 3: A relatively deeper pool will form at the tip of an abutment. This hypothesis explored the secondary effect of the abutment on the stream morphology. Artificial groynes, which protrude into the flow, are a well established bank protection technique [*Jackson*, 1935; *Przedwojski*, 1995]. We hypothesized that root-plate abutments would have the same effect, creating scour around their tip [*Shields et al.*, 1995] and producing 'gyres' between abutments that would encourage deposition and protect the

adjacent bank from erosion [*Przedwojski*, 1995; *Shields et al.*, 1995; *Sukhodolov et al.*, 2002].

When we surveyed the bank profiles at the abutments, we continued the surveys out into the stream bed. At all sites there was a distinct deepening of the bank at the tip of the abutment. At Site 1 (Figure 11), the bed was 0.6 m (approximately 30% of the bank height) lower at the abutment tip as compared to the bed up and downstream. The zone of deepening is associated with both the densest root zone and the widest section of the abutment (Figure 11).

In terms of bank protection, we looked for evidence of deposition between adjacent abutments, or less migration of the bank face downstream of an abutment. Such asymmetry would have indicated a lower erosion rate on the downstream

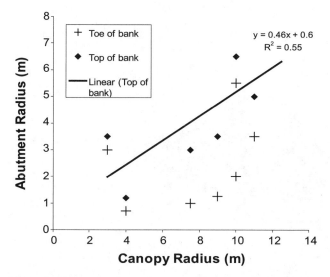

Figure 9. Abutment radius at the top and toe of the bank relative to the radius of the associated tree canopy. (Note that two trees have canopies of 20 m and abutments of 13 m.)

bank of the abutment, as we would have expected if the abutment was producing flow separation and recirculation. Abutments at Sites 2, 3, and 4 were slightly skewed, with more erosion and deeper undercutting on the upstream side of the abutment (Figures 5 to 8). However, Site 6 showed more erosion on the downstream side, and the other three sites were symmetrical. Thus, there is no clear evidence that isolated abutments alter general erosion rates up or downstream, but they do appear to increase depth at the abutment tip.

5. DISCUSSION

5.1. Dimensions and Distribution of Abutments

Abutments formed around the root plate of trees are rare on the Acheron River. Despite inspecting hundreds of trees along the river, we were only able to find seven abutments formed around isolated trees. These abutments were found in the upland and lowland reaches of the stream, in gravel or silty-clay banks 1 to 5 m high, and in four different tree species.

The dimensions of a root-plate abutment (in planform and in section) are an indirect indicator of the magnitude of the effect of a single tree on bank erosion processes. Despite the small sample size, we are confident that the root plate of a tree will begin to materially reduce bank erosion rates when the bank has eroded within about 50% of the canopy radius from the trunk. Thus, a typical abutment on the Acheron River is likely to be half the radius of the tree canopy at the bank top, and perhaps 25% of the radius at the bank toe. Whilst the abutment will probably lead to deepening at its

tip, it is not clear that the abutment will alter erosion rates on the bend as a whole.

5.2. Erosion Mechanisms Around Abutments

We have not measured erosion processes around abutments, but we can make some inference about erosion mechanisms from the abutments themselves. *Davis and Gregory* [1994] describe erosion undercutting the root plate of a small ash tree in the Highland Water, southern England. A log jam trapped against the abutment produced a rise in the water surface that drove the hydraulic sapping of gravels from around the root plate. The abutments on the Acheron River were formed as a result of increased resistance provided by the tree, rather than the flow deflection described by *Davis and Gregory*. This resistance could be produced by roots increasing the apparent cohesion of the banks, by surcharge from the weight of the tree, or from many other mechanisms [*Simon and Collison*, 2002].

Ubiquitous undercutting beneath the root plate is associated with a decrease in root density (Figure 11). At the upstream sites, where the bed and lower banks were gravel, the undercutting took place entirely in the basal gravels. In the two downstream sites (6 and 7), the undercutting took place in uniform material.

The main explanation for the presence of the abutments is that roots support the overhanging block as well as increasing the resistance to scour. Clearly, the roots also strengthen the toe of the bank because in all but one of the abutments (Site 4), the toe of the bank deviated from the curve of the bank profile, implying some increased resistance at the toe.

In some cases, it is surprising that the material in the undercut could be removed. At Site 2 (Figure 6), cobbles up to 200 mm in diameter have been removed from an undercut that is nearly 2 m deep but only 0.4 m high. We would have assumed that flow velocities at the back of the undercut would be insufficient to entrain these particles. It is likely that the removal process is related more to undercutting of the bank toe and collapse of individual gravel structures rather than to direct fluvial entrainment.

Deepening at the tip of the abutment corresponds with predictions made by model studies of spur dikes [*Kuhnle et al.*, 1999], despite the abutments encroaching on less than 50% of the stream width as was simulated in the model study. The increase in the local bank height at the scour hole could be expected to encourage slump failures, although we saw no examples. In fact, trees along the banks tend to fail by toppling into the river as the undercut collapses.

Further downstream where the toe of the bank is composed of finer material, the presence of very fine roots in the undercuts demonstrates that erosion is by the failure of small

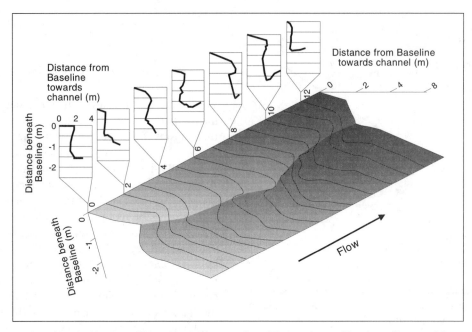

Figure 10. A series of vertical bank profiles at Site 1. The gray dotted line on the profiles shows the top of the gravel layer.

(<100 m wide) saturated slumps, or slurry flows, that slide away leaving the roots intact. Whilst roots initially add support to the fine material, it is theorized that this support allows the sediment to become saturated. Rather than failing in a block, such as a cantilever, the root mass is strong enough to hold the soil until it becomes a slurry that flows out between the roots.

5.3. The Role of Trees in Altering Bank Erosion Rates

A general conclusion from our observations is that isolated trees, of any size, on the Acheron River, will not alter the long-term migration rate of a bend. Individual abutments were smaller than we had expected (thus influencing less bank length), even for large trees over 20 m tall. Ultimately, erosion rates are controlled by removal of material from the toe of the bank. The ubiquity of undercutting beneath abutments, even in shallow sections of stream, indicates that it is erosion at this point that reflects the true impact of trees on bank processes. This should be born in mind when designing river rehabilitation projects.

Isolated abutments at the scale observed in the Acheron River do increase depth at their tips but do not obviously protect the downstream bank from erosion. This is not surprising because the effectiveness of engineered groynes is strongly related to their close spacing. However, there is no doubt that if large trees lined the top of the bank at spacings where their canopies merged and where their root plates

extended to the toe of the bank, then trees would provide some protection from erosion.

6. CONCLUSIONS

Stream bank erosion often isolates the root plate of a riparian tree on a pedestal of sediment jutting out from the stream bank. Such root-plate abutments are a distinct fluvial feature, yet they are a transitory landform produced as a result of greater erosion resistance provided by trees. The morphology of abutments integrates the many effects of isolated trees on erosion rates. From measuring seven abutments formed along the Acheron River in southeastern Australia, we conclude the following. (1) Roots from a single tree increase the resistance of impinging banks in a semi circle centered on the trunk. The abutment has a radius that is always smaller than (usually less than half) the canopy radius. This relationship holds for four dominant riparian tree species along the Acheron River situated on gravel and sandy-loam banks that are 1 m to 4 m high. (2) All abutments are deeply undercut, with most of the abutment forming a 0.5 m to 1 m thick overhanging plate of finer sediments reinforced by its roots. However, the deviation of the bank curve at its toe indicates that trees also provide some strengthening at the bank toe, even when the bank is nearly 4 m high. This strengthening is not enough to materially alter the migration rate of a meander bend. Abutments fail by toppling. (3) The bed is deepened

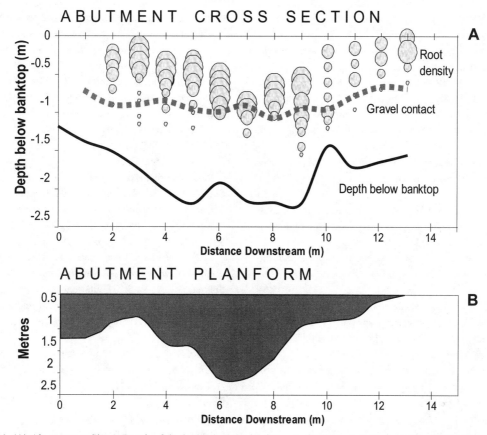

Figure 11. (A) Abutment at Site 1. Depth of the bed below the bank top (solid line) plotted against the distribution of root densities (circles graded from 1 smallest to 5 largest), position of the gravel contact (dotted line). (B) Planform of the abutment (dashed line at the bottom). Note the relationship between higher depth, longer abutment overhang, and denser roots.

at the tip of the abutment by as much as 30% of the bank height. Thus the abutments have a secondary effect on channel morphology.

Our observation of abutments on the Acheron River suggests the following processes that clearly have a bearing on erosion around trees but have been researched little: (1) the effect of trees (particularly roots and surcharge from the weight of trees) on erosion of gravels, (2) hydraulics and erosion mechanisms in undercut banks, and (3) interaction between abutments and the adjacent bed.

Of particular interest is the potential use of abutments as a tool for integrating the numerous scaled interactions between rivers and trees. That is, river morphology varies in a regular, if not predictable, pattern downstream (e.g. hydraulic geometry), whilst the types and characteristics of trees also change downstream for ecological reasons. Understanding these interactions improves our understanding of river channel change and vegetation [*Hupp*, 1990; *Abernethy and Rutherfurd*, 1998].

Acknowledgements. This research was made possible by a grant to the University of Melbourne from Land and Water Australia as part of the second Riparian Zone Research Program. The second author was supported by a post-doctoral fellowship from the Royal Society. Reviews by Natasha Pollen and an anonymous reviewer substantially improved the manuscript. We would like to thank Chris Parkinson and Subhadra Jha for assistance in the field, and Brett Anderson for figure preparation.

REFERENCES

Abernethy, B., and I. D. Rutherfurd, Where along a river's length will vegetation most effectively stabilise stream banks?, *Geomorphology*, 23, 55–75, 1998.

Abernethy, B., and I. D. Rutherfurd, The effect of riparian tree roots on the mass-stability of riverbanks, *Earth Surface Processes and Landforms* 25, 1–17, 2000.

Abernethy, B., and I. D. Rutherfurd, The distribution and strength of riparian tree roots in relation to riverbank reinforcement, *Hydological Processes*, 15, 63–79, 2001.

Adams, J., Gravel size analysis from photographs, *Journal of the Hydraulics Division, ASCE*, 105, 1247–1255, 1979.

Beeson, C. E., and P. F. Doyle, Comparison of bank erosion at vegetated and non-vegetated channel bends, *Water Resources Bulletin*, 31, 983–990, 1996.

Carbon, B. A., A. M. Bartle, A. M. Murray, and D. K. MacPherson, The distribution of root length and the limits of flow of soil water to roots in dry sclerophyll forest, *Forest Science*, 26, 656–64, 1980.

Davis, R. J., and K. J. Gregory, A new distinct mechanism of river bank erosion in a forested catchment, *Journal of Hydrology*, 157, 1–11, 1994.

Gordon, N. D., *The Hydraulic Geometry of the Acheron River, Victoria, Australia*, Department of Civil and Environmental Engineering, University of Melbourne, 1996.

Gray, D. H., and R. B. Sotir, *Biotechnical and Soil Bioengineering Slope Stabilisation: A Practical Guide for Erosion Control*, New York, Wiley, 1996.

Hey, R. D., and C. R. Thorne, Stable channels with mobile gravel beds, *Journal of Hydraulic Engineering*, 112, 671–689, 1986.

Hickin, E. J. and G. C. Nanson, Lateral migration rates of river bends, *Journal Hydraulic Engineering* 110(11), 1557–1567, 1984.

Hudson, H. R., A field technique to directly measure riverbank erosion, *Canadian Journal of Earth Science*, 9, 381–383, 1982.

Hupp, C. R., Vegetation patterns in relation to basin hydrogeomorphology, in *Vegetation and Erosion*, edited by J. B. Thornes, pp. 217–237, John Wiley and Sons, Chichester, 1990.

Jackson, R. B., J. Canadell, J. R. Ehleringer, H. A. Mooney, O. E. Sala, and E. D. Schulze, A global analysis of root distributions for terrestrial biomes, *Oecologia*, 108, 389–411, 1996.

Jackson, T. H., *Bank Protection on Mississippi and Missouri Rivers*, U.S. Army Corps of Engineers, 1935.

Kuhnle, R. A., C. V. Alonso, and F. D. Shields, Geometry of scour holes associated with 90° spur dikes, *Journal of Hydraulic Engineering*, 125, 972–978, 1999.

Laclau, J., M. Arnaud, J. Bouillet, and J. Ranger, Spatial distribution of Eucalyptus roots in a deep sandy soil in the Congo: Relationships with the ability of the stand to take up water and nutrients, *Tree Physiology*, 21, 129–136, 2001.

Micheli, E. R., and J. W. Kirchner, Effects of wet meadow riparian vegetation on streambank erosion, 2. Measurements of vegetated bank strength and consequences for failure mechanisms, *Earth Surface Processes and Landforms*, 27, 687–697, 2002a.

Micheli, E. R., and J. W. Kirchner, Effects of wet meadow riparian vegetation on streambank erosion, 1. Remote sensing measurements on stream bank migration and erodibility, *Earth Surface Processes and Landforms*, 27, 627–639, 2002b.

Pizzuto, J. E., Bank erodibility of shallow sandbed streams, *Earth Surface Processes and Landforms*, 9, 113–124, 1984.

Przedwojski, B., Bed topography and local scour in rivers with banks protected by groynes, *Journal of Hydraulic Research*, 33, 257–273, 1995.

Shields, F. D., and D. H. Gray, Effect of woody vegetation on sandy levee integrity, *Water Resources Bulletin*, 28, 917–931, 1992.

Shields, F. D., S. S. Knight, and C. M. Cooper, Incised stream physical habitat restoration with stone weirs, *Regulated Rivers: Research and Management*, 10, 181–198, 1995.

Simon, A., and A. J. C. Collison, Quantifying the mechanical and hydrologic effects of riparian vegetation on streambank stability, *Earth Surface Processes and Landforms*, 27, 527–546, 2002.

Sukhodolov, A., W. S. J. Uijttewaal, and C. Engelhardt, On the correspondence between morphological and hydrodynamical patterns of groyne fields, *Earth Surface Processes Landforms*, 27, 289–305, 2002.

Thorne, C. R., Processes and mechanisms of river bank erosion, in *Gravel Bed Rivers*, edited by R. D. Hey, J. C. Bathurst, and C. R. Thorne, pp. 227–271, Wiley, Chichester, 1982.

Thorne, C. R., Effects of vegetation on riverbank erosion and stability, in *Vegetation and Erosion*, pp. 125–144, John Wiley and Sons, Chichester, 1990.

Tufekcioglu A., J. W. Raich, T. M. Isenhart, and R. C. Schultz, Fine root dynamics, coarse soil biomass, root distribution and soil respiration in a multispecies riparian buffer in Central Iowa, USA, *Agroforestry Systems*, 44, 163–174, 1999.

Walker, M., and I. D. Rutherfurd, An approach to predicting rates of bend migration in meandering alluvial rivers, in *Second Australian Stream Management Conference*, Adelaide, South Australia, Vol 2, 659–66, 1999.

James R. Grove and Ian D. Rutherfurd, School of Anthropology, Geography and Environmental Studies, University of Melbourne, and Cooperative Research Center for Catchment Hydrology, Victoria, Australia, 3010. idruth@unimelb.edu.au.

Effects of Riparian Vegetation on Stream Channel Stability and Sediment Budgets

Stanley W. Trimble

Department of Geography, University of California Los Angeles, Los Angeles, California.

Does grass or forest promote greater stream channel stability? The question is important because (a) many sodded streambanks are naturally reverting to forest and (b) many governmental agencies in the U.S. are actively promoting forested streambanks and are even financially subsidizing the transition in some cases. Some previous studies suggest that stream channels are narrower, deeper, and smaller under sod cover than under forest cover. Their explanation was that grass promoted bank accretion while large woody debris (LWD) from forest destabilized the streams. This study examines four reaches of Coon Creek, Wisconsin, each with a grassed and forested subreach. Grassed reaches had smaller average bankfull cross-sectional area, baseflow width, baseflow cross-sectional area and width-depth ratio ($p < 0.001$). However, there was less or no statistical difference for bankfull width, average baseflow depth and maximum baseflow depth. These results were consonant with much of the literature and suggest that sediment budgets may be managed, in part, by managing riparian vegetation. Other implications are that riparian vegetation may affect channel size in relation to streamflow regime and that stream stability is not directly related to biomass. The differential effects of LWD on stream stability in different climates suggest that hydroclimatology plays an important role.

1. INTRODUCTION

A long-unresolved question is the relative size, shape, and stability of stream channels ("channel," as used here, includes channel and banks) as influenced by riparian vegetation [e.g. *Hadley*, 1961; *Parsons*, 1963; *Smith*, 1967; *Heede*, 1972; *Graf*, 1978, 1979; *Grissinger and Murphey*, 1983; *Hickin*, 1984; *Pizzuto*, 1984; *Volny*, 1984; *Shine*, 1985; *Hey and Thorne*, 1986; *Nanson and Hickin*, 1986; *Gregory and Gurnell*, 1988; *Elmore and Beschta*, 1989; *Coppin and Richards*, 1990; *Thorne*, 1990; *Shields*, 1992; *Malanson*, 1993; *Dunaway et al.*, 1994; *Fetherston et al.*, 1995; *McKenne et al.*, 1995; *Gurnell and Gregory*, 1995; *Shields et al.*, 1995; *Davies-*

Colley, 1997; *Montgomery*, 1997; *Piegay and Gurnell*, 1997; *Stott*, 1997; *Trimble*, 1997; *Lyons et al.*, 2000; *Millar*, 2000; *Brooks and Brierley*, 2002; *Simon and Collison*, 2002; *Hession et al.*, in press]. More specifically, does grass or does forest promote greater stream channel stability? A provocative and perhaps prescient hypothesis was offered by *Zimmerman et al.* [1967] who found that headwater streams were significantly narrower and smaller under sod cover than under forest cover. Their explanation was that forests supplied much woody debris to channels, destabilizing them, while sod banks accreted with sediment deposition and encroached on the stream. However, this effect appeared to decrease downstream.

At the same time the *Zimmerman et al.* study appeared, a trout management field manual was published by Wisconsin Department of Natural Resources [*White and Brynildson*, 1967]. Although that study was a pragmatic management guide based on long field experience rather than "scientific" inquiry, the authors gave basically the same conclusions as those of *Zimmerman et al.* regarding forested versus sodded

Riparian Vegetation and Fluvial Geomorphology
Water Science and Application 8
10.1029/008WSA12

channels. Moreover, they showed, for their region of the north-central U.S., what would happen when small, grassy channels revert to woodlands (Figure 1).

The question of forest versus grass is more than academic. Collectively, large areas of bottomlands in the United States have reverted from cropland or pasture to woodland over the past 60 years or so. This process continues, often abetted by government policies. During the 1930s and 1940s, the U.S. Civilian Conservation Corps planted millions of willows and other trees along eroding stream channels. For the last few decades, riverine rights-of-way have been acquired by state agencies which generally allow the land to revert to forest and have actually afforested the areas in some cases. Some scientists have recently advocated wholesale reforestation of stream-banks [*Welsch*, 1991; *Sweeney*, 1993]. Since 1991 the U.S. Federal government has encouraged, by subsidies through the Conservation Reserve Program, the planting of riverine areas to forest [*USDA-SCS*, 1991]. A similar scenario as above has also evolved in parts of France with trees being the riparian vegetation of choice [*Piegay et al.*, 1997] and it thus appears reasonable to assume that it is a widespread phenomenon. Given this systematic and continuing conversion of riverine zone from grass to forest, it is important to know the potential effects on stream channels. While the subject has received increasing interest in the scientific literature over the past decade, it received little attention in a recent National Research Council (NRC) report on riparian management [*NRC*, 2002].

2. GENERAL (THEORETICAL) CONSIDERATIONS

2.1. Forest

Trees might appear to be more effective in establishing and holding a stream bank against hydraulic scour. They have dense root systems that often extend below the water level and may extend several meters into the bank. Thus, a bank protected by riverine forest might enjoy the protection of an interlaced root system. The large roots provide reinforcement, tying the soil mass together, increasing tensile strength and reducing mass movement, while the fine roots provide protection against direct hydraulic scour. The effectiveness of such root systems has been suggested by earlier studies [*Hadley*, 1961; *Smith*, 1967; *Graf*, 1978; 1979; *Hickin*, 1984; *Pizzuto*, 1984; *Beschta and Platts*, 1986; *Gregory and Gurnell*, 1988; *Coppin and Richards*, 1990; *Thorne*, 1990; *Sweeney*, 1993; *Trimble*, 1994], but detailed, sophisticated work by *Abernethy and Rutherfurd*, [2000a, 2001] and *Simon and Collison* [2002] shows that tree root effectiveness is far more complicated.

Yet another apparent advantage of trees is that they can keep streambanks drier during the growing season by their higher rates of water use. *Wolman* [1959] and *Hooke* [1979]

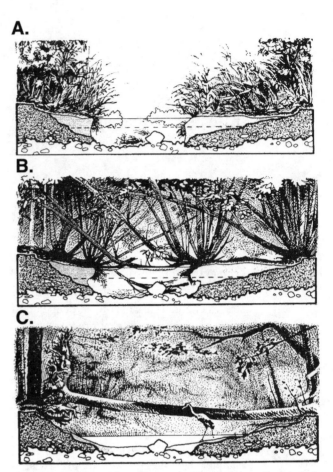

Figure 1. Transition of small, narrow, deep streams to broad, shallow streams as a result of afforestation. (a) Narrow, deep stream resulting from grass and small woody vegetation. (b) 10 to 20 years later, woody vegetation predominates, shading out grass on banks. Large woody debris begins to clog channel. (c) Many years later. Mature forest with dense shade and few plants on banks which have eroded away. Large woody debris clogs channel. Modified from White and Brynildson [1967].

have shown that wet banks are much more vulnerable to erosion and mass movement. With trees, the combination of high transpiration demand and extensive root systems can significantly reduce soil moisture in the banks [*Thorne*, 1990; *Smith*, 1992; *Simon and Collison*, 2002]. Bank failure is thereby theoretically reduced. Of course, this comes at the cost of increased water consumption, an important consideration in many regions [*Smith*, 1992].

Frost can also contribute significantly to bank erosion [*Lawler*, 1993]. *Stott* [1997] found that forested reaches had only half the incidence of frost as compared to grassy reaches.

There are, however, at least four destabilizing characteristics of mature forest on stream channels. While forested banks

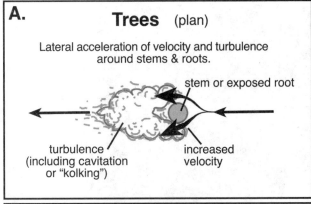

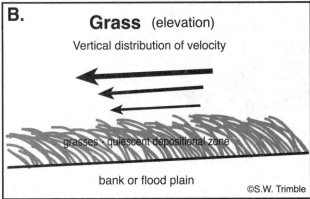

Figure 2. Comparison of velocity profiles for trees and grass.

and floodplains may more effectively decrease the mean velocity of streamflow because of hydraulic friction, there is the hydraulic scour created by the acceleration of water around rigid trunks and roots [*Parsons*, 1963; *Zimmerman et al.*, 1967; *Thorne*, 1990; *Trimble*, 1994; Figure 2A]. Thus, local bank and floodplain erosion can be effective at high stream discharges. As scour occurs around a root mass, and exposes more rigid roots, there is positive feedback that can accelerate the erosion [*Thorne*, 1990; *Trimble*, 1994].

One result of mature trees along streams is that they eventually find their way into the stream as large woody debris or LWD [*Heede*, 1972; *Swanson et al.*, 1976; *Swanson and Leinkaemper*, 1978; *Keller and Swanson*, 1979; *Keller and Tally*, 1979; *Mosley*, 1981; *Marston*, 1982; *Gregory et al.*, 1985; *Harmon et al.*, 1986; *Duijsings*, 1987; *Malanson and Butler*, 1991; *Gregory and Davis*, 1992; *Gregory et al.*, 1993; *Smith et al.*, 1993; *Nanson et al.*, 1995; *Montgomery*, 1997; *Piegay and Gurnell*, 1997; *Stott*, 1997; *Trimble*, 1997; *Piegay et al.*, 1999; *Lyons et al.*, 2000; *Manga and Kirchner*, 2000]. This occurs on a sustained basis from old timber but disease and storms can also supply a surge of LWD. The maximum residence time of LWD may be on the order of centuries [*Keller and Tally*, 1979; *Keller and Swanson*, 1979], so any effects are of long duration.

In contrast to the concept that LWD destabilizes stream channels, several studies suggest that LWD can create effective dams, storing sediment and perhaps even increasing stream stability [*Bilby and Likens*, 1980; *Marston*, 1982; *Hickin*, 1984; *Beschta and Platts*, 1986 *Duijsings*, 1987; *Gregory and Gurnell*, 1988; *Montgomery*, 1997; *Brooks and Brierley*, 2002]. However, a close reading of these studies shows that most are set in regions with mild hydrologic regimes such as those with low gradients in marine west coast climates (such as the lowland UK), or in glaciated areas with interrupted drainage and suppressed stormflow peaks, or in areas receiving equitable streamflow from mountain snowmelt. Conversely, during large floods and more commonly in flashier streams as found in humid tropical, sub-tropical, and continental climates, scour typically develops around the LWD so that the bed and banks may be severely eroded (Figure 3). In turn, this can undermine other streamside trees that eventually join the LWD. Thus, the effects of LWD may be event and climate specific, a point to which we will later return.

These processes can be aided by the third liability of trees, which is their mass along with their inherently unstable weight distribution. The moment of a tall tree, whether acted on by the forces of gravity, wind, or flood waters, is often enough to rip out all or part of the root mass [*Abernethy and Rutherfurd*, 1998]. The void left in the bank may not only be erodible, but it may create turbulence in high flows, further destabilizing the bank [*Shields and Gray*, 1992; *Trimble*, 1994; *Keller and McDonald*, 1995]. The effect of mass alone is still open to question and may be site specific, but *Abernethy and Rutherfurd* [2000b] found that the weight of trees alone did not destabilize a stream in Australia.

Yet a fourth problem with tree cover is that the banks and floodplain are shaded thus limiting the growth of grass and other ground cover. This results in increased surface erosion during high flows [*Hunt*, 1979; *White and Brynildson*, 1967; *Murgatroyd and Ternan*, 1983; *Nanson and Hickin*, 1986; *Peterson*, 1993; *Trimble*, 1994].

2.2. Grass

Thick, well-rooted grass can protect channels and banks from rapidly moving water [*Parsons*, 1963; *White and Brynildson*, 1967; *Dunaway et al.*, 1994]. Long grass, in particular, can lie over with the current and form a "thatch" which can prevent scour. Indeed, such a thatch can create a relatively thin quiescent environment directly above the ground that induces sediment accretion (Figure 2B). *White and Brynildson* [1967] and *Clary and Webster* [1990] suggest that ungrazed, grass-covered channel banks can accrete not only vertically, but also laterally so that they may extend over the water, forcing the stream into a deep and narrow, almost tunnel-like channel.

Figure 3. Accumulation of large woody debris (LWD) with resulting scour of bed and banks. This is Range 4 of Reach 1 Forest, this study. See Figure 7 for profile.

Grass has at least two major liabilities in regard to channel stability. First, on high steep banks, the grass roots rarely extend down to or below the water level as they often do with trees [*Thorne*, 1990]. Such banks are especially vulnerable to undercutting by stream erosion. Second, some grass roots are too shallow and weak to reinforce a bank so it may be susceptible to slumping under conditions of high moisture with high pore pressure [*Wolman*, 1959; *Hooke*, 1979]. During the extremely wet summer of 1993 in the northcentral U.S., streambank slumping was widespread on grass banks (Figure 4) but comparatively absent along forested banks.

Thus, it may be seen that trees and grass each have their theoretical advantages and disadvantages. It is not immediately clear that either enjoys a clear edge and indeed, any advantages might also depend on other factors such as existing channel morphology and slope, bank slope, height and strength, bank and floodplain material, species of trees and grass involved, stream order, management of trees and grass, and climate. Of the factors listed above, gravel-bed streams appear to be particularly problematic [*McKenney et al.*, 1995; *Jacobson and Gran*, 1999]. *Hey and Thorne* [1986] give a set of equations that show trees to be associated with narrower, gravel-bed channels. Another uncertainty is stream order, or more precisely, stream width. It appears that if streams are wider than the length of riparian trees, the effects of LWD are reduced [*Zimmerman*, 1967; *Davies-Colley*, 1997], presumably because the debris can float away downstream and is less likely to be

caught in logjams. The best way to establish the factors involved is by controlled experiments in which streams evolve over time under different vegetation, but only one has been done. *Hunt* [1979] removed all woody bank vegetation from a reach of three trout streams in Wisconsin. Only two of the three streams narrowed as expected and net trout increased in only one. However, the experiment ran only 3 1/2 years (1974 to 77), a period marked by highly erratic streamflow, so the results appear indeterminate.

An imperfect but quicker method than controlled experiments is by means of empirical measurements as used by *Zimmerman et al.,* [1967]. Four subsequent studies have used this approach. The first, in Dartmoor, UK, showed that channels under coniferous forest were significantly wider and larger than other local channels under grass cover [*Murgatroyd and Ternan*, 1983]. A similar study is the unpublished thesis of *Shine* [1985]. By examining contiguous grass and forested reaches along two small streams in Wisconsin, Shine was able to show that forested reaches had statistically significant greater average bankfull width, wetted perimeter and channel cross-sectional area than sodded reaches, but that there was no difference in water depths. In the Pennsylvania Piedmont, *Sweeney* [1993] found that forested streams were much wider than non-forested ones, but the difference decreased downstream, findings very similar to those of *Zimmerman et al.* [1967]. In headwater trout streams of upstate New York, *Peterson* [1993] showed that forested reaches were significantly

A.

B.

Figure 4. Bank slumping under grass, July 1993: (A) a tributary of Coon Creek just upstream of confluence with Reach 3 of this study, and (B) Coon Creek about 2 km upstream of Reach 3.

wider and shallower than reaches which had been deforested for electric transmission rights-of-way. The significance of this study was that the transition to a narrower and deeper channel had more clearly occurred as a result of vegetation change to grass. Later empirical studies are those of *Davis-Colley* [1997], *Stott* [1997], *Trimble* [1997, more details given later], and *Hession et al.* [in press]. Of these four studies, only *Stott* found narrower channels under forest and he was comparing two different streams.

Thus, the logic and the literature are ambivalent but it appears that grass has a slight advantage. Following are the results of a study from the northcentral U.S.

3. AN EXAMPLE FROM THE NORTHCENTRAL U.S.

Prior to European settlement, many valley bottoms of the northcentral U.S. appear to have been covered with grass. For the Driftless Area (Figure 5), there are two strong lines of evidence for this: (1) reports of early explorers [*Trimble and Lund*, 1982] and (2) the widespread presence of floodplain Prairie soils (Mollisols), now buried under the modern sediment [*McKelvey*, 1939]. The presence of these Mollisols not only indicates grass cover in periods before European settlement, it also suggests environmental stabil-

ity: streams were undergoing very slow lateral erosion and floodplain accretion rates were low. Streams were everywhere clear and Brook Trout abounded [*Trimble and Lund*, 1982; *Trimble and Crosson*, 2000]. These floodplain grasslands were maintained by fire from Native Americans and, to a lesser degree, by large grazing animals such as Buffalo [*Lyons et al.*, 2000].

Over twenty years of observation in streams of the region suggested to me that forested stream reaches were wider than grassy ones. The question became more pressing as more stream reaches, once in grass pasture, were allowed to revert to forest. This process has continued for many reasons but prominent among them were (1) the decline of agriculture in general, especially dairy farming, (2) bottomlands in some areas becoming wetter because of continuing stream sedimentation, making grass untenable [*Trimble and Lund*, 1982; *Trimble*, 1993] and (3) purchase of riverine rights of way by governmental agencies.

The region has a long history of sediment problems [*McKelvey*, 1939; *Happ et al.*, 1940; *Trimble and Lund*, 1982]. In particular, the storage of sediment and its reentrainment have been problematic [*Trimble*, 1975, 1981, 1983, 1999; *Knox*, 1987] and management of sediment has been of great concern over the past two decades [*Trimble*, 1993, 1999]. Thus, the containment or loss of sediment within and from stream channels is of major interest.

The methodology for the present study was to compare contiguous or near-contiguous reaches of grassy and forested stream channels in Coon Creek, Wisconsin, a stream which has received extensive study in the past [e.g. *McKelvey*, 1939; *Trimble*, 1975, 1981, 1983, 1993, 1999; *Trimble and Lund*, 1982]. The zone of the basin selected for study is the lower main valley [Figure 5; *Trimble*, 1983]. This zone, a long, graded reach of about 25 km, has aggraded about 3 to 5 meters since European settlement and the banks are formed of historical alluvium and are uniformly about 1 to 1.5 meters above baseflow water level. In most places, the floodplain (actually, broad natural levees) level is easily discerned, so measuring bankfull cross-sectional area (to the level of the levee tops) was relatively simple. Bank and channel materials are generally fine sand with generous admixtures of silt in the banks.

Sediment accretion continues in this zone so that channels and banks (but not necessarily floodplains) have aggraded several centimeters over the past 20 years or so [*Trimble*, 1993, 1999]. There are no stream discharge stations in the reach so one cannot easily compare bankfull discharge capacity to flow frequency, but the high frequency of overbank flows suggests that these channels are perhaps undersized. A summary of this study has already appeared [*Trimble*, 1997] but only a fraction of the data and limited applications could be presented.

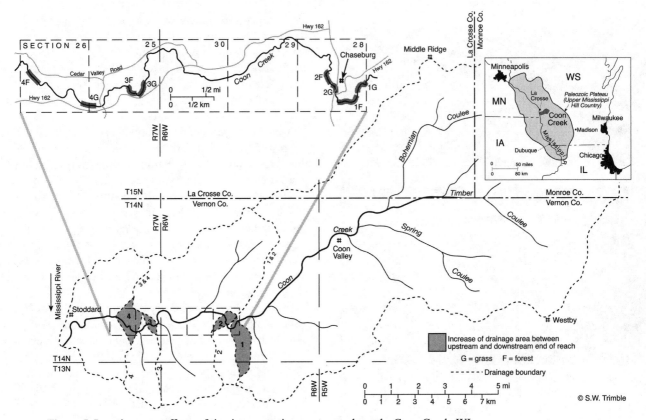

Figure 5. Location map, effects of riparian vegetation on stream channels, Coon Creek, WI.

4. SAMPLE REACHES

Sample reaches containing forest and grass sub-reaches were selected using several criteria. First, the subreaches had to be distinctly grass or forest and had to have been in that state for the 50 or so years previous, a condition verified by aerial photography starting in 1934. Secondly, in order to decrease the variable of stream discharge, there had to be a minimal increase of basin drainage area over the span of the reach. That is, no major tributaries could enter within the study reach. Third, the grass reach had to be upstream of the forested reach because LWD sometimes affects channels some distance downstream from the forested reach. Because of this "plume effect," a grass reach for this study could begin only with a buffer zone of 500 m or more downstream from forests. These stringent criteria severely limited possible sample sites, especially regarding grassy reaches. Four study reaches were found which generally fit the criteria (Figure 5, Table 1).

Additionally, it was desirable that "forest" and "grass" be uniform among study reaches. This was the case for forest, with mature stands of box elder (*Acer negundo*), silver maple (*Acer saccharinum L.*), willow (*Salix* spp.) and cottonwood (*Populus deltoides*) but grass was mixed domestic and native species.

It was also desirable that all grassland be ungrazed to eliminate the additional impacts of cattle on streambanks [*Trimble*, 1994; *Trimble and Mendel*, 1995]. However, Reach 1G (grass) was heavily grazed with obvious impacts such as eroding and slumping banks (Figure 6). Also, the left bank of Reach 3G was grazed, but not as heavily as Reach 1G. Reach 2G, while ungrazed, was partially in residential lawns in the village of Chaseburg but grass on the steeper banks was rarely mowed or clipped. Additionally, much of Reach 2 had been historically one of two branches of Coon Creek flowing through the village of Chaseburg. The other channel had gradually occluded and the present channel has carried the entire flow of Coon Creek for the past 15 or so years. Reach 2 also begins directly

Table 1. Drainage area at each reach.

Reach	Upstream km^2	Downstream km^2	PI
1	169.9	171.5	0.9
2	174.4	174.8	0.3
3	209.0	209.3	0.2
4	214.9	216.5	0.7

PI – Percent increase over reach

downstream from a forested reach but a low bridge directly upstream (Wis. Hwy. 162) blocks most LWD from floating down into Reach 2G. Additionally, the village maintenance person quickly removes any such LWD because it would increase local flooding. In Reaches 1, 2 and 3, less than 100 m separated the grass reach and the downstream forested reach. In Reach 4, however, problems of land access dictated that the sub-reaches be separated by 1500 m. Even so, the increase of drainage area was fortuitously minimal and indeed, the maximum areal increase over any study reach was only 0.9% at Reach 1 (Table 1).

Channel cross-sections were profiled by precise (3rd order) instrumented surveys orthogonal to the current at 30 m intervals. Anomalies (e.g., trees along a grass reach and one bridge) were avoided. More profiles would have been desirable from the standpoint of statistical analysis, but it appeared that a smaller interval between profiles might greatly lessen the independence of each profile. Thus, the need for statistical independence of each profile, along with the relatively short available grassy reaches, limited the number of possible sample profiles in each reach. All profiles were surveyed during a 10-day period in June 1995. There was no significant rain during the period and baseflow was high and uniform because of a wet spring.

The surveyed profiles allowed the measurement of several dimensions. First, bankfull cross-sectional area was defined as that of the stream flowing up to the level of the floodplain lying along each bank (Figure 7). In most cases, the flood-

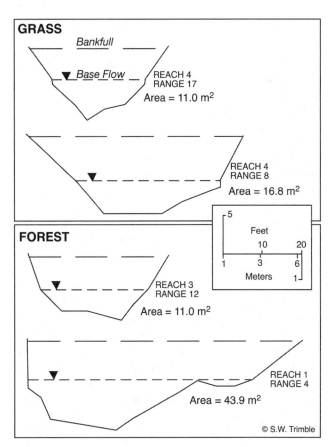

Figure 7. Contrasts in channel sizes, grass and forest

Figure 6. Reach 1G at Hwy. K bridge just upstream from Chaseburg, WI. View is downstream. Note battered and eroded stream banks with slumping (arrows), the result of heavy grazing.

plain (natural levee) on either side of the stream lies at the same level. The second measurement was the width of the channel at bankfull flow, measured from top of bank (edge of floodplain) to top of bank (Figure 7). These measurements are important because their differences indicate to what degree channels might be expected to increase in size if converted from grass to forest. The next four measurements are for baseflow only and are the width, cross-sectional area, average depth (cross-sectional area ÷ width) and maximum depths. Aside from pure geomorphologic questions, the latter four measurements also relate to fish habitat; a narrower, deeper stream is usually considered to be better [*White and Brynildson*, 1967; *Hunt*, 1979; *Peterson*, 1993; *Lyons et al.*, 2000]. In the same category is width-depth ratio (baseflow width ÷ baseflow average depth), a dimensionless value which gives a generalized geometry of the channel. Finally, maximum baseflow depth gives another index of fish habitat.

5. FINDINGS AND DISCUSSION

The collected data for all profiles are given in Tables 2-9, and the analysis is summarized in Table 10. Because variance could be caused not only by the primary factor of vegetation, but also by local and systematic factors of stream reach, two-way analysis of variance (ANOVA) was used for the primary analysis. Significance was indicated by the F-test. The two-way ANOVA shows (a) the main effects of, by turn, vegetation and reach, and (b) interactive effects of vegetation and reach (Table 10). To provide further clarification, a secondary analysis examined differences of means by reach (Table 10). The hypothesis tested is that the average forested channel dimension is larger than that of the grassed channel. Although the samples were small, the Kolmogorov-Smirnov test indicated that all of the samples compared in Table 10 have approximately normal distributions (p = 0.95). Following is a discussion of each morphological category examined.

5.1. Bankfull Channel Size

In the main, the forested channel was larger than the grassed channel (p < 0.001; Tables 2-10, Figure 8A). Differences ranged from 2.1 m^2 in Reach 4 to 8.8 m^2 in Reach 2, but reach did not significantly affect the results. In the context of other studies [Hunt, 1967; Smith, 1992; Murgatroyd and Ternan, 1993; Peterson, 1993; Davies-Colley, 1997; Lyons et al., 2000], this seems ample evidence to suggest that grassy chan-

Table 2. Effects of vegetation on stream channels by reach and profile, Coon Creek: Reach 1G, grass, heavily grazed.

R	BA m^2	BW m	W m	A m^2	D m	W/D	MD m
1	24.5	16.8	10.7	4.5	0.4	25	0.9
2	24.9	16.8	9.1	4.5	0.5	19	0.9
3	23.2	18.3	10.1	5.0	0.5	21	0.8
4	26.6	16.5	11.9	5.0	0.4	28	1.2
5	20.1	14.0	8.5	5.6	0.7	13	1.2
6	20.4	14.6	8.2	4.8	0.6	14	0.9
7	24.7	19.2	10.1	5.9	0.6	17	0.8
8	31.2	21.3	11.6	5.9	0.5	22	1.1
9	25.7	15.2	11.0	6.1	0.6	20	1.0
10	21.7	14.0	8.8	5.4	0.6	15	0.8
11	19.1	13.7	8.8	5.0	0.6	15	0.6
12	23.8	16.8	9.1	5.0	0.5	17	0.6
13	19.9	16.8	9.1	5.4	0.6	16	0.9
14	23.0	15.2	10.1	5.6	0.6	18	1.1
15	22.7	14.3	9.8	5.9	0.6	16	0.8
16	18.6	16.8	8.8	6.1	0.7	13	0.8
17	26.4	16.8	9.1	5.4	0.6	16	0.6
18	24.9	18.3	9.4	5.9	0.6	15	0.9
19	19.7	19.2	11.9	6.7	0.6	22	0.5
20	22.7	19.8	11.3	5.9	0.5	22	0.6
Ave.	23.2	16.7	9.9	5.5	0.6	18	0.9
SD	3.1	2.1	1.2	0.6	0.1	4.1	0.2

R – range; BA – bankfull cross-sectional area; BW – bankflow width; W – baseflow width; A – baseflow cross-sectional area; D – baseflow average depth; W/D – baseflow width-depth ratio; MD – baseflow maximum depth.

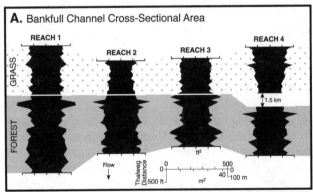

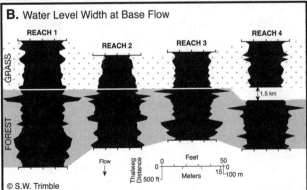

Figure 8. Channel size related to vegetation, grass versus forest, Coon Creek, WI.

nels would become larger by reverting to forest. The magnitude of this channel erosion is suggested by the range of values. The data from Reach 4 indicates that sediment loss would be 2.1 m^3/m of channel or 2100 m^3/km (2.67 ac-ft/mile) of channel. Expressed another way, the grassed channel is storing 2100 m^3/km of more sediment than is the forested channel. Assuming that it took 50 years for the forested reaches to assume their present size, the average annual rate of channel erosion (sediment export) would be about 42 m^3/km of channel. Assuming a bulk specific gravity of 1.4 [Trimble and Lund, 1982] this would be about 59 Mg/km-yr.

Looking at Reach 2, the reach with the greatest average disparity between grassy and forested (8.8 m^2), the total lost of afforested channel would be about 8800 m^2/km (11.4 ac-ft/mile). Again at a bulk specific gravity of 1.4 and an afforestation period of 50 years, the average sediment export would be about 245 Mg/km (434 tons/mi-yr). However, 30 years of

Table 3. Effects of vegetation on stream channels by reach and profile, Coon Creek: Reach 1F, forest.

R	BA m^2	BW m	W m	A m^2	D m	W/D	MD m
1	23.4	12.2	10.7	8.7	0.8	13	0.9
2	20.6	13.7	10.1	6.1	0.6	17	0.8
3	43.3	21.9	19.8	15.1	0.8	26	1.5
4	43.9	21.0	17.7	16.0	0.9	19	1.9
5	27.1	16.5	11.3	6.5	0.6	19	1.0
6	20.8	14.3	9.4	5.6	0.6	16	0.7
7	27.3	17.7	10.7	7.1	0.7	16	1.1
8	28.4	16.8	11.3	7.4	0.7	17	0.8
9	21.7	13.7	9.1	7.6	0.8	11	0.9
10	22.9	15.8	10.7	6.5	0.6	18	0.9
11	19.7	14.3	10.4	4.5	0.4	24	0.8
12	16.0	11.0	9.4	6.5	0.7	13	1.0
13	29.2	16.8	12.8	7.1	0.6	23	0.9
14	29.4	19.8	9.1	6.7	0.7	13	1.1
15	26.0	17.1	10.1	6.5	0.6	16	0.9
16	22.5	13.7	11.0	5.6	0.5	21	0.6
17	26.4	16.2	11.3	6.5	0.6	19	0.8
18	27.1	16.5	11.9	7.2	0.6	20	1.0
19	27.7	15.2	14.6	7.1	0.5	30	0.9
20	28.1	16.2	14.0	6.9	0.5	29	0.9
21	30.5	15.8	13.7	7.8	0.6	24	1.2
22	29.0	16.8	13.4	6.1	0.5	29	0.5
23	25.1	15.5	12.2	6.3	0.5	24	0.7
24	26.0	13.7	11.3	6.1	0.5	21	0.9
25	29.6	16.5	12.5	7.1	0.6	22	0.7
26	32.2	16.2	12.8	9.3	0.7	18	0.9
27	26.8	14.6	11.0	6.3	0.6	19	0.6
28	24.5	14.6	10.4	5.9	0.6	18	0.9
29	20.4	14.6	99.8	4.6	0.5	20	1.1
30	23.4	14.3	10.1	5.6	0.6	18	0.9
Ave.	26.6	15.8	11.8	7.2	0.6	20	0.9
SD	5.9	2.3	2.4	2.5	0.1	4.8	0.3

observation within the region suggest that forested channels, while having both erosion and accretion, eventually attain a reach equilibrium so that net channel erosion ceases. That is, channels afforested for long periods seem to have assumed stable proportions on average. Thus, the foregoing observations would apply only to newly (≤ 30 years) forested reaches. Because few data exist, we must speculate on actual export rates over the 50-year period and those would decrease if the period were longer. Drawing on the general principles given earlier and a study by *Shields et al.,* [1995], it appears that afforesting channels might remain stable or even gain sediment

during the early stages of forest growth when small trunks are supple and root growth adds bank strength. After about two decades trees would have become large enough to add substantial LWD to streams, thus beginning the destabilization process described earlier. Many riverine trees tend to be short-lived, so after 2 to 3 decades there should begin a surge of LWD. Given the consequent bank erosion with collapse of mature trees into the stream augmented with windthrow and other chance delivery of LWD to the stream, bank and channel erosion should be at a maximum and erosion rates could be conceivably several times the average rates cited earlier. However, once the channel reaches equilibrium where cut and fill are roughly equal over some reach, then presumably there is no further net sediment loss. The latter point deserves emphasis: "no net loss" from an "equilibrium" forested reach may disguise significant fluxes with channels locally eroding in some places but accreting in others. Such within-reach fluxes would presumably be greater than in grassed reaches. While the foregoing processes and changes are partially hypothetical, they appear to be consonant with all the literature available thus far.

Much of the literature suggests that geomorphologic work or flux is inversely correlated with phytomass [e.g., *Langbein and Schumm*, 1958; *Graf*, 1979; *Kirkby*, 1980]. However, the accumulated evidence suggests that the relationship along stream channels is more complex (Figure 9). As would be intuitively expected, bare banks allow rapid erosion [e.g.,

Table 4. Effects of vegetation on stream channels by reach and profile, Coon Creek: Reach 2G, grass.

R	BA m^2	BW m	W m	A m^2	D m	W/D	MD m
1	20.1	12.5	7.0	6.7	1.0	7	1.4
2	25.3	12.8	6.4	5.6	0.9	7	1.2
3	17.5	11.6	7.3	5.8	0.8	9	1.0
4	16.9	11.6	7.6	5.6	0.7	10	0.9
5	15.6	11.0	7.9	5.6	0.7	11	0.9
6	17.7	11.6	7.9	5.8	0.7	11	0.9
7	18.2	10.7	7.6	5.9	0.8	10	0.9
8	15.2	10.7	7.6	5.8	0.8	10	1.0
9	21.4	11.9	8.5	5.9	0.7	12	0.8
10	19.7	13.1	9.4	6.1	0.6	15	0.9
11	18.8	13.7	9.8	4.8	0.5	20	0.9
12	18.2	13.7	9.1	5.8	0.6	14	1.0
13	21.0	16.8	9.8	6.5	0.7	15	1.1
14	21.0	13.4	10.1	6.3	0.6	16	0.9
Ave.	19	12.5	8.3	5.9	0.7	12	1.0
SD	2.6	1.6	1.2	0.5	0.1	3.5	0.2

Table 5. Effects of vegetation on stream channels by reach and profile, Coon Creek: Reach 2F, forest.

R	BA m²	BW m	W m	A m²	D m	W/D	MD m
1	22.9	13.4	10.4	6.5	0.6	16	1.0
2	25.5	13.7	9.1	8.0	0.9	10	1.3
3	44.6	20.7	17.1	14.5	0.8	20	1.9
4	24.5	14.6	10.4	7.4	0.7	14	0.9
5	34.8	15.2	14.6	9.5	0.7	23	1.0
6	27.3	18.3	12.2	9.3	0.8	16	1.5
7	27.5	16.8	11.3	6.9	0.6	19	1.2
8	24.9	15.2	11.3	7.21	0.6	18	0.9
9	27.1	16.2	11.9	6.5	0.5	22	0.7
10	28.1	16.8	11.6	7.1	0.6	19	0.9
11	32.9	19.8	14.9	7.8	0.5	29	0.9
12	25.7	15.8	14.3	7.4	0.5	28	1.2
13	23.8	13.4	10.1	5.8	0.6	17	0.8
14	21.7	15.2	9.8	5.6	0.6	17	0.9
15	26.8	18.3	10.1	7.1	0.7	14	0.9
16	24.9	15.2	10.1	6.5	0.6	16	1.2
17	24.7	15.2	12.2	6.7	0.5	22	0.9
18	30.7	17.4	11.3	7.6	0.7	17	1.2
19	30.7	18.3	10.7	8.6	0.8	13	1.0
20	21.6	13.4	9.8	6.9	0.7	14	0.9
21	30.3	16.8	11.9	8.6	0.7	16	1.4
22	31.4	16.2	13.4	7.8	0.6	23	0.7
23	27.0	14.9	11.9	7.2	0.6	20	0.4
Ave.	27.8	16.1	11.8	7.7	0.6	18	1.0
SD	5	2	2	1.8	0.1	4.5	0.3

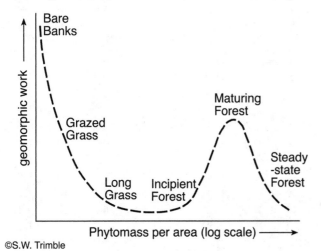

Figure 9. Relation of unit riparian phytomass to net geographic work along stream channels (schematic)

Keller, 1975], but bank erosion would be expected to decrease with increasing grass cover. The best protection for banks, at least based on most studies so far, appears to be long grass or perhaps long grass with small woody plants. As the woody plants reach maturity, and then senility, gross bank erosion might be expected to reach a maximum. This relationship is shown as a continuous function in the schematic model (Figure 9) but the transition from grass to trees may, in fact, be a step function. The model is more applicable to humid areas; in arid regions, there is inadequate rain for grass cover and phreatophytic woody plants would predominate.

Even in humid climates, there may be significant regional differences in the role of LWD depending on stormflow frequency-magnitude relationships. This study was in the humid continental climate which features intense precipitation events. As mentioned earlier, other studies done in areas of milder hydroclimatology such as marine west coast [Koeppen; Cbf] suggest that LWD may act as a dam and help stabilize streams [e.g., *Marston*, 1982; *Montgomery*, 1997; *Brooks and Brierley*, 2002] For example, one-year, 2 to 10 hour storms are 2 to 3 times greater in the Driftless Area (Daf) as compared to the Puget Sound region (Cbf, Figure 10). Similar differences of stormflow magnitude between the two regions are given by the U.S. Flash Flood Index [*Kochel*, 1988]. Thus a moderate flow might allow LWD to act as a dam and stabilize the

Table 6. Effects of vegetation on stream channels by reach and profile, Coon Creek: Reach 3G, grass.

R	BA m²	BW m	W m	A m²	D m	W/D	MD m
1	24.0	15.2	10.1	6.1	0.6	17	0.9
2	20.3	0.9	8.5	6.3	0.7	12	1.1
3	23.4	14.9	8.2	6.9	0.8	10	1.2
4	25.5	15.2	9.8	5.8	0.6	17	0.9
5	23.4	13.7	9.4	6.7	0.7	13	1.1
6	23.4	12.5	8.5	6.3	0.7	12	1.1
7	21.6	11.6	7.9	6.1	0.8	10	1.1
8	23.6	13.7	8.8	5.9	0.7	13	1.0
9	22.1	14.3	7.9	5.9	0.7	10	0.9
10	25.1	17.4	9.1	10.6	1.2	8	2.1
11	20.3	12.2	7.9	4.8	0.6	13	0.9
12	22.3	12.2	8.8	6.3	0.7	13	0.9
13	23.6	14.6	9.8	6.9	0.7	14	1.2
14	20.3	12.2	8.8	6.9	0.8	11	0.9
15	28.6	16.8	10.1	8.7	0.9	12	1.2
Ave.	23.2	13.2	8.9	6.7	0.7	12	1.1
SD	2.2	3.8	0.8	1.4	0.2	2.4	0.3

Table 7. Effects of vegetation on stream channels by reach and profile, Coon Creek: Reach 3F, forest.

R	BA m²	BW m	W m	A m²	D m	W/D	MD m
1	40.5	22.9	19.2	11.3	0.6	33	1.0
2	21.2	12.8	9.1	7.4	0.8	11	1.5
3	34.4	18.3	14.6	7.8	0.5	27	1.1
4	26.6	14.0	11.0	7.8	0.7	16	1.1
5	19.3	12.2	10.7	8.7	0.8	13	1.2
6	20.3	12.2	10.1	5.4	0.5	18	0.9
7	20.8	12.2	9.1	6.7	0.7	13	0.9
8	24.7	15.5	9.8	6.5	0.7	15	1.2
9	20.3	11.3	8.8	5.9	0.7	13	1.2
10	24.2	12.8	10.1	8.4	0.8	12	1.3
11	18.6	11.9	8.2	6.1	0.7	11	1.2
12	18.0	11.0	8.5	6.7	0.8	11	1.2
13	28.1	18.3	15.8	7.4	0.5	35	1.1
14	37.0	18.6	15.2	11.3	0.7	21	0.9
15	40.3	18.3	13.4	14.5	1.1	13	1.9
16	27.3	15.2	11.3	9.3	0.8	14	1.5
17	32.7	17.4	13.1	8.6	0.7	20	0.9
18	25.7	14.6	11.6	7.4	0.6	18	0.8
19	19.7	11.6	8.2	6.5	0.8	10	1.1
20	27.3	14.6	9.1	11.2	1.2	8	1.8
Ave.	26.4	14.8	11.3	8.2	0.7	17	1.2
SD	7.2	3.2	3	2.3	0.2	7.4	0.3

channel whereas the much greater and more rapid flows expected elsewhere might cause LWD to induce stream channel erosion (Figure 2).

Given that forested channels are larger on average in some regions, it might be surmised that they have the advantage in transporting more streamflow and permitting less flooding. Although the LWD increases Manning's n values up to two times at baseflow, *Shields and Gippel* [1995] found that removal of LWD in Obion River, Tennessee, increased bankfull conveyance only 22%. However, many forested cross-section sizes are comparable to the small sections of grassed channels (Figure 7) and such small cross-sections would be locally limiting (bottlenecks) so that higher flows might locally be forced overbank. Because stream velocity would be accelerated through these smaller cross-sections, they should be subject to higher shear stresses.

A final point in this section is that the average bankfull channel size within the grassed sub-reach of Reach 2 is anomalously low. I attribute this to the fact that much of Reach 2 is located on a channel that, up to the 1970s, carried only about half the flow of Coon Creek, and the bankfull channel size has not yet adjusted to the greater flow.

5.2. Bankfull Width

This measure is a surrogate for bankfull area. Additionally, the difference between grass and forest gives a general idea of how much land area would be eroded from the floodplain by conversion to forest. The data (Tables 2–10) are mixed but, overall, forested reaches are wider than grassed ones (p < 0.02).

There is a strong downstream trend of decreasing bankfull channel width (p < 0.001), but this may be highly influenced by anomalously wide channels in Reach 1G. Indeed, the two-way ANOVA shows that Reach 1G is anomalously wide, being 0.9 m wider than Reach 1F. This may be explained by the fact that Reach 1G was heavily grazed (Figure 6), so that bank crests had been trampled and eroded away from the stream, making the channel wider. Reach 1F and 2F are also anomalously wide, but no explanation is readily evident.

Noted earlier was the observation that some grassed streambanks underwent significant slumping during the extremely wet summer of 1993 (Figure 4). The effect appeared to be almost exclusively on grazed banks but that may be due to the fact that grazed banks are more easily observed. Such slumping would have left grassed channels larger and wider than normal so that differences between grassed and forested channels, as considered in this study, would be less than normal. That is, failed banks of grassed channels would presumably recover over a number of years, further restricting channel cross-section and making differences between forested and grassed channels even more significant.

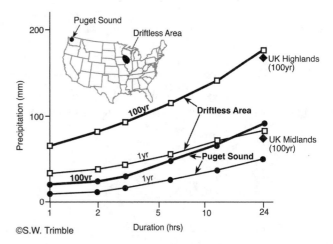

Figure 10. Frequency-magnitude relationships for precipitation events, Driftless Area humid continental) and Puget Sound and UK (marine west coast).

5.3. Baseflow Width

The channel width at baseflow was significantly (p < 0.001) narrower in grass (Tables 2–10, Figures 6, 8B, and 10). This strongly suggests that channel widening under forest is a real and strong tendency that is readily apparent to the observer (Figure 7). The average difference ranges from 1.9 m for Reach 1 to 3.5 m for Reach 2.

The main effect of reach on baseflow width is insignificant, but Reach 1G is anomalously wide. Again, this is attributed to heavy grazing of Reach 1G, which had eroded and compacted the wetted edge of the stream. Note that its average width, 9.9 m is 10.6% wider than Reach 3G, the next widest grass reach. Perhaps significantly, the left bank of Reach 3G is also grazed, but not so heavily as Reach 1G.

5.4. Baseflow Cross-Sectional Area

Again, all reaches showed that forests had significantly larger baseflow channels (p < 0.001). Given the equation for continuity of flow (Q = AV), one might conclude that

Figure 11. A narrow channel under high grass. This is the down-stream part of Reach 4G (Ranges 17 to 19, see Figure 6) where width-depth ratios range from 8 to 12. Slumping is not observable here but may be hidden by the high grass.

Table 8. Effects of vegetation on stream channels by reach and profile, Coon Creek: Reach 4G, grass.

R	BA m^2	BW m	W m	A m^2	D m	W/D	MD m
1	33.1	19.8	10.1	6.7	0.7	15	1.0
2	25.5	13.1	8.2	5.8	0.7	12	1.0
3	26.0	14.3	8.2	5.9	0.7	11	1.1
4	25.1	12.8	8.5	6.9	0.8	11	1.1
5	22.3	13.7	9.1	5.2	0.6	16	0.7
6	27.0	15.8	11.9	8.4	0.7	17	1.2
7	20.1	11.6	7.6	6.5	0.9	9	1.2
8	34.2	16.8	11.3	9.9	0.9	13	1.2
9	24.3	11.6	7.9	7.8	1.0	8	1.3
10	23.0	11.6	7.3	6.1	0.8	9	1.2
11	22.1	11.6	8.5	6.3	0.7	12	1.2
12	23.4	12.5	9.1	5.8	0.6	14	0.9
13	23.2	13.7	8.8	6.5	0.7	12	0.9
14	19.7	11.9	8.2	5.9	0.7	11	1.5
15	17.8	11.3	7.9	7.4	0.9	8	1.3
16	19.9	14.0	9.8	8.7	0.9	11	1.5
17	17.5	11.0	7.3	6.5	0.9	8	1.5
18	17.5	10.1	7.6	5.9	0.8	10	1.2
19	19.7	11.9	9.1	7.1	0.8	12	0.9
Ave.	23.2	13.1	8.8	6.8	11.0	11	1.2
SD	4.7	2.3	1.3	1.2	2.6	2.6	0.2

forested channels, being large, would have lower baseflow velocities. However, probably half or more of LWD in Coon Creek was below water level. Not only does this reduce the effective channel cross-sectional area; it also causes flow to accelerate around the debris so that velocities are highly varied in and around LWD. The result is channel and bank erosion where velocities are high, and sediment infills in quiescent water. Such infill is usually noncompacted, easily-eroded silty material.

Baseflow cross-sectional areas also increase downstream for both grassed and forested conditions. This is expected because streams of the region are highly effluent so that base-flow increases downstream. Additionally, slope decreases so that a larger cross-section might be expected.

5.5. Average Baseflow Depth

Because grassed streams are visibly narrower at baseflow (Figure 11), it would seem intuitive that they would be deeper because the flow appears to be laterally constrained. Also, some qualified observers have suggested that grassed channels are deeper [*White and Brynildson*, 1967; *Peterson*, 1993]. However, the data for this study do not support that idea (Tables 2–10). In fact, there is almost no difference, a result consonant with *Shine* [1985]. However, baseflow average depth does increase downstream, a result that is expected from previous discussion, showing the baseflow cross-section increasing downstream but with width remaining fairly uniform. Downstream increases of depth were greater for grassed than for forested reaches, and this may relate to the downstream decrease of width-depth ratio discussed next.

Table 9. Effects of vegetation on stream channels by reach and profile, Coon Creek: Reach 4F, forest.

R	BA m²	BW m	W m	A m²	D m	W/D	MD m
1	24.2	11.6	10.1	9.9	1.0	10	1.7
2	27.5	15.5	11.9	9.9	0.8	14	1.5
3	19.3	11.3	9.4	7.4	0.8	12	1.0
4	21.9	11.9	9.4	5.6	0.6	16	1.0
5	26.2	13.1	9.8	9.5	1.0	10	1.4
6	33.6	16.8	14.9	12.1	0.8	18	2.1
7	20.8	11.6	9.1	7.1	0.8	12	1.3
8	21.4	12.5	10.4	7.8	0.8	14	1.1
9	26.4	15.2	11.9	7.4	0.6	19	1.5
10	25.3	13.4	10.1	10.4	1.0	10	1.6
11	24.3	13.7	11.3	8.2	0.7	15	1.0
12	23.6	14.3	11.3	7.8	0.7	16	1.0
13	25.7	15.2	11.6	8.2	0.7	17	0.9
14	29.9	14.9	13.4	9.3	0.7	19	0.8
15	23.2	12.8	11.3	8.7	0.8	15	0.9
16	28.4	16.8	11.6	12.1	1.0	11	1.4
17	43.1	21.3	20.4	9.9	0.5	42	1.9
18	19.9	11.3	10.1	6.3	0.6	16	0.9
19	20.6	12.2	10.1	5.9	0.6	17	0.9
20	20.1	10.4	9.1	7.1	0.8	12	1.1
Ave.	25.3	13.8	11.4	8.5	0.8	16	1.3
SD	5.6	2.6	2.6	1.8	0.1	6.8	0.4

5.6. Baseflow Maximum Depth

Again one might invoke the reasoning above to suggest that grassed channels would have the greater maximum depths at baseflow. Countering this, however, is the fact that the locally high water velocities flowing under and around LWD can erode deep holes as discussed above and, indeed, some of these holes were deep enough to make the surveys more difficult. In fact, every forested reach had greater average maximum depths and these are quite evident in Reaches 3 and 4 (Tables 2–10) but the differences were not very significant ($p < 0.1$). Maximum depths tend to increase downstream, and this may relate to the downstream decrease of width-depth ratio, discussed next.

5.7. Baseflow Width-Depth Ratio

This dimensionless measure shows the relationship between stream width and average depth at baseflow. Not surprisingly, grassed reaches have average W/D ratios that are only 67 to 72% of their forested counterparts and overall

are statistically very significant ($p < 0.001$). The exception is Reach 1 where the heavy grazing presumably affected this configuration so that the W/D ratio of Reach 1G was 90% of 1F (Tables, 2, 3, and 10). The W/D ratios for both grass and forest decrease in a downstream direction, a tendency which accords with that of *Knox* [1987] who found that stream channels in the Driftless Area became narrower downstream, and may relate to downstream fining of channel and bank sediments.

6. CONCLUSIONS AND IMPLICATIONS

I conclude that there are significant and reasonably consistent differences between several morphological characteristics of wooded and grassed channels in Coon Creek. A primary question was one of channel sediment budgets: would a grassed reach become a sediment source if reversion to forest occurs? Based on this study and the literature reviewed, the answer seems to be yes, and the potential sediment loss appears considerable. By implication and based on the other studies

Table 10. Summary of average channel dimensions differences under grass (G) and forest (F) vegetation cover [from *Trimble*, 1997].

R	BA m²	BW m	W m	A m²	D m	W/D	MD m
1G	23.2	16.7	9.9	5.5	0.6	18	0.9
1F	226.2	15.8	11.8	7.2	0.6	20	0.9
Diff.	3.4	-0.9	1.9	1.7	0	2	0
Prob.	<u>0.025</u>	0.1	**0.001**	N.S.	N.S.	0.1	0.15
2G	19	12.5	8.3	5.9	0.7	12	1
2F	27.8	16.1	11.8	7.7	0.6	18	1
Diff.	8.8	3.6	3.5	1.8	-0.1	6	0
Prob.	**0.0005**	**0.0005**	**0.0005**	**0.0005**	**0.0025**	**0.0005**	> 0.25
3G	23.2	13.2	8.9	6.7	0.7	12	1.1
3F	26.4	14.8	11.3	8.2	0.7	17	1.2
Diff.	3.2	1.6	2.4	1.5	0	5	0.1
Prob.	<u>0.05</u>	0.1	**0.001**	**0.01**	> 0.25	**0.005**	0.2
4G	23.2	13.1	8.8	6.8	0.8	11	1.2
4F	25.3	13.8	11.4	8.5	0.8	16	1.3
Diff.	2.1	0.7	2.6	1.7	0	5	0.1
Prob.	0.15	0.2	**0.0005**	**0.001**	N.S.	**0.0025**	0.2

Diff. – Difference between Reach F and G; for probability (Prob.), underlined values mean statistical significance at P < 0.05; bold values mean statistical significance at P < 0.01; N.S. – sample distribution not suitable for comparison by difference of means.

reviewed here [e.g., *Hunt*, 1975; *Peterson*, 1993] it also appears likely that forested reaches converted to grass would become sediment sinks. If so, conversion from forest to grass would create a sediment sink in channels of the same magnitude already discussed for sediment sources. Thus, sediment budgets may be managed in part by managing riparian vegetation. I do not argue these results to be universal, but believe them to be correct for the Driftless Area and by implication, in most alluvial channels composed of silt to fine sand. Indeed the studies reviewed here suggest that similar results are widespread.

Another implication of the effect of riparian vegetation on channel size is that it affords one explanation for the variance of channel size from flow regime. That is, the failure of channel size or bankfull discharge capacity to accord with the presumed channel-forming discharge [e.g., *Williams*, 1978] might be in part attributed to the effects of riparian vegetation. Such studies should henceforth specify vegetation, both for reaches and for individual profiles. Such an approach (examining the variance from size predicted by regime) might be another way of more precisely quantifying the role of riparian vegetation on channel size.

One control on the effects of LWD on stream size and stability appears to be stormflow regime and particularly regional hydroclimatology. If so, would this affect not only channel size, but also form? LWD has been shown to affect pool spacing [*Montgomery et al.*, 1995], channel pattern [*Piegay and Gurnell*, 1997; *Millar*, 2000], and river evolution [*Brooks and Brierley*, 2002]. How might these processes and forms differ by hydroclimtalogical regime? Certainly, the climate classification (e.g., Koeppen) should be specified for any study. In the UK, stormflow regimes are much more severe for the highland areas as opposed to the lowlands [*Faulkner*, 1999] so that location even within the UK is important (Figure 10).

A long-standing tenet of geomorphology is a positive correlation between biomass and geomorphic stability (e.g., low rates of erosion). Most studies of riparian vegetation, however, have indicated a negative correlation in that forested banks often appear to cause more bank erosion than grass. This study suggests that the relationship may be far more complex and appears to be time specific.

Assuming that the physical characteristics exhibited in this and other studies by grass-lined channels are found to be the goals of stream management in some regions, how can grass cover be maintained? That is, how can grass be encouraged and trees be eliminated? Probably the only effective and cost efficient method would be some combination of fire, mechanical cutting, and perhaps controlled grazing at optimum times. Otherwise, grass reaches would quickly become woodland [*Lyons et al.*, 2000].

Removal of LWD does not appear to be a management alternative. Aside from great expense, experimental clearance of LWD from streams elsewhere has generally been disastrous with at least short-term channel destabilization and downstream movement of sediment (*Robison and Beschta*, 1990; *Gippel et al*, 1992; *Gregory*, 1992; *Keller and McDonald*, 1995] although it is not clear this result would be universal.

Acknowlegements. I thank the Wisconsin Department of Natural Resources for support, my daughter, Jennie Trimble, for field assistance, Alex Mendel for research and statistical assistance, and Chase Langford for the illustrations. Critical readings by John Lyons (Wis. Dept. of Nat. Resources), George Malanson (Univ. of Iowa), Timothy Diehl (USGS), David Sear (Univ. of Southampton) and one anonymous reviewer greatly improved the paper. In particular, Dr. Diehl had a profound influence on Figure 9. Fellow members of the Committee on Watershed Management, National Research Council, provided stimulating discussions related to this topic during the period 1996–99.

REFERENCES

Abernethy, B., and I. Rutherfurd, Where along a river's length will vegetation most effectively stabilize stream banks? *Geomorphology*, 23, 55–75, 1998.

Abernethy, B., and I. Rutherfurd, The effect of riparian trees on the mass-stability of riverbanks, *Earth Surface Processes and Landforms*, 25, 921–937, 2000a.

Abernethy, B., and I. Rutherfurd, Does the weight of riparian trees destabilize riverbanks, *Regulated Rivers: Research and Management*, 16, 565–576, 2000b.

Abernethy, B., and I. Rutherfurd, The distribution and strength of riparian tree roots in relation to river bank reinforcement, *Hydrological Processes*, 15, 63–79, 2001.

Bilby, R. E., and G. E. Likens, Importance of organic debris dams in the structure and function of stream ecosystems, *Ecology*, 61, 1107–1113, 1980.

Brooks, A., and G. Brierley, Mediated equilibrium: the influence of riparian vegetation and wood on the long-term evolution and behaviour of a near-pristine river, *Earth Surface Processes and Landforms*, 27, 343–367, 2002.

Clary, W. P., and B. F. Webster, Riparian grazing guidelines for the intermountain region, *Rangelands*, 12, 209–212, 1990.

Coppin, N. J., and I. G. Richards, (Eds.), *Use of Vegetation in Civil Engineering*, 292 pp., Butterworths, London, 1990.

Couper, P., T. Stott, and I. Maddock, Insights into river bank erosion derived from analysis of negative erosion-pin recordings: observations from three recent UK studies, *Earth Surface Processes and Landforms*, 27, 59–79, 2002.

Davies-Colley, R. J., Stream channels are narrower in pasture than in forest, *New Zealand Journal of Marine and Freshwater Research*, 31, 599–608, 1997.

Duijsings, J. J., A sediment budget for a forested catchment in Luxembourg and its implications for channel development, *Earth Surface Processes and Landforms*, 12, 173–184, 1987.

Dunaway, D., S. R. Swanson, J. Wendel, and W. Clary, The effect of herbaceous plant communities and soil textures on particle erosion of alluvial streambanks, *Geomorphology*, 9, 47–56, 1994.

Elmore, W., and R. L. Beschta, The fallacy of structure and the fortitude of vegetation, USDA Forest Service *GTR PSW-110*, 116–119, 1989.

Faulkner, D., Rainfall frequency estimation, *Flood Estimation Handbook, Wallingford*, UK, Institute of Hydrology, v. 2, 2000.

Fetherston, K. L., R. J. Naiman, and R. E. Bilby, Large woody debris, physical process and riparian forest development in montane river networks of the Pacific Northwest, *Geomorphology*, 13, 133–144, 1995.

Gippel, C. J., I. C. O'Neill, and B. L. Finlayson, *The Hydraulic Basis of Snag Management*, Centre for Applied Environmental Hydrology, 116 pp., University of Melbourne, 1992.

Graf, W. L., Fluvial adjustments to the spread of tamarisk in the Colorado Plateau Region, *Bulletin Geological Society of America*, 89, 1491–1501, 1978.

Graf, W. L., Mining and channel response, *Annals of the Association of American Geographers*, 69, 262–275, 1979.

Gregory, K. J., Vegetation and river channel process interactions, in *River Conservation and Management*, edited by P. J. Boon, P. Calaw, and G. E. Petts, p. 255–269, Wiley, Chichester, 1992.

Gregory, K. J., and R. J. Davis, Coarse woody debris in stream channels in relation to river channel management in woodland areas, *Regulated Rivers: Research and Management* 7, 117–136, 1992.

Gregory, K. J., and A. M. Gurnell, Vegetation and river channel form and process, in *Biogeomorphology*, edited by H. Viles, p. 11–42, Blackwell, Oxford, 1988.

Gregory, K. J., A. M. Gurnell, and C. T. Hill, The permanence of debris dams related to river channel processes, *Hydrological Science Journal*, 30, 371–381, 1985.

Gregory, K. J., R. J. Davis, and S. Tooth, Spatial distribution of coarse woody debris dams in the Lymington Basin, Hampshire, U.K., *Geomorphology*, 6, 207–224, 1993.

Grissinger, E. H., and J. B. Murphey, Present channel stability and late Quaternary valley deposits in northern Mississippi, *Special Publication of the International Association of Sedimentologists*, 6, 241–250, 1983.

Gurnell, A. M., and K. J. Gregory, Interactions between semi-natural vegetation and hydrogeomorphological processes, *Geomorphology*, 13, 49–69, 1995.

Hadley, R. F., Influence of riparian vegetation on channel shape, Northeastern Arizona, *USGS Professional Paper* 424-C, p. 30–31, Washington D.C., 1961.

Happ, S. C., G. Rittenhouse, and G. C. Dobson, Some principles of accelerated stream and valley sedimentation, USDA *Technical Bulletin*, 695, 133 pp., 1940.

Harmon, M. E., J. F. Franklin, J. F. Swanson, P. Sollins, S. V. Gregory, J. D. Lattin, N. H. Anderson, S. P. Cline, N. G. Aumen, J. P. Sedell, G. W. Lienkaemper, K. Cromack, and K. W. Cummins, Ecology of coarse woody debris in temperate ecosystems, *Advances in Ecological Research*, 15, 133–302, 1986.

Heede, B. H., Influences of a forest on the hydraulic geometry of two mountain streams, *Water Research Bulletin*, 8, 523–530, 1972.

Hession, W. C., J. E. Pizzutto, T. Johnson, and R. Horwitz, Influence of bank vegetation on channel morphology in rural and urban watersheds, *Geology*, in press.

Hey, R. D., and C. R. Thorne, Stable channels with mobile gravel beds, *Journal of Hydraulic Engineering*, 112, 671–689, 1986.

Hickin, E. J., Vegetation and river channel dynamics, *Canadian Geographer*, 28, 111–126, 1984.

Hooke, J. M., An analysis of the processes of river bank erosion, *Journal of Hydrology*, 42, 39–62, 1979.

Hunt, R. L., Removal of woody streambank vegetation to improve trout habitat, Wisconsin Department of Natural Resources *Technical Bulletin*, 115, 36 pp., 1979.

Jacobson, R. B., and K. Gran, Gravel sediment routing from widespread, low-intensity landscape disturbance, Current River basin, Missouri, *Earth Surface Processes and Landforms*, 24, 897–917, 1999.

Keller, E. A., Channelization: a search for a better way, *Geology*, 3, 246–248, 1975.

Keller, E. A., and F. J. Swanson, Effects of large organic material on channel form and fluvial processes, *Earth Surface Processes*, 4, 361–380, 1979.

Keller, E. A., and T. Tally, Effects of large organic debris on channel form and process in the coastal redwood environment, in *Adjustment of the Fluvial System*, edited by D. D. Rhodes, and G. P. Williams, p. 169–198, Kendall Hunt Publications, Dubuque, 1979.

Keller, E. A., and A. McDonald, River channel change: the role of large woody debris, in *Changing River Channels*, edited by A. Gurnell and G. Petts, pp. 217–235, Wiley, Chichester, 1995.

Kirkby, M. J., The problem, in *Soil Erosion*, edited by M. J. Kirkby and R. P. C. Morgan, pp. 1–12, John Wiley, New York, 1980.

Knox, J. C., Historical valley floor sedimentation in the upper Mississippi Valley, *Annals of the Association of American Geographers* 77, 224–244, 1987.

Kochel, R. C., Geomorphic impact of large floods: review and new perspectives on magnitude and frequency, in *Flood Geomorphology*, edited by V. R. Baker, R. C. Kochel and P.C. Patton, pp.169–188, Wiley, New York, 1988.

Lawler, D. M., Needle ice processes and sediment mobilization on river banks; the River Ilston, West Glamorgan, UK, *Journal of Hydrology*, 15, 81–114, 1993.

Langbein, W. B., and S. A. Schumm, Yield of sediment in relation to mean annual precipitation, *American Geophysical Union Transactions*, 39, 1076–1084, 1958.

Lyons, J., S. W. Trimble, and L. K. Paine, Grass versus trees: managing riparian areas to benefit streams of central North America, *Journal of the American Water Resources Association*, 36, 919–930, 2000.

Malanson, G. P., *Riparian Landscapes*, 287 pp., University Press, Cambridge, 1993.

Malanson, G. P., and D. R. Butler, Woody debris, sediment, and riparian vegetation of a subalpine river, Montana, USA, *Arctic and Alpine Research*, 22, 183–194, 1991.

Manga, M., and J. Kirchner, Stress partitioning in streams by large woody debris, *Water Resources Research*, 36, 2373-2379, 2000.

Marston, R. A., The geomorphic significance of log steps in forest streams, *Annals of the Association of American Geographers*, 72, 99–108, 1982.

McKelvey, V. E., *Stream and valley sedimentation in the Coon Creek drainage basin*, unpublished M.A. thesis, 122 pp., University of Wisconsin, Madison, 1939.

McKenney, R., R. B. Jacobson, and R. C. Wertheimer, Woody vegetation and channel morphogenesis in low-gradient, gravel-bed streams in the Ozark Plateaus, Missouri and Arkansas, *Geomorphology*, 15, 175–198, 1995.

Millar, R. G., Influence of bank vegetation on alluvial channel patterns, *Water Resources Research*, 36, 1109–1118, 2000.

Montgomery, D. R., What's best on banks? *Nature*, 188, 328–29, 1997.

Montgomery, D. R., J. Buffington, R. Smith, K. Schmidt, and G. Pess, Pool spacing in forest channels, *Water Resources Research*, 31, 1097–1105, 1995.

Mosley, P. M., The influence of organic debris on channel morphology and bedload transport in a New Zealand forest stream, *Earth Surface Processes and Landforms*, 6, 571–579, 1981.

Murgatroyd, A. L., and J. L. Ternan, The impact of afforestation on stream bank erosion and channel form, *Earth Surface Processes and Landforms*, 8, 357–370, 1983.

Nanson, G. C., and E. J. Hickin, A statistical analysis of bank erosion and channel migration in western Canada, *Bulletin of Geological Society of America*, 97, 497–504, 1986.

Nanson, G. C., M. Barbetti, and G. Taylor, River stabilization due to changing climate and vegetation during the late Quaternary in western Tasmania, Australia, *Geomorphology*, 15, 145–158, 1995.

National Research Council (NRC), *Riparian Areas: Functions and Strategies for Management*, 386 pp., National Academy Press, Washington, D.C., 2002.

Parsons, D. A., Vegetative control of streambank erosion, USDA *Misc. Pub.*, 970, p. 130–136, 1963.

Peterson, A. M., Effects of electric transmission rights-of-way on trout in forested headwater streams in New York, *North American Journal of Fisheries Management*, 13, 581–585, 1993.

Piegay, H., and A. M. Gurnell, Large woody debris and river geomorphological pattern: examples from S. E. France and S. England, *Geomorphology*, 19, 99–116, 1997.

Piegay, H., A. Thevenet, and A. Citterio, Input, storage and distribution of large woody debris along a mountain river continuum, the Drome River, France, *Catena*, 35, 19–39, 1999.

Pizzuto, J. E., Bank erodibility on shallow sandbed streams, *Earth Surface Processes and Landforms*, 9, 113–124, 1984.

Robison, E. G., and Beschta, Coarse woody debris and channel morphology interactions for undisturbed streams in southeast Alaska, USA, *Earth Surface Processes and Landforms*, 15, 149–156, 1990.

Shields, F. D., Effects of woody vegetation on sandy levee integrity, *Water Resources Bulletin*, 28, 917–931, 1992.

Shields, F. D., A. J. Bowie, and C. M. Cooper, Control of streambank erosion due to bed degradation with vegetation and structure, *Water Resources Bulletin*, 31, 475–489, 1995.

Shields, F. D. and C. J. Gippel, Prediction of effects of woody debris, *Journal of Hydraulic Engineering*, 121, 341–354, 1995.

Shine, M. J., *Effect of riparian vegetation on stream morphology*, unpublished M.S. thesis, 82 pp., University of Wisconsin-Madison, 1985.

Smith, C. M., Riparian afforestation effects on water yields and water quality in pasture catchments, *Journal of Environmental Quality*, 21, 257–245, 1992.

Smith, D. G., Effect of vegetation on lateral migration of anastomosed channels of a glacial meltwater river, in *Proceedings of the Symposium on River Morphology*, pp. 255–275, International Association of Scientific Hydrology, Pub. 75, 1967.

Smith, R. D., R. C. Sidle, P. E. Porter, and J. R. Noel, Effects of experimental removal of woody debris on the channel morphology of a gravel-bed stream, *Journal of Hydrology*, 152, 153–178, 1993.

Simon, A., and A. J. Collison, Quantifying the mechanical and hydrologic effects of riparian vegetation on streambank stability, *Earth Surface Processes and Landforms*, 27, 527–546, 2002.

Stott, T., A comparison of stream bank erosion processes on forested and moorland streams in the Balquhidder catchments, central Scotland, *Earth Surface Processes and Landforms*, 22, 383–399, 1997.

Swanson, F. J., and G. W. Leinkaemper, Physical consequences of large organic debris in Pacific Northwest streams, USDA Forest Service, *GTR PNW-69*, 12 pp., 1978.

Swanson, F. J., G. W. Leinkaemper, and J. R. Sedell, History, physical effects and management inputations of large organic debris in western Oregon streams, USDA Forest Service, *GTR PNW-56*, 15 pp., 1976.

Sweeney, B. W., Effects of streamside vegetation on macroinvertebrate communities of White Clay Creek in Eastern North America, *Proceedings of the Academy of Natural Sciences of Philadelphia*, 144, 291–340, 1993.

Thorne, C. R., Effects of vegetation on riverbank erosion and stability, in *Vegetation and Erosion*, edited by J. B. Thornes, pp. 123–144, Wiley, Chichester, 1990.

Trimble, S. W., Response of Coon Creek, Wisconsin, to soil conservation measures, in *Landscapes of Wisconsin: A Field Guide*, edited by B. Borowieki, pp. 24–29, Association of American Geographers, Washington, 1975.

Trimble, S.W., Changes in sediment storage in the Coon Creek basin, Driftless Area, Wisconsin, 1853 to 1975, *Science*, 214, 181–183, 1981.

Trimble, S. W., A sediment budget for Coon Creek basin in the Driftless Area, Wisconsin, 1853-1977, *American Journal of Science*, 283, 454–474, 1983.

Trimble, S. W., The distributed sediment budget model and watershed management in the Paleozoic Plateau of the upper midwestern United States, *Physical Geography*, 14, 285–303, 1993.

Trimble, S.W., Erosional effects of cattle on streambanks in Tennessee, USA, *Earth Surface Processes and Landforms*, 19, 451–464, 1994.

Trimble, S.W., Stream channel erosion and change resulting from riparian forests, *Geology*, 25, 467–469, 1997.

Trimble, S. W., Decreased rates of alluvial sediment storage in the Coon Creek Basin, Wisconsin, 1975-93, *Science*, 285, 1244–1246, 1999.

Trimble, S.W., and S. W. Lund, Soil conservation and the reduction of erosion and sedimentation in the Coon Creek Basin, Wisconsin, *USGS Professional Paper*, 1234, 35 pp., 1982.

Trimble, S. W., and A. C. Mendel, The cow as geomorphic agent—a critical review, *Geomorphology*, 13, 233–253, 1995.

Trimble, S. W., and P. Crosson, U.S. soil erosion rates-myth and reality, *Science*, 289, 248–250, 2000.

USDA-SCS, *National Manual for Assisting ASCS Cost Sharing Programs, Part 539, National Bulletin* No. 300-1–4, Washington, 35 pp., 1991.

Volny, S., Riparian stands, in *Developments in Agricultural and Managed Forest Ecology*, edited by O. Riedl, and D. Zachar, pp. 423–453, Elsevier, Amsterdam, 1984.

Welsch, D., Riparian forest buffers: function and design for protection and enhancement of water resources, USDA Forest Service *Report No.* NA-PR-07-91, 1991.

White, R. J., and O. M. Brynildson, Guideline for management of trout habitat in Wisconsin, Wisconsin Department of Natural Resources *Technical Bulletin*, 39, 65 pp., 1967.

Williams, G. P., Bankfull discharge of rivers, *Water Resources Research*, 14, 1141–58, 1978.

Wolman, M. G., Factors influencing erosion of a cohesive river bank, *American Journal of Science*, 257, 204–216, 1959.

Zimmerman, R. C., J. C. Goodlett, and G. H. Comer, The influence of vegetation on channel form of small streams, in *Symposium on River Morphology*, pp. 255–275, Int. Assoc. Sci. Hydrol. Pub., 75, 1967.

Stanley W. Trimble, Department of Geography, University of California Los Angeles, 310/825–1071, 1314, Los Angeles, CA.

Flow, Sediment, and Nutrient Transport in a Riparian Mangrove

S. Ikeda and Y. Toda

Department of Civil Engineering, Tokyo Institute of Technology, Tokyo, Japan.

Y. Akamatsu

Interdisciplinary Graduate School of Science and Engineering, Tokyo Institute of Technology, Kanagawa, Japan.

Riparian mangroves have been known to play an important role in the trapping of sediment and nutrients transported from upstream areas. The present work examines flow, sediment, and nutrient transport in a mangrove area on Ryukyu Island, Japan. Field observations made during neap tides show that nutrient supply from the flood plains with mangroves to the river was controlled by subsurface flow. At spring tide when water inundated the mangrove swamps, nutrients were transported from the flood plains to the river directly by surface flowing water. The temporal asymmetry of the flow field induced by the existence of mangrove trees, as well as sediment transport and deposition within the mangrove forest, were reproduced well by a numerical simulation employing a depth-averaged zero-equation turbulence model.

1. INTRODUCTION

Riparian mangrove forests usually grow in subtropical and tropical estuarine zones, and they control the movement of estuarine water, supply nutrients to the coastal zones, and serve as good habitat for mangrove biota. A mangrove does not indicate a kind of tree but rather special kinds of trees that grow along estuarine rivers and coastal zones. Mangrove trees usually grow in calm, saline shallow-water zones with muddy soils. Mangrove forests retard flow and induce deposition of fine sediment. Since nutrients such as phosphorous and nitrogen are attached to fine sediments, mangrove forests control the movement of such materials derived from upstream catchment areas. Thus, mangrove forests work as a buffering interface between the land and sea.

In subtropical and tropical zones, riparian mangroves constitute a typical estuarine ecosystem. However, they are decreasing rapidly in size due to land development. In the southwest islands of Japan, for example, mangroves have been reduced in size by bridge construction and channel works at river mouths. Mangrove forests are a major source of nutrients for estuaries and coral, and they also serve as good fish habitat [*Macnae*, 1974; *Mann*, 1982; *Robertson et al.*, 1988; *Jansson*, 1988; *Alongi et al.*, 1989]. Therefore, conservation of mangroves is urgently required. Many synthetic studies have examined mangrove ecosystems [*Snedaker and Snedaker*, 1984; *Robertson and Alongi*, 1992]. But these studies are not very comprehensive because too few measurements were obtained of the tidal flow and sediment and nutrient transport to fully understand the dynamics of mangrove ecosystems.

Tidal flow in mangroves has been studied using numerical analysis [*Wolanski et al.*, 1980; *Kanazawa and Mazda*, 1994; *Nakatsuji et al.*, 1994]. However, mangroves are complex systems, in which water movement, geomorphology, sediment transport, and ecology all interact. There is a clear need to understand the mangrove ecosystem from the viewpoint of material cycling, which contains physical, biological, and chemical processes. Mass balance of nutrients and organic material between mangrove and coastal zones has been studied in several mangroves [*Boto and Bunt*, 1981; *Twilly*, 1985; *Woodroffe*, 1985]. However, existing knowledge of flow and nutrient transport processes is still limited.

Riparian Vegetation and Fluvial Geomorphology
Water Science and Application 8
Copyright 2004 by the American Geophysical Union
10.1029/008WSA13

In this paper, the characteristics of flow at a mangrove zone and the transport of organic materials and nutrients are studied, treating neap and spring tides separately. At the neap tide, water inundates only small creeks existing in mangrove swamps and the transport through these small creeks is examined. At the spring tide, during which the whole mangrove swamp is inundated, mass balance of organic materials and nutrients is described and discussed.

2. OUTLINE OF THE OBSERVATIONS

2.1. Location and Terms for Observations

Field observations were conducted along the Nagura River on Ishigaki Island, Japan (Figure 1a), which is situated at the west part of Ryukyu Islands close to Taiwan. The Nagura River has a length of 4.5 km and a catchment area of 16.1 km². The river possesses mangrove forests near the river mouth and a large lagoon at the coast (Figure 1b).

Field observations were conducted during five time periods: January 11 to 13, 2000 during a neap tide (TERM 1), June 28 to July 4, 2000 during a spring tide (TERM 2), December 17 to 23, 2000 during a near-neap tide (TERM 3), July 20 to 22, 2001 for a neap tide (TERM 4), and September 1 to November 5, 2002 for a pluvious season (TERM 5). The observations are still on-going, and sediment yield is still being monitored from sugar cane, rice, and grass fields because these fields are the source areas for much of the sediment and nutrients.

2.2. Characteristics of the Observation Zone

The observation zone extends from the river mouth (close to Station B) to Station A near Kanda Bridge, which is situated at about 600 m upstream of the river mouth (Figure 1c). The dominant species of trees in the 16-ha mangrove forest are *Bruguria gymnorhiza* and *Rhizophora stylosa* [*Nakasuka et al.*, 1974]. The characteristics of mangrove forests on the river banks, such as number of trees, mean breast height diameter (BHD), and litter fall rate, are described in Table 1. Small creeks exist in the mangrove swamps and connect with the main channel. These creeks are important agents for the exchange of materials between the main channel and the mangrove forest.

Table 1. Characteristic of mangrove trees.

	Right bank	Left bank
Number of trees per hectare	6330	2133
Mean BDH (m)	0.228	0.321
Litter fall rate (kg/ha/day)	15	16.5

The estuary zone affected by tides extends to the Kanda Bridge as seen in the Figures 1b and 1c. The contour lines of elevations measured using ordinary optical levels and a differential GPS are depicted in Figure 2. A large island-like stable sandbar exists near the river mouth. The main channel and the mangrove swamps constitute a two-stage channel, and the difference in elevation is typically 1 m. At neap tide, the water only enters in the small creeks and the mangrove swamps are not completely inundated. At spring tide, the swamps are completely inundated.

Bank protection works were constructed near Station A between 450 m and 600 m from the river mouth. This work has increased the difference in elevations between the main channel and the mangrove swamp. The mangrove swamps in this area are no longer inundated even at spring tide.

There is an intake gate that supplies water to crop fields at a location of 2.5 km from the river mouth. There is a channel between the Kanda Bridge and the intake gate. The areas close to the channel are rice fields, whereas those fields far from the channel are mainly grass to feed cows and sugarcane.

The lagoon connects to the sea under the Nagura-Oohashi Bridge (Figure 1b). The intake gate and the Nagura Bridge are employed as the boundary conditions for numerical simulations described later.

3. METHODS

3.1. Flow

The longitudinal flow velocity was measured using two electromagnetic current meters (Alec Electronics Co. Ltd.) at Stations A and B (Figure 1c). Since the water was well-mixed vertically, the sensors were placed at 0.3 m from the river bed. The temporal variations of surface water elevations were also measured at the same locations using pressure-type water elevation meters (Daikirika Co. Ltd.). The probes were used to estimate the fluxes of material between the mangrove swamps and the main channel.

Since subsurface flow plays an important role in transporting dissolved nutrients and organic material, eight wells with an inner diameter of 0.1 m were dug in the mangrove swamps along a transverse at Station B. During TERM 3, ground water elevations were measured electronically every one hour. The ground water was also sampled at the wells at every two hours.

3.2. Water Quality and Materials

During TERM 1 when a neap tide occurred, field observations were conducted to determine the characteristics of transport of organic material in the main channel and the

124°

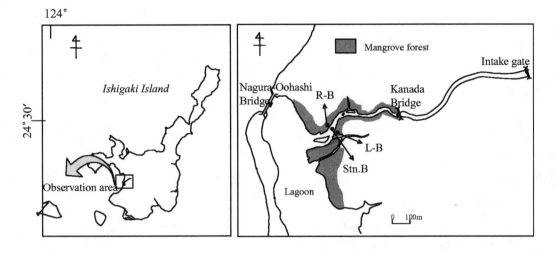

(a) Map of Isigaki Island.   (b) Map of the mangrove estuary of Nagura river

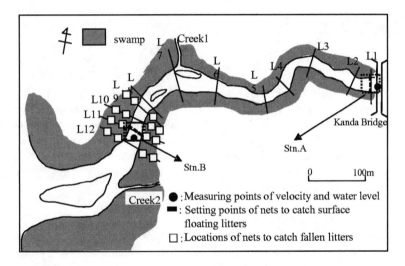

(c) Locations of the observations

Figure 1. Maps of the location areas.

creeks of mangrove swamps. Water sampling was carried out at sections L7, L10, and Creek 1 at both high and low tides. The concentrations of dissolved oxygen and salinity were measured using a portable water-quality meter (Horiba Co. Ltd.). Sampling was conducted for only one vertical point in a section because the flow was well mixed. Water samples packed in cool boxes were sent to the Hydraulics Laboratory of Tokyo Institute of Technology, where the concentration of dissolved organic carbon (DOC) was measured (Shimadzu Co. Ltd.).

During TERM 2 when a spring tide occurred, longitudinal transport of surface-floating litter (fallen leaves), dissolved materials, and particulate materials were observed. During one tidal cycle (7:30~19:30, July 2, 2000), the surface-float-

ing litter was collected at Stations A and B using 2 m-long meshed nets. The amount of the surface-floating litter trapped by the nets was measured every two hours. At the same locations with the same interval, water sampling and monitoring of dissolved oxygen and salinity were conducted. For the water samples, the following parameters were measured: ignition loss (IL, which is a measure of the content of carbonic materials), dissolved organic carbon (DOC), and nutrients (orthophosphate PO_4-P, total phosphorus T-P, ammonia NH_4-N, nitrate NO_3-N, nitrite NO_2-N, and total nitrogen T-N). The concentrations of nutrients were determined using a standard colorimetric method. The concentration of IL was determined by igniting the dried residuals filtered from the sampled water at 700° C for two hours.

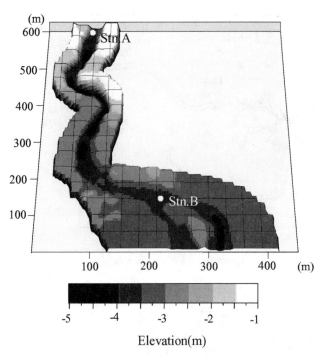

Figure 2. Contour map of the location area. Elevation is the height above the average sea level of Port Naha.

During TERM 5, water sampling was also carried out every two hours during floods at the intake gate and the Kanda Bridge (Figure 1b) using automatic water samplers (ISCO Co. Ltd.). The sampled water was sent to the laboratory of Tokyo Institute of Technology, and the concentration of suspended solids (SS) was determined. The size distribution of suspended solid was also measured employing a laser diffraction particle size analyzer (Shimadzu Co. Ltd.).

During TERM 3, 15 nets (1 m × 1 m) were placed horizontally in the mangrove forests to catch the amount of fallen litter. The location of the nets is shown as dark squares on Figure 1c.

3.3 Soils

During TERM 2, the surface soils at sections L1 to L12 were sampled both at the river bed and in the swamps. These samples were analyzed for IL, T-P, and T-N, and their grain size distributions were determined. The contents of T-P and T-N were measured using colorimetric method as described previously.

A list of these measurements, terms, and the methods are summarized in Table 2.

4. NUMERICAL SIMULATIONS

4.1. Surface Flow

The present estuarine zone was well-mixed vertically, and therefore a depth-averaged flow model was employed for numerical simulation. For describing the depth-averaged flow, the Reynolds equation of zero-order closure model was used. The equations were written in arbitrary coordinate system. The bed flow resistance was expressed using Manning's formula, and the mangrove tree flow resistance was described by a function of a resistance coefficient to flow, depth-averaged flow velocity, and the density of trees [*Chen and Ikeda*, 1997].

4.2 Sub-Surface Flow

From several field observations, the underground subsurface flow was found to play an important role in transporting dissolved materials from soil layers in swamps. Ground-water

Table 2. Measured parameters and applied methods.

	Items	Date	Method
Water quality:	SM	January 11 to 13, 2000	Filtration
	PO$_4$-P, T-P, NH$_4$-N, NO$_2$-N, T-N, NO$_3$-N	January 11 to 13, 2000	Colorimetric method
	DOC	June 28 to July 4, 2000	TOC-control
	Salinity, DO	June 28 to July 4, 2000	Portable water-quality meter
Vegetation:	Number of trees	July 2 to August 18, 2000	
	BDH	July 2 to August 18, 2000	
	Litter fall	July 2 to August 18, 2000	
Soil:	Size distribution	June 28 to July 4, 2000	Sieves, Laser diffraction analyzer
	T-N, T-P, IL	June 28 to July 4, 2000	Colorimetric method
Flow:	Velocity and water level	January 11 to 13, 2000	Electromagnetic current meter and
		June 28 to July 4, 2000	water level meter
		December 17 to 23, 2000	
		July 20 to 22, 2001	

flow was treated by employing the Richards equation, which is valid for unsaturated seepage flow.

4.3. The Governing Equations

The governing equations for surface flow are as follows:

$$\frac{\partial h}{\partial t} + \frac{\partial (uh)}{\partial x} + \frac{\partial (vh)}{\partial y} = 0 \tag{1}$$

$$\frac{\partial (uh)}{\partial t} + \frac{\partial (u^2 h)}{\partial x} + \frac{\partial (uvh)}{\partial y} =$$
$$-gh\frac{\partial z_s}{\partial x} - \frac{\tau_{bx}}{\rho} - \frac{F_x}{\rho} + \frac{\partial}{\partial x}\left(-\overline{u'^2}h\right) + \frac{\partial}{\partial y}\left(-\overline{u'v'}h\right) \tag{2}$$

$$\frac{\partial (uh)}{\partial t} + \frac{\partial (uvh)}{\partial x} + \frac{\partial (v^2 h)}{\partial y} =$$
$$-gh\frac{\partial z_s}{\partial y} - \frac{\tau_{by}}{\rho} - \frac{F_y}{\rho} + \frac{\partial}{\partial x}\left(-\overline{u'v'}h\right) + \frac{\partial}{\partial y}\left(-\overline{v'^2}h\right) \tag{3}$$

where h is local depth of flow, u and v are depth-averaged velocity components in x and y directions, respectively, g is gravitational acceleration, τ_{bx} and τ_{by} are bottom shear stresses in x and y directions, respectively, F_x and F_y are components of drag due to vegetation in x and y directions, respectively, and $-\overline{u'^2}$, $-\overline{u'v'}$, and $-\overline{v'^2}$ are the depth-averaged Reynolds stresses.

The governing equation for ground water flow is as follows:

$$[C_w(\psi) + \beta_0 S]\frac{\partial \psi}{\partial t} = \frac{\partial}{\partial x}\left(K_x \frac{\partial \psi}{\partial x}\right)$$
$$+ \frac{\partial}{\partial y}\left(K_y \frac{\partial \psi}{\partial y}\right) + \frac{\partial}{\partial z}\left[K_z\left(\frac{\partial \psi}{\partial z} + \frac{\rho}{\rho_f}\right)\right] \tag{4}$$

where Ψ is the pressure potential, K_x, K_y, and K_z are the hydraulic conductivities in x, y, and z directions, ρ is depth-averaged density of saline water, ρ_f is depth-averaged density of fresh water, $C_w(\Psi)$ is slope of the soil water retention curve, S is specific storage, and β_0 is a numerical coefficient. Assuming that the porosity of soils is independent of pressure potential, β_0 is expressed as

$$\beta_0 = \begin{cases} 0 & \text{for unsaturated conditions} \\ 1 & \text{for saturated conditions} \end{cases} \tag{5}$$

The transport equation for depth-averaged concentration of suspended sediment and the equation for the temporal variation of the amount of sediment are described by

$$h\left(\frac{\partial C}{\partial t} + u\frac{\partial C}{\partial x} + v\frac{\partial C}{\partial y}\right) =$$
$$h\left[\frac{\partial}{\partial x}\left(\varepsilon_{sh}\frac{\partial C}{\partial x}\right) + \frac{\partial}{\partial y}\left(\varepsilon_{sh}\frac{\partial C}{\partial y}\right)\right] + Er - De \tag{6}$$

$$\frac{d(sed)}{dt} = De - Er \tag{7}$$

where C is the depth-averaged concentration of suspended sediment, ε_{sh} is the lateral diffusivity of suspended sediment, Er is the erosion rate from the bottom, De is the deposition rate, and sed is the amount of sediment stored in the surface layer. The lateral diffusivity is assumed to be $\varepsilon_{sh} = D_h$, and the erosion rate is taken as

$$Er = 6.7 \times 10^{-5}\left(\frac{u_*}{V_s}\right)^2 V_s \tag{8}$$

where V_s is the settling velocity of suspended sediment. The deposition rate of suspended sediment is given by

$$De = C_b V_s = \alpha_s C V_s \tag{9}$$

where C_b is the concentration of suspended sediment near the bottom and α_s is a coefficient correlating C_b with C. Assuming an exponential distribution for sediment concentration in the vertical direction, α_s is expressed by [Ikeda et al., 1992]

$$\alpha_s = \frac{V_s h}{\varepsilon_z\left[1 - \exp\left(-\dfrac{V_s}{\varepsilon_z}h\right)\right]_s} \tag{10}$$

where the ε_z is the vertical diffusivity of suspended sediment. The value of ε_z is assumed to be $\varepsilon_z = 0.077\, u_* h$ [Ikeda and Izumi, 1991].

For the surface flow, the flow discharge based on the observed values was the inlet condition. The fraction of flow discharge in each grid at the upstream end was determined in proportion to $h^{5/3}$ according to Manning's formula. At the downstream end, the water depth was given, and the longitudinal gradient of the mass flux was taken to be zero.

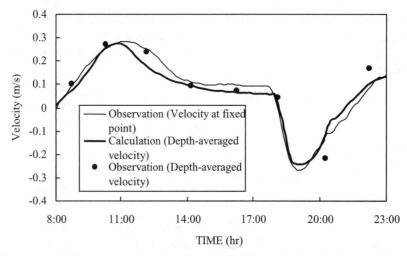

Figure 3. Comparison of the temporal variation of longitudinal flow velocity calculated and those observed.

For the ground water flow, the pressure potential $\Psi = z_0 \rho / \rho_f$, where z_0 is the vertical distance from the soil surface, was given as the horizontal boundary condition. The vertical boundary condition was a no-flux condition.

The concentration of suspended sediment obtained from observations was the initial upstream boundary condition. In calculating the transport of suspended sediment, a sediment size d of 10 μm was used because the size of sediment discharged from the catchment area during floods was of this order. At the downstream end, the water depth was given and the downstream gradients of the mass flux and the concentration of suspended sediment in each grid were taken to be zero. However, the observed concentration of suspended sediment was given at the downstream end when the direction of flow is toward upstream during high tide.

5. RESULTS

As described previously, the behavior of flow was different for the neap tide compared to the spring tide. The results are described for both cases separately.

5.1. Surface Flow

Numerical simulations were performed for the spring tide of TERM 4. Figure 3 describes the temporal variation of longitudinal flow velocity in the main channel at Station B (downstream direction is taken to be positive), where the thick line indicates the predicted depth-averaged flow velocity and dark circles are the measured values. This figure shows that the temporal variation of longitudinal flow velocity was not sinu-

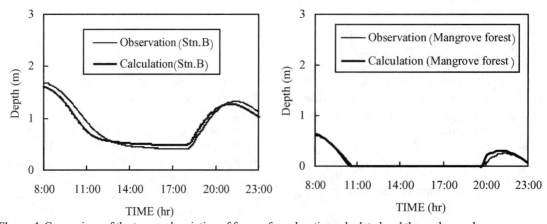

Figure 4. Comparison of the temporal variation of free surface elevation calculated and those observed.

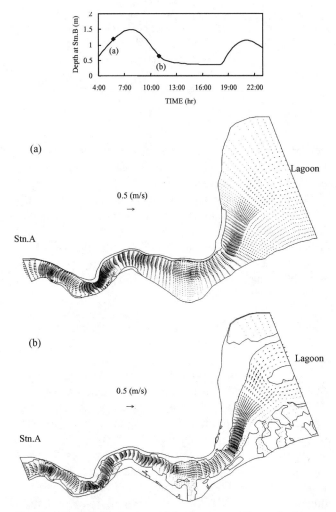

(a)

(b)

Figure 5. Spatial distributions of depth-averaged flow velocities at flood tide and ebb tide.

soidal but skewed in time, though the temporal variation of tide at the downstream boundary was almost sinusoidal. This skewness was induced by the drag due to mangrove trees.

The temporal variations of free surface elevation for the main channel and the left bank swamp are shown in Figure 4. The numerical predictions agree reasonably well with the measurements, and the inundation of the swamp at high tide is reproduced well by the numerical simulation. The two-dimensional spatial distributions of depth-averaged flow velocities at flood tide (rising stage) and at ebb tide are depicted in Figures 5a and 5b, respectively.

5.2. Sub-Surface Flow

Since the sub-surface flow is expected to be one of the major agents in transporting dissolved materials at neap tide, the observations made at TERM 3 were employed for a comparison with the numerical simulation. Figure 6 shows the temporal and spatial variations at a transverse section of Station B during 11:00 to 23:00, December 21, 2000. The dashed lines are the results of numerical simulations made by employing the Richards equation, and the symbols are the measurements. The numerical calculations based on unsaturated flow model describe the actual data reasonably well. As discussed later, seepage flow played an important role in transporting nutrients from the swamps to the main channel at neap tide.

5.3. Transport of Organic Material at Neap Tide

The observation during January 13 to 15, 2000 (TERM 1) was made at neap tide. Figure 7 shows DOC concentrations of river water at sections L7, L10, and Creek 1 at the peak high tide and the peak low tide on January 11, 2000. At high tide, there was no significant difference in the concentration of DOC anywhere. At low tide, however, the concentrations of DOC in Creek 1 and at its outlet were higher than in the main channel. The soils in the swamps contained larger amounts of organic material than in the river bed (Figure 8). However, near the outlet of Creek 1, the soil in the river bed included a fairly large amount of organic material, which was probably discharged from Creek 1 and deposited there. These results

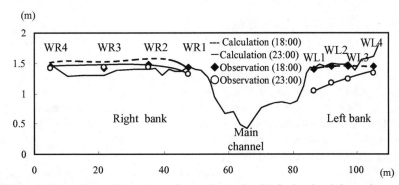

Figure 6. Comparison of the values of groundwater level calculated and those observed.

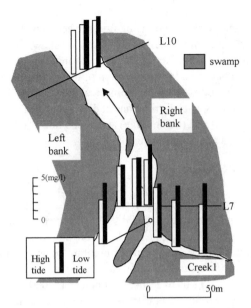

Figure 7. Concentration of dissolved organic carbon in river water (at neap tide).

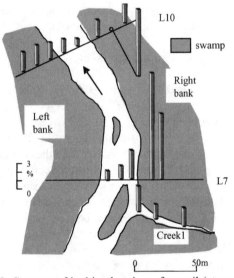

Figure 8. Contents of ignition loss in surface soil (at neap tide).

indicate that during low tide Creek 1 transported dissolved organic material toward the main channel.

Table 3 represents the comparison of the amount of nutrient increase in the river and the export fluxes of nutrients by seepage flow in the swamps between Stations A and B during the low tide of TERM 3. Since the bank protection works were made, the geographical feature near Station A was different from that along Line A, and water does not cover the swamps near Station A. However, the mangrove swamp downstream extended for more than 50 m, similar to Line A. The nutrient fluxes by the seepage water flow in the swamps

Table 3. Comparison of nutrient fluxes by seepage water flow and amount of increase of nutrients in river (18:00 to 23:00, December 21, 2000).

Nutrient	Groundwater flux estimated		Amount of increase in river (g/5hr)
	WL1 (g/50m/5hr)	WR1 (g/50m/5hr)	
PO$_4$-P	6.0	0.3	10.1
DOC	5.8 x 10^3	0.4 x 10^3	27.4 x 10^3

between Stations A and B were estimated by multiplying the nutrient flux per unit width along a transversal line at Station B and 50 m. The amount of nutrient increase in the river between Stations A and B was also estimated by subtracting the import flux at Station A from the export flux at Station B. It was found that the flux of nutrients by the seepage water flow in the swamps was the same order of magnitude as the amount of increase along the river course. Thus, the major increase of nutrients along river course was supplied by the seepage flow at low tide.

5.4. Transport of Organic Materials and Nutrients at Spring Tide

5.4.1 Litter transport. Figure 9 shows the fluxes of surface-floating litter, trapped by the three nets at Station A and B for TERM 2 (spring tide), where the flux to the downstream direction is taken to be positive. The temporal variation of longitudinal flow velocity at Station B is also shown in Figure 9. At spring tide, large amounts of litter were transported toward the lagoon after the peak high tide, and most returned to the river before the next peak high tide. The observation shows that only a small amount of litter was provided from the swamps upstream of Station A. Since pure mangrove forests do not exist upstream of Station A, the litter was provided by the mangrove forests existing between Stations A and B.

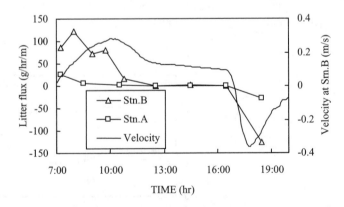

Figure 9. Temporal variation of surface-floating litter fluxes at Stations A and B (at spring tide)

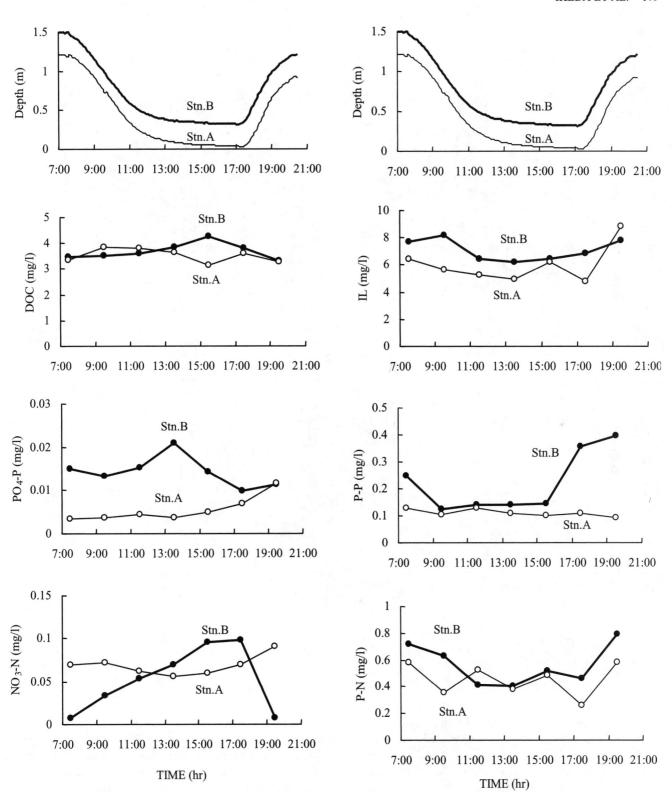

Figure 10. Temporal variation of dissolved materials at Stations A and B (at spring tide)

Figure 11. Temporal variation of particulate materials at Stations A and B (at spring tide).

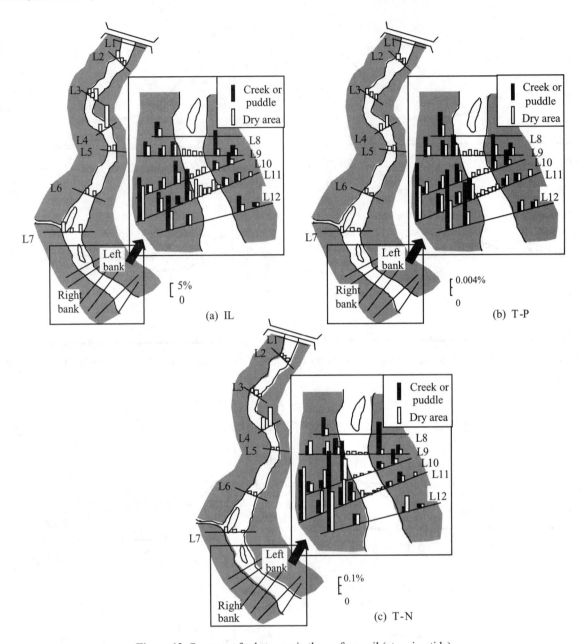

Figure 12. Contents of substances in the surface soil (at spring tide).

5.4.2 Transport of dissolved material. Figure 10 shows the temporal variations of the concentrations of DOC, NO_3-N and PO_4-P at Stations A and B during TERM 2. At Station A, the concentrations of these parameters did not vary much with time. However, concentrations at Station B increased during low tide and were larger than those at Station A around peak low tide. Thus, dissolved organic materials and dissolved nutrients were exported from the swamps to the main channel during low tide. For NO_3-N, this tendency was most pronounced. The concentration of PO_4-P at Station B was larger than at Station A not only during low tide but also during high tide. It is possible that orthophosphate detached easier from the suspended materials at higher values of salinity.

5.4.3 Transport of particulate material. The temporal variations of particulate phosphorus (P-P), particulate nitrogen

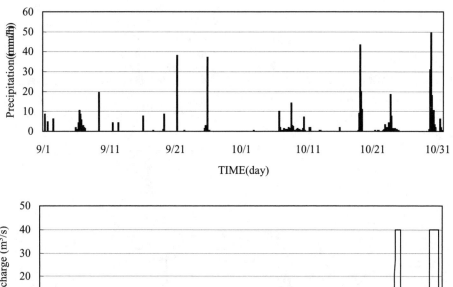

Figure 13. Precipitation and discharge of the river.

(P-N), and IL at Stations A and B for during TERM 2 are shown in Figure 11. The value of IL represents the amount of organic particles. The concentration of P-P was calculated by subtracting PO_4-P from T-P, and the concentration of P-N was calculated by subtracting NO_3-N, NO_2-N, and NH_4-N from T-P. The maximum concentration of all particulate material occurred around peak high tide, except for the concentration of P-P at Station A. The concentrations of particulate material exceeded those of dissolved material by more than ten times, which indicates that large amounts of particulate material existed in mangrove zones.

5.4.4 Organic material and nutrients contained in the surface soil. Swamps are a major source of nutrients in mangrove zones where fallen leaves and fruits are decomposed by crabs, insects, and bacteria. Figure 12 shows the fractions of T-P, T-N, and IL contained in the surface soil of the swamps and river bed. In the swamps, the sampling of soil was conducted at a small creek and at a dry area during low tide. There were higher concentrations of organic material and nutrients in the swamps than in the river bed. In addition more organic material and nutrients were contained in the surface soils in small creeks than in dry areas. This suggests that these materials were moved from the small creeks to the main chan-

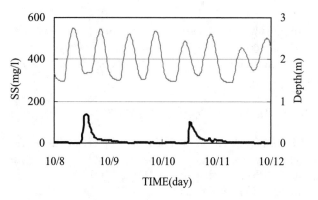

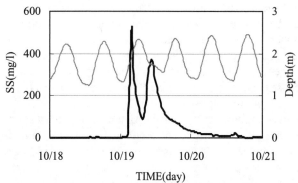

Figure 14. Temporal variations of suspended solids at Kanda Bridge and depth at Nagura-Oohasi Bridge.

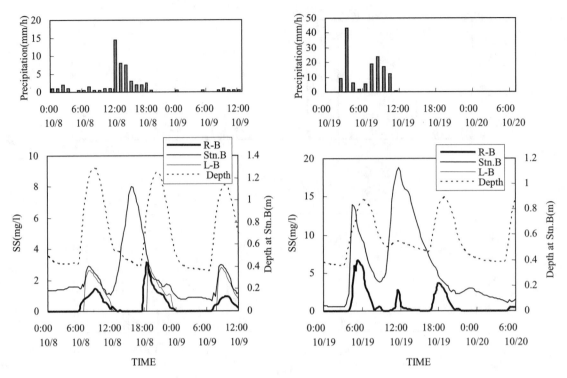

Figure 15. Temporal variation of suspended solids during floods.

nel during low tide. Near the right bank of L4, organic material (IL) and nutrients (T-N, T-P) were higher here than at other points in the main channel, presumably the result of litter deposition.

5.5. Transport of Suspended Sediment During Flood

5.5.1 Field observation. Figure 13 depicts the temporal variation of precipitation and discharge at the intake gate during TERM 5. On September 21 to 25, heavy rains caused the increase in river discharge from 5.7 to 7.5 m³/s. The river discharge intermittently exceeded 5 m³/s during October 7 to 11 because of continuous precipitation. A heavy rain on October 19 with a maximum intensity of 43 mm/h increased river discharge to 10 m³/s. During October 23 to 25 and October 30 to 31, water was released from the intake gate because the discharge exceeded 40 m³/s, and these discharge data are unavailable.

Figure 14 shows the temporal variation of suspended solids (SS) at the Kanda Bridge and the flow depth at the Nagura-Oohasi Bridge during October 8 to 12 and October 18 to 21, 2002. In the flood during October 8 to 12, where the concentration of SS at the Kanda Bridge increased, the tide was low and SS was immediately flushed to the lagoon and the coastal area. The flood during October 18 to 21 occurred at high tide,

and the concentration of SS decreased around the peak high tide because of the inundation of seawater. Thus, the tidal condition has a large influence on the transport of SS during flood flows in the estuary.

5.5.2 Numerical simulation. Figure 15 shows the temporal variations of flow depth at Station B and the concentration of SS at R-B, Station B, and R-L (Figure 1b) from 0:00 on October 8 to 24:00 on October 9, 2002. The precipitation observed for the term is also depicted in Figure 15. Although the concentration of SS at Station B increased with rainfall, the concentration of SS at R-B and L-B increased only at high tide except for the daytime low tide on October 19. The spatial distributions of SS at the low tide (October 8, 13:45) and at the high tide (October 8, 20:30) are depicted in Figures 16a and 16b. The inundation of seawater into mangrove forest, which contained the suspended sediment, occurred only at high tide. Thus suspended sediment flushed to the lagoon and the coastal area during the flood returned to the mangrove area

Table 4. Content of litter.

Litter	C/DW (%)	N/DW (%)	P/DW (%)	IL (%)
Floating	40–42	0.86–0.95	0.035–0.045	75–85
Fallen	46–51	1.03–1.12	0.026–0.031	88–92

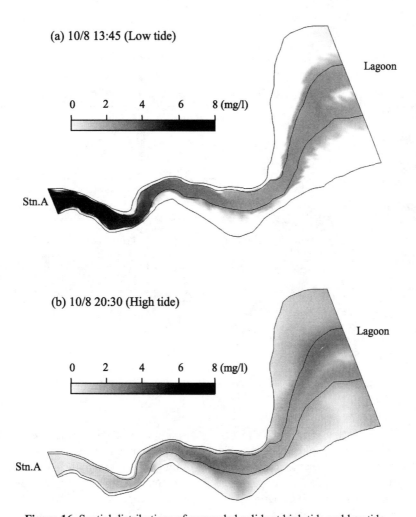

Figure 16. Spatial distributions of suspended solids at high tide and low tide.

by way of the tide, and the sediment was transported toward the mangrove forest by way of the inundation of seawater.

5.6. Mass Balance of Organic Material and Nutrients in the Mangrove Area

Figure 17 shows the net flux of the organic material and the nutrients between Stations A and B during one tidal period (7:00 to 16:30, July 2, 2000). The contents of C, T-P, T-N, and IL in the surface-floating litter and in the fallen litter were estimated by using the percentage of C, T-P, T-N, and IL per unit dry weight as shown in Table 2. The litter fall rate was estimated by averaging the values of 15 locations (Figure 1c). The contents of T-P, T-N, and IL in the surface soil of the swamps per unit area were estimated by the averaged values of the sampling points in the swamps, assuming that the porosity of the soil was 0.4 [Koumura, 1982].

Comparing the total import flux at Station A with the total export flux at Station B in Figure 17, it is found that the mangrove forests between these stations provided a large amount of organic material and nutrients as litter, dissolved material and particulate material. Referring to the export flux of Station B, it is seen that most of the organic material and the nutrients were exported as dissolved or particulate material. In addition most of the fallen litter was trapped in the swamps, and moved back and forth between the lagoon and the main channel. Finally after decomposition by crabs, insects, and bacteria, these materials were available as dissolved or particulate material to the coastal area.

The export flux of POC at the river mouth (Station B), which amounted to 11 to 13 kg C/ha/tide, can be compared with the results of *Boto and Bunt* [1981] who estimated that 2.2 kg C/ha/tide was exported from a mangrove forest in Missonary Bay, Hinchinbrook Island. The difference between these

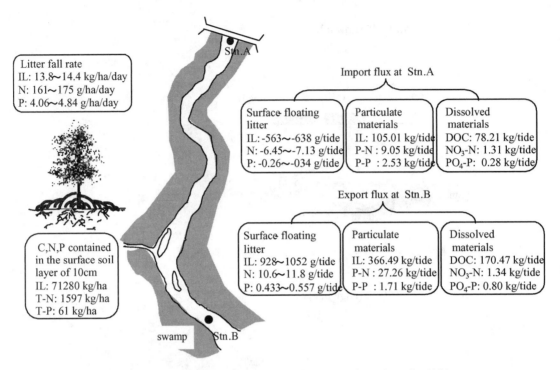

Figure 17. Mass balance of organic material and nutrients (at spring tide).

sites is due to the existence of a lagoon at the river mouth. In the mangrove forest of Missonary Bay, which is directly connected to an open sea, most of the litter was flushed to the coastal area, hence the effect of grazing and decomposition into particulate material was small. This result suggests that the mangrove forest in a semi-closed lagoon is an effective source of nutrients for the coastal area as it provides more nutrients as dissolved or particulate material.

6. CONCLUSIONS

A mangrove forest along the Nagura River on Ishigaki Island, Japan grows in a semi-closed lagoon. In this mangrove area, dissolved organic material was efficiently transported through creeks at neap tide when water did not inundate the swamps. Employing field measurements for ground-water elevations and numerical simulation, the amount of nutrients supplied from the flood plains to the main channel at neap tide were estimated. This amount was found to be the same order of magnitude as the increase of nutrients along the river course, indicating that nutrient flux due to the ground water flow played an important role at neap tide. At spring tide, when water inundates the swamps, litter provided by the mangrove forests was flushed from the swamps during low tide. This litter moved back and forth between the lagoon and the river, and most was transported to the coastal area only after

it was decomposed into dissolved or particulate material. The remaining litter trapped in the swamp provided a large amount of organic material to the surface soils in the swamps. These soils contained a larger amount of organic material and nutrients as compared to the river beds. Moreover, the suspended sediment flushed to the lagoon and the coastal area during floods returned to the mangrove area by the inundation of seawater.

Acknowledgments. The present study has been conducted under the financial support of Grant-in-Aid for Scientific Research, the Ministry of Education, and Culture of Japan (Grant No.11305035, No. 11875104). The Kajima Research Foundation has also supported this program.

REFERENCES

Alongi, D. M., K. G. Boto, and F. Tirendi, Effect of exported mangrove litter on bacterial productivity and dissolved organic carbon fluxes in adjacent tropical nearshore sediments, *Marine Ecology Progress Series*, 56, 133–144, 1989.

Boto, K. G., and J. S. Bunt, Tidal export of particulate organic matter from a northern Australian mangrove system, *Estuarine, Coastal and Shelf Science*, 13, 247–255, 1981.

Chen, F. Y., and S. Ikeda, Horizontal separation flows in shallow open channels with spur dikes, *Journal of Hydroscience and Hydraulic Engineering, JSCE*, 15, 1–14, 1997.

Ikeda, S., and N. Izumi, Effect of pile dikes on flow retardation and sediment transport, *Journal of Hydraulic Engineering*, 117, 1459–1478, 1991.

Ikeda, S., K. Ohta, and H. Hasegawa, Effect of bank vegetation of flow and sediment deposition, *Journal of Hydraulic, Coastal and Environmental Engineering*, 447, 25–34, 1992 (in Japanese).

Jansson, B. O., *Coastal-offshore Ecosystem Interactions*, Lecture notes on Coastal and Estuarine Studies, 22, 367 pp., Springer-Verlag, Berlin, 1988.

Kanazawa, N., and Y. Mazda, Tidal flow asymmetry in mangrove estuaries, *Umi no Kenkyuu*, 3, 1–11, 1994 (in Japanese).

Koumura, J., *Sedimentation Hydraulics 1*, Morikita Publishing Co., Tokyo, 1982 (in Japanese).

Macnae, W., Mangrove forest and fisheries, FAO/UNDP Indian Ocean Fishery Programme, *Indian Ocean Fishery Commission, Publication IOFC Dev/74/73*, 35 pp., 1974.

Mann, K. H., *Ecology of Coastal Waters: A Systems Approach*, 322 pp., University of California Press, Berkeley, 1982.

Nakasuka, T., Y. Ooyama, and M. Hruki, Study on mangrove 1, Distribution of mangroves in Japan, *Journal of Japan Ecosystem Society*, 24, 237–246, 1974 (in Japanese).

Nakatsuji, K., A. Itoh, K. Muraoka, and R. A. Falconer, Hydraulic characteristics of tropical mangrove lagoon, *Proceedings of Coastal Engineering*, 41, 1126–1130, 1994 (in Japanese).

Robertson, A. I. and D. M. Alongi, *Tropical Mangrove Ecosystems*, 329 pp., American Geophysical Union, Washington, DC, 1992.

Robertson, A. I., D. M. Alongi, P. Daniel, and K. G. Boto, How much mangrove detritus enters the Great Barrier Reef lagoon, in *Proceedings of the 6th International Symposium on Coral Reefs*, Townsville 2, pp. 601–606, 1988.

Snedaker, S. C., and J. G. Snedaker, *The Mangrove Ecosystem: Research Methods*, 251 pp., UNESCO, 1984.

Twilley, R. R., The exchange of organic carbon in basin mangrove forests in a southwest Florida estuary, *Estuarine, Coastal and Shelf Science*, 20, 543–557, 1985.

Wolanski, E. M., M. Jones, and J. S. Bunt., Hydrodynamics of tidal creek-mangrove swamp system, *Aust. J. Mar. Freshwater Res.*, 31, 431–450, 1980.

Woodroffe, C. D., Studies of a mangrove basin, Tuff Crater, New Zealand, *Estuarine, Coastal and Shelf Science*, 20, 447–46, 1985.

S. Ikeda and Y. Toda, Department of Civil Engineering, Tokyo Institute of Technology, 2-12-1, O-okayama, Meguro, Tokyo, Japan.

Y. Akamatsu, Interdisciplinary Graduate School of Science and Engineering, Tokyo Institute of Technology, 4259, Nagatsuda-tyou, Midori-ku, Yokohama-shi, Kanagawa, 226-8502, Japan.

Sedimentation in Floodplains of Selected Tributaries of the Chesapeake Bay

K. M. Ross[1,2] C. R. Hupp[2], and A. D. Howard[1]

[1]University of Virginia, Department of Environmental Sciences, Charlottesville, Virginia

[2]U.S. Geological Survey, Water Resources Discipline, Reston, Virginia

Coastal Plain floodplains in the Mid-Atlantic United States are the last place for sediment storage before reaching sensitive estuarine and marine environments. Field monitoring of sedimentation rates and patterns, suspended-sediment concentrations, substrate characteristics, and woody vegetation has been conducted in eight 1-ha floodplain sites during 2000–02 to describe the main controls on sedimentation and to determine the effects of channelization and urbanization on floodplain fluvial geomorphic processes and sedimentation patterns along three tributaries to the Chesapeake Bay. These tributaries include the Chickahominy River, the Pocomoke River, and Dragon Run. Site-scale deposition patterns are spatially highly variable. Multiple regression analyses were used to determine that inundation frequency and duration, and lateral distance from the channel and sloughs jointly explain most of this variation (between 65 and 85%). Channelized floodplain reaches are flooded less frequently, have lower sedimentation rates (0.19 mm/yr to 3.2 mm/yr), less storage (700 metric t-sediment/yr to 1,090 metric t-sediment/yr) and less diversity than the non-channelized reaches. Inundation (from 27 days/yr to 97 days/yr), sedimentation rates (3 mm/yr to 7.2 mm/yr), and diversity (0.36 to 0.84 on the Shannon-Weiner diversity index) increase with distance from urbanization along the Chickahominy River. These results suggest that understanding floodplain sediment and community dynamics and geomorphic processes with respect to dominant watershed land-use patterns is critical for effective water-quality management and restoration efforts.

1. INTRODUCTION

Coastal Plain floodplains in the Mid-Atlantic States are the last place for sediment and contaminant storage from continental runoff before reaching critical estuarine and marine habitats of the Chesapeake Bay (Figure 1). The bay was listed as an impaired water body under the Clean Water Act in 1999 due in part to the elevated sediment and nutrient loads delivered by the tributaries. Rivers connect the terrestrial and

Riparian Vegetation and Fluvial Geomorphology
Water Science and Application 8

marine ecosystems. Floodplains are an integral component within the watershed sediment budget. Much of the current Chesapeake Bay policy towards the forested floodplain has been focused on buffers that function as 'filters' for laterally moving overland flows from the adjacent agricultural fields and urban areas. However, these floodplains also affect the rate of sediment transport longitudinally downstream by trapping and storing (retention) overbank deposits. In addition, nutrients, metals, and contaminants, which influence water-quality and habitat conditions in the Bay, readily adsorb to fine-grained minerals. The fluvial geomorphic processes that control floodplain-sediment patterns and storage are not well known, despite being extremely sensitive to environmental

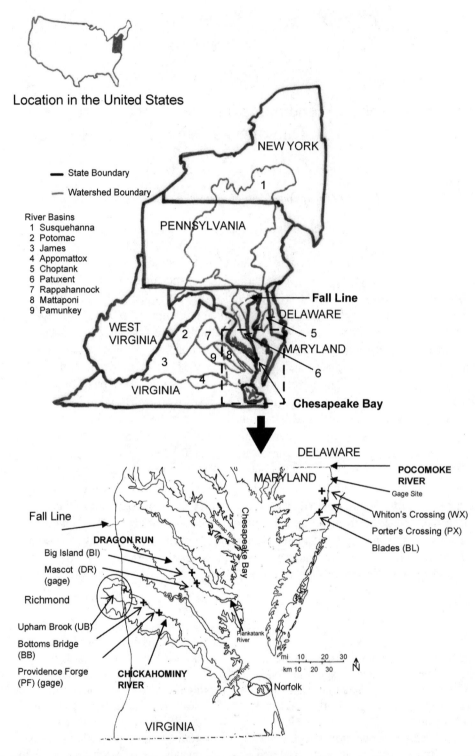

Figure 1. Locations of the eight study sites (+) along the Chickahominy and Pocomoke Rivers and Dragon Run in Virginia, Maryland, and Delaware. The Fall Line (hatched line) is at the boundary between the Piedmont (to the west) and the Coastal Plain (to the east) physiographic provinces.

changes within the watershed [*Hupp*, 2000; *Jones et al.*, 2001; *Phillips*, 1991]. Lastly, the river continuum and the transition between the upland and aquatic zones create a dynamic physical environment that promotes a diverse and rich ecosystem.

1.1. Coastal Plain Floodplains

Vertical accretion, the 'slow' accumulation of overbank fines without appreciable lateral channel migration, is the primary process by which lowland floodplains develop, such as Coastal Plain floodplains [*Middelkoop and VanDerPerk*, 1998; *Nanson and Croke*, 1992; *Walling and He*, 1998]. In the Coastal Plain Physiographic Province of the Mid-Atlantic States, the frequency of events capable of causing substantial lateral migration is limited by low valley gradients and broad, flat floodplains; many of which may be underfit [*Hupp*, 2000]. Even rare, catastrophic flows in the headwaters dissipate and lose the potential to erode once the rivers cross the Fall Line onto the Coastal Plain, as was observed during the Rapidan flood in June 1995 [*Ross*, 1999]. From about 50 to 90 percent of suspended sediment is estimated to be transported during discharges that occur 10 percent of the time [*Meade*, 1982]. With minimal erosion caused by lateral migration and little remobilization and export of floodplain sediments, particulate storage could be long (years) in the Coastal Plain [*Meade*, 1982; *Raymond and Bauer*, 2001; *Walling et al.*, 1996].

Two types of rivers occur on the Coastal Plain—alluvial rivers that originate in the uplands and Piedmont, and blackwater rivers that head on the Coastal Plain [*Hupp*, 2000]. The origin of the river has a strong effect on the nature and quantity of the suspended sediment. Alluvial streams tend to be larger and have higher suspended sediment loads, typically with a considerable mineral fraction. These larger rivers contribute a significant portion of the particulates and nutrients to the Chesapeake Bay [*Langland et al.*, 2000]. Blackwater rivers are smaller, with poorly drained and less developed floodplains. Suspended sediments, often with high organic content, are fine-grained (silt to clay-sized). Tannins released from the decaying organic matter cause the characteristically tea-colored water. Soils within the floodplain are highly porous with low bulk densities (0.2 to 0.3 g/cm^3) and organic contents often exceed 20% [*Hupp*, 2000; *Hupp and Osterkamp*, 1996; *Messina and Connor*, 1998; *Wharton et al.*, 1982].

Bank sediments along Coastal Plain rivers are typically fine-grained and cohesive, a combination that further limits active channel migration. The banks usually are low and once overbank flows occur, the inundated width extends across the entire floodplain, significantly limiting flow competence. Natural levees, usually composed of fine- to coarse-grained sands, frequently form adjacent to the channel as suspended load sediments are deposited [*Hupp*, 2000; *Pizzuto*, 1987]. Eleva-

tions vary only 1 to 2 m within the floodplain, such that small differences in flood stage or ground-water elevation, can appreciably affect inundation frequency and duration at various locations. The floodplains are typically inundated multiple times a year, often for extended durations, particularly during the winter and spring. A combination of low evapotranspiration rates, low intensity but long-lasting rainfall events that cause high water tables, and limited floodplain drainage during typical winter and spring seasons, causes large portions of the floodplain to be ponded for extended periods (up to months). This ponded water is accumulated from either rainwater or hyporheic waters, and consequently contains very little suspended sediment. In contrast, the precipitation events during the summer and fall typically are infrequent, but locally intense thunderstorms, often with high rainfall totals. Channel stage typically rises quickly as a result of these storms. The water table usually is deep and there is little ponding on the floodplain, even in the lowest elevations, prior to the entry of sediment-laden channel flow into the floodplain.

The Coastal Plain floodplains of the Eastern United States historically have supported expansive forested wetlands. Less than 25 percent of the pre-colonial coverage remains, initially converted primarily to agricultural land and, more recently, to urban land. These watershed land-cover changes have contributed to alterations in the terrestrially derived sediment and nutrient loadings delivered to the tidal Chesapeake Bay [*Cooper and Brush*, 1993; *Jones et al.*, 2001; *Langland et al.*, 2000; *Sheridan and Hubbard*, 1987]. Localized Chesapeake Bay sedimentation rates have increased from an average of 0.5 mm/yr to greater than 2.2 mm/yr since the 19th century when 80 percent of the watershed was cleared for agriculture [*Brush*, 1989; *Cooper and Brush*, 1993]. However, a large portion of the sediment that was mobilized as early as the 1600s, and particularly during the 1800s, is believed to still be stored as valley fill deposits that can compose the top 1–5 meters of floodplains and channels of the Piedmont and Coastal Plain (up to 90%) [*Costa*, 1975; *Happ*, 1945; *Meade*, 1982; *Phillips*, 1989; *Trimble*, 1975].

Contemporary processes in river corridors, such as channel incision and sediment storage, reflect the influence of past land-use changes, as well as the more current urban- growth patterns. Early stages of urbanization (land clearing and construction) increase sediment generation [*Dauer et al.*, 2000; *Hupp*, 1999; *Wolman*, 1967], alter flood hydrographs by changing watershed-runoff patterns (frequency and magnitude) and by lowering the water table [*Hollis*, 1975], and alter hydraulic conditions by simultaneously increasing flow velocities and reducing the floodplain hydroperiod (the frequency and duration of inundation). This combination can reduce the storage function of the floodplains, because velocities may be too rapid for deposition.

Table 1: Characteristics of the investigated rivers and floodplain sections. The study site abbreviations for the Chickahominy River are Upham Brook (UB), Bottom's Bridge (BB), and Providence Forge. The Pocomoke River sites are Whiton's Crossing (WX), Porter's Crossing (PX), and Blades (BL). The Dragon Run sites are Big Island (BI) and Mascot (DR). Slopes were differentiated along the Chickahominy River to show the change with downstream distance.

River	Downstream distance from headwaters (km)	Drainage Area (km²)	Stream flow data record	Mean Q (Stage (m)) Winter/Spring	Mean Q (Stage (m)) Summer/Fall	Maximum Q (Stage (m)) Date	Slope
Chickahominy	790			10.2 (1.7)	4.4 (1.4)	216 (3.6) Aug 1955	
Upham Brook[c]	A	--	1990-1994				5.6×10^{-4}
Bottom's Bridge[c]	87	--	1942-present				9.4×10^{-5}
Providence Forge[c]	109	--					
Pocomoke	158			4.77 (1.6)	1.93 (1.2)	72 (4.09) Aug 1989	8×10^{-5}
Whiton's Crossing[b,c]	23	--	1950-present				
Porter's Crossing[c]	28	--					
Blades[d]	39.5	--					
Dragon	280			2.96 (1.76)	1.13 (1.27)	78 (2.7) Feb 1998	7×10^{-4}
Big Island[d]	34	--					
Mascot[c]	40	--	1981-present				

[a]: Upham Brook is a small tributary of the Chickahominy River and is 27 km upstream of Bottom's Bridge.

[b]: USGS streamflow-gaging station is located 9 km upstream from Whiton's Crossing

[c]: Sites with complex topography

[d]: Sites with uniform topography

USGS: U.S. Geological Survey

Q: Discharge in m³/s

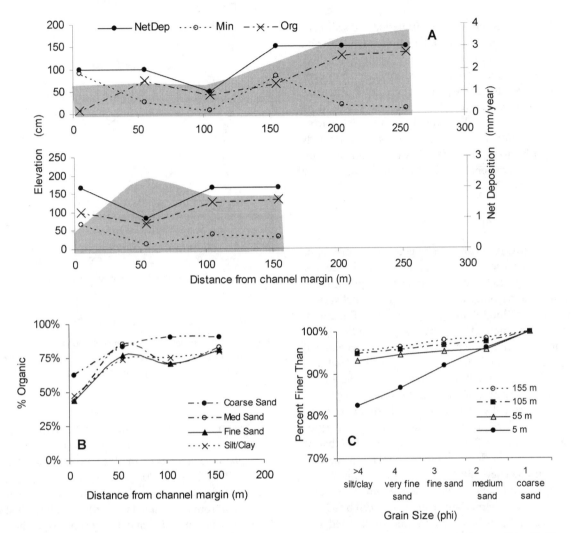

Figure 2. Annual net deposition rates (total, and mineral and organic fraction), from clay pad measurements (A) with distance from the channel margin (m) and elevation (cm) (THICK DARK LINE) along two sample transects at Blades, Pocomoke River. Substrate properties, organic content (percent) (B) and grain size distribution (phi) (C) along the same transects.

Channelization is extensive in portions of the Chesapeake Bay watershed, which restricts the accumulation of sediments by vertical accretion, the predominant floodplain-building process within Coastal Plain floodplains, to all but the largest floods because levees usually increase bank heights by more than a meter [*Hupp and Bazemore*, 1993; *Simon*, 1995]. The connectivity between the channel and floodplain is reduced severely by levees. Deposition potential within the floodplains immediately downstream of channelization is high because flow velocities would be considerably reduced when the routed flows initially enter the floodplains through breaches and sloughs or overtop the banks.

The floodplain vegetation community is diverse and productive [*Hupp*, 2000; *Messina and Connor*, 1998; *Mitsch and Gosselink*, 1993; *Wharton et al.*, 1982], driven by variations in hydroperiod, elevation, and in sediment deposition and erosion. Community characteristics, including diversity and richness, are affected by the local hydroperiod and adaptive ecological strategies, such as the regeneration mechanisms of the dominant species. The frequency and duration of flooding at any given place on the floodplain is linked closely to surface topography, and, therefore, to the fluvial geomorphic processes and sediment deposition patterns. Greater topographic heterogeneity within the floodplain equates to a larger range of 'wetness zones' for different species to establish and thrive. Physical changes are complicated by commensurate changes in the forested vegetation community impacted by the changing fluvial environment within urban and channelized watersheds. In particular, the hydroperiod and sedimentation processes are altered. The links between land use and sediment yields in the bay have been conceptually established. However, few studies have evaluated and connected how contemporary land-use changes are altering the floodplain sedimentation processes and development, which would be expected to affect further the diversity of the rich riparian vegetation community. Restoration of the forested wetlands along the bay tributaries is one of the key management strategies for managing sediment and adsorbed particulate delivery into the Chesapeake Bay.

1.2. Sediment Deposition

Floodplain deposition is affected by geomorphic processes and watershed properties that vary in relevance to the scale of interest. Within a given floodplain segment of approximately 1 ha, deposition and storage are controlled by the delivery of sediment onto the floodplain, the hydroperiod, floodplain topography and geometry, the connectivity of the floodplain to the channel, and sediment characteristics. At regional and watershed scales, channel gradient, drainage area, land use, downstream distance, and basin geology also may be important explanatory variables of sediment loads and storage.

Sediment can be delivered to and transported within the floodplain by advection, diffusion, and by sloughs. Advective transport involves a flow component normal to that of the channel flow; water flows into the floodplain. Advection typically occurs by overbank flows when maximum differences in stage result between the channel and floodplain. Crevasse splays are a geomorphic feature produced by advection. In the Coastal Plain, this feature would be expected during flows in the summer/fall when water tables are low. However, when the floodplain is already inundated by an elevated water table or rainwater, as is characteristic during the winter/spring, the pressure gradient would be low, and transport by advection would be limited. Under these conditions, strong concentration and velocity gradients would be expected between the channel water (high) and the antecedent water (low), and diffusive transport would dominate. As channel water overtops the banks during flood stage, particles are deposited with distance from the channel as a function of settling and flow velocities. Natural levees of comparatively coarser sediments frequently form adjacent to the channel [*Hupp*, 2000; *Pizzuto*, 1987]. Along fairly flat floodplain surfaces with little variation in inundation within the floodplain, deposition rates tend to exponentially decrease with distance from the active channel [*Pizzuto*, 1987; *Walling et al.*, 1996]. As the floodplain topography increases in complexity with sloughs and depressions (which is typical of Coastal Plain floodplains), the hydraulic conditions become more complex.

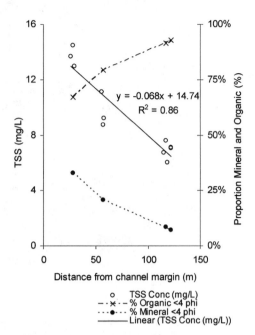

Figure 3. Total suspended solid concentration (TSS) (mg/L) with distance from the channel margin during a March 2001 overbank flow event with mineral and organic fractions along a Blades transect, Pocomoke River.

During moderate channel flows, in the absence of overbank flows, sediment and associated particles can be delivered to floodplain via sloughs or small depression channels [*Dunne et al.*, 1998; *Hupp*, 2000; *Patterson et al.*, 1985]. Sediment is deposited as the sloughs dissipate within the floodplain (50 to 100s m from the channel margin). The effect of sloughs or levee breaches in sediment delivery is probably the least documented transport mechanism, but could be important in development of Coastal Plain floodplains. It is possible that the contributing effect of sloughs in sediment delivery varies temporally and spatially depending on the flow magnitude and antecedent water-table levels. As flows recede, deposition would be expected within the slough networks. However, slough morphometry appears to be stable. Fines that are deposited within the sloughs as the flows recede may be entrained and transported onto the floodplain surface during subsequent higher velocity flows. The concentrations are typically lower during the smaller events, but depending on the frequencies compared to overbank events, the contribution to overall sediment delivery and storage could be substantial.

Sediment processes are closely tied to the hydroperiod, which is driven by seasonal precipitation trends. In winter and spring, low elevations initially are inundated by a rising water table, hyporheic flow, direct precipitation, or flow through sloughs and levee breaches. As the channel stage rises with larger events, water continues to enter the floodplain through the sloughs. Once the channel stage exceeds bank height, sediment-laden flows rapidly enters the swamp [*Mertes*, 1997]. Mixing of these two water sources might be expected. However, *Mertes* [1997] and *Dietrich et al.* [1999] have suggested that they may not readily mix, potentially altering expected

flow and deposition patterns. During the summer and fall months, the water table is low and flooding is driven by locally intense precipitation events. Flows during moderate events enter through the sloughs, and during larger events can overtop the banks into a dry floodplain. These temporal and spa-

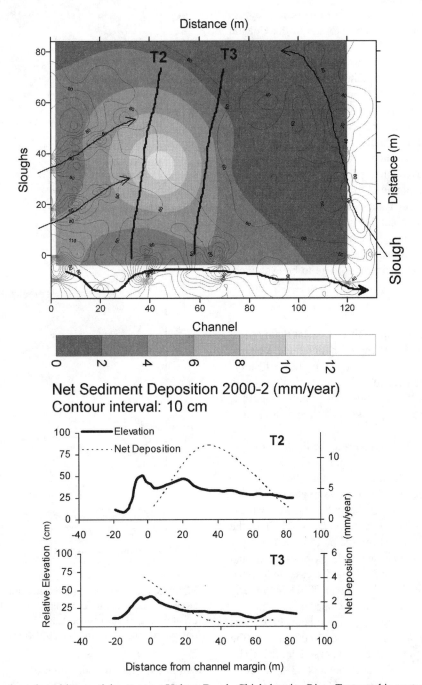

Figure 4. Annual net deposition spatial patterns at Upham Brook, Chickahominy River. Topographic contour intervals are 15 cm. The transect locations (T2 and T3) are shown on theupper image of the site. The channel and slough locations are shown with the arrows. Net deposition along the two transects are shown in the lower figure.

tial differences in the antecedent water level on the floodplain would be expected to have important implications in Coastal Plain floodplain development. Despite the importance of these hydrologic factors, few studies have evaluated the relationship between hydroperiod and sediment delivery and transport within Coastal Plain floodplains.

In this chapter, we present results from work in progress on vertical accretion processes and storage in the Coastal Plain. The main objectives are to describe sedimentation patterns and to relate these patterns to topography and floodplain geometry, hydroperiod, and sediment-transport processes. Patterns will be described at the floodplain scale (1 ha) and watershed scale. Eight study sites were investigated along three representative Coastal Plain rivers tributary to the Chesapeake Bay. A combination of landscape features and fluvial geomorphic processes, including elevation, distance from the channel, distance from a slough (if present), flooding duration, flooding frequency, and potential for deposition, were investigated that may offer significant explanation for annual and long-term sediment floodplain deposition rates and patterns. The selected watersheds have substantially different land uses (channelized-agriculture/urbanized/forested), thereby allowing land-cover effects on sediment storage to be evaluated.

1.3. Study Area

The study was conducted along three Coastal Plain rivers, including the Chickahominy River, near Richmond, Virginia (urban); the Pocomoke River on the Eastern Shore of Maryland and Delaware (channelized and heavy agriculture); and Dragon Run (forested), Western Shore of Virginia (Figure 1, Table 1). Total annual precipitation within the region is approximately 100 cm.

The Chickahominy River originates in the Piedmont Physiographic Province near Richmond, Virginia and flows through the Coastal Plain into the James River, which flows into the Chesapeake Bay. Appreciable wetland loss has resulted in the upper third of the watershed since the 1920's due to urbanization [Hupp et al., 1993; Syphard and Garcia, 2001]. Since the 1960s, population growth has increased by 155% in the upper third of the watershed, predominately in the Richmond area, and is expected to continue with concomitant decreases in forested and agricultural lands [Syphard and Garcia, 2001]. In the early 1990s, 17.5% of the watershed was developed urban area with 623 people per mi^2, 18.9% was agricultural land, 49.9% was forested land, and 11.6% was wetlands. In comparison, the middle and lower thirds of the watershed are still dominated by forest (64.4%), with very little development (1.3%), with 63 people per mi^2 [Chesapeake Bay Program, 2002]. Study reaches were established at Upham Brook (UB), Bottoms Bridge (BB), and Providence Forge (PF) (from upstream to downstream)

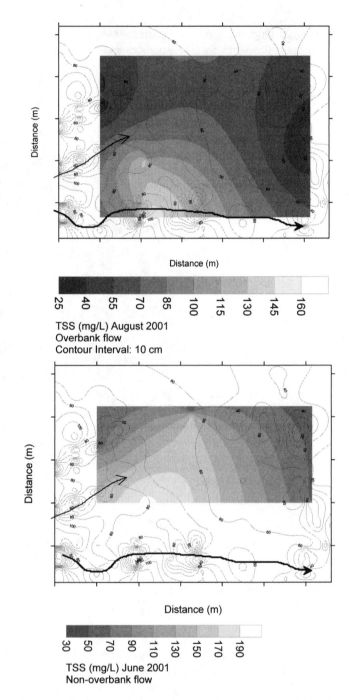

Figure 5. Total suspended solid concentration patterns (TSS) (mg/L) at Upham Brook, Chickahominy River, during an overbank flow event in August 2001 (upper) and flow through the sloughs that did not overtop the banks in June 2001 (lower).

within the Coastal Plain (Figure 1). Upham Brook is a tributary to the Chickahominy River within the upper section of the watershed. The study site was established immediately upstream of its confluence with the Chickahominy River.

Channelization, agriculture, and poultry industries are common in the Pocomoke River watershed on the Eastern Shore of Maryland and Delaware. Channelization in this area began as early as the 1600s to facilitate farming on the floodplains and continued to accelerate into the 20th century. Channelization of the upstream section of the Pocomoke River was completed in the early 1940s. Over 40% of the stream miles have been channelized and the majority of the agricultural fields are ditched extensively. These ditches provide a conduit for runoff from the fields to travel rapidly into the mainstem of the Pocomoke River. Agricultural and forested lands comprise about 80% of the landscape, with less than 2% of the land developed [*Chesapeake Bay Program,* 2002]. This watershed was not well known until the 1990s when *Pfiesteria piscicida* and the massive fish kills thought to be caused by nutrient enrichment from land-use practices were publicized widely. Three intensive study sites were established, including Whiton's Crossing (WX), Porter's Crossing (PX) and Blades (BL) (listed in downstream order). Deposition measurements also were made at two sites within floodplains along the channelized section of the Pocomoke River.

Dragon Run Swamp is a small, forested blackwater river that flows into the Piankatank River. It is the third largest swamp in Virginia, flows within the largest forested area in the Chesapeake Bay watershed, and is one of the most pristine swamps in the Eastern U.S. It has the second highest ecological value within the Chesapeake watershed (Friends of Dragon Run, personal communication [2001], [*The Nature Conservancy,* 2002]). Two study reaches were established, including Big Island (BI) and Mascot (DR).

The bottomlands along the three rivers support a diverse vegetation community, including *Acer rubra, Carpinus caroliniana, Liquidambar styraciflua, Fraxinus pennsylvanica, Quercus sp.,* and *Taxodium distichum.* With downstream distance, the dominance of *Taxodium distichum* and *Nyssa aquatica* increases. Herbaceous species include *Anelima, Leersia, Polygonum, Rhus radicans, Sagittaria, Suararus,* and *Smilax,* and common shrubs include *Ilex opaca.*

2. METHODS

2.1. Field and Laboratory Data Collection

2.1.1. Floodplain topography. Floodplain surfaces were surveyed at 10 m grid intervals supplemented by measurements at breaks in slope. Elevations were interpolated with Kriging and mapped by use of three-dimensional mapping software (Surfer 8.0). Temporary benchmarks (TBM) were established by use of permanent markers to enable repeated surveying.

2.1.2. Suspended sediment concentration. Floodplain suspended sediment samples were collected after inundation events from June 2000 through June 2002. Triplicate arrays of single stage sediment samplers [*Edwards and Glysson,* 1999], were installed and elevations surveyed in June through August 2000 at various locations (at least 5/reach) on the floodplains (for example, on the levees/banks, in/near sloughs, levee crevasses, and in the midswamp and backswamp). The sediment samplers collect approximately 400 mL of water and suspended solids during the rising limb of the hydrograph at two discrete water stages (about 20 and 80 cm above the ground surface). Thus, the samplers collect water at specific water levels on the floodplain, providing information on the spatial variability of suspended sediment concentrations. This sampler type is in contrast to other types of samplers that collect at particular time intervals throughout the duration of an event. The filled sample bottles were collected after each sampling event and returned to the laboratory for suspended sediment concentration analyses. Concentrations were determined by filtering the samples through a pre-combusted and weighed 1 μm glass-fiber filter. Each sample was dried at 60°C and ashed at 450°C for at least 24 hours to determine organic and mineral fractions [*Brower et al.,* 1989].

2.1.3. Sediment deposition and erosion. Net annual (short-term) and long-term deposition rates (up to 100 years) were measured using clay pad horizon markers and dendrogeomorphic techniques, respectively, between 1998 and 2002. Annual floodplain deposition rates and patterns were determined with feldspar clay pads, which were initially placed at ground level at a minimum density of 12 per site. The pads were laid in a grid arrangement, with three or four transects extending into the floodplain from the channel margin [*Hupp and Bazemore,* 1993]. The pads were revisited each year and the amount of net sediment accumulation or erosion was measured. Long-term net floodplain sediment accumulation was determined using dendrogeomorphic techniques, in which rates are estimated by dividing the root burial depth by the tree age [*Hupp and Morris,* 1990]. At least four trees surrounding each clay pad were measured. For clarification, annual deposition rates refer to measurements made using clays pads and long-term rates include those made utilizing the dendrogeomorphic technique. The pad elevations were surveyed along with the site topography and the distances of each from the channel margin and sloughs (if present) were measured in the field and from the surface maps.

2.1.4. Grain-size analysis and bulk density. Soil-surface samples for grain-size analyses were collected from the top 2 cm near each clay pad. Care was taken to sample only sediment that had been deposited within the last year. Samples

Table 2: Pearson correlation matrix for Upham Brook on the Chickahominy River, a site with complex topography. Grain-size parameters (size and sorting) are in phi units.

		Hydrology			Geomorphology			Substrate			
		Flood Frequency	Duration	Event Frequency	Distance Channel	Distance Slough	Elevation	OM	Sand: Clay	Mean Grain Size	Sort
Deposition	ST Dep	-0.30	-0.33	-0.46	-0.27	-0.24	0.39	-0.14	**0.96**	**-0.86**	0.49
	LT Dep	*-0.66*	**-0.71**	**-0.75**	-0.13	0.12	**0.69**	0.21	0.17	*-0.62*	*0.61*
Substrate	OM	-0.2	-0.07	-0.05	0.29	0.37	-0.14				
	Sand:Clay	-0.32	-0.38	-0.45	-0.33	-0.22	0.44				
	Mean Grain Size	*0.50*	*0.57*	**0.69**	**0.60**	-0.05	**-0.68**				
	Sort	**-0.75**	**-0.75**	**-0.84**	**-0.61**	0.28	**0.65**				
Hydrology	Duration	**0.97**									
	Event Frequency	**0.89**	**0.92**								
	Distance Channel	0.51	*0.57*	**0.69**							
	Distance Slough	-0.25	-0.14	-0.20	-0.20						
	Elevation	**-0.69**	**-0.82**	**-0.89**	**-0.74**	0.12					

Significance levels
p<0.05
p<0.10

Abbreviations

OM: Organic %
Mean Grain Size: Mean mineral grain size (phi)
Flood Frequency: Total number of inundated days per year
Event: Number of events per year (Event Frequency)
STDep: Annual net deposition (mm/yr)
LTDep: Long-term net deposition (mm/yr)

Table 3: Best model fit results for short and long-term net deposition rates for combined data from Mascot (DR) and Upham Brook (UB). Coefficient and p-values for each variable are provided, and the success of the model (Model P-value).

Independent Variable	DR STDep	DR LTDep	UB LTDep
		Dependent Variable	
Elevation	-0.59 (p=0.10)		1.06 (p=0.02)
Flood Frequency	ns	-.60 (p=0.02)	-0.35 (p=0.20)
Distance Slough	-0.46 (p=0.09)	ns	ns
Distance Channel		-.23 (p=0.39)	0.83 (p=0.02)
Event Frequency	0.89 (p=0.02)	0.18 (p=0.39)	ns
Intercept		-0.03	0.23
Total Variability Explained	65%	85%	85%
F statistic	4.98	15.22	9.55
Model P-value	0.03	0.004	0.02

Abbreviations

STDep: Annual net deposition (mm/yr)
LTDep: Long-term net deposition (mm/yr)
DR: Mascot site on Dragon Run
UB: Upham Brook site on the Chickahominy River

were stored on ice in a cooler and in a refrigerator prior to analyses. Organic content was determined from a subsample, which was dried at 60°C for 24 hours, weighed, ashed at 450°C, and reweighed. Grain-size distributions were determined using wet sieving techniques of disaggregated samples at 1 phi (0.50 mm) (coarse sand); 2 phi (0.25 mm) (medium sand); 3 phi (0.125 mm) (fine sand); 4 phi (0.0625 mm) (very fine sand); and smaller than 4 phi (0.0625 mm) (silt and clay) size fractions. Each size fraction was dried at 60°C for 24 hours and weighed. The size fractions were then ashed at 450°C for 24 hours and weighed to determine organic content. Surface samples also were collected for bulk density analyses at six to eight locations within each site. Bulk-density samples were collected by cutting a square into the top 2 cm of the floodplain surface, and immediately measuring the dimensions. Due to the high clay content of the soil, it was possible to cut and remove an intact sample. Care was taken in sample collection to ensure accuracy of volume measurements. The sample was transferred to a preweighed plastic bag, reweighed, dried at 60°C for 24 hours, weighed, ashed at 450°C for 24 hours, and re-weighed [Brower et al., 1989].

2.1.5. Calculation of sediment storage. Sediment storage over a given period along a floodplain reach can be computed from net deposition rates measured within the study sites:

$$S(t)= \sum_{i=1}^{n} \left(\frac{R_i + R_{i+1}}{2} \right) \left(\frac{L_i + L_{i+1}}{2} \right) Dk \qquad (1)$$

where S is annual storage within the river reach (ton sediment /yr); R (g cm^{-2}/yr) is the average deposition rates measured within a floodplain site at i and the next downstream site, $i+1$; L (km) is the average width of the floodplain within each floodplain site (i and $i+1$); D (km) is the distance between the two study sites; and k is a scaling factor [Owens et al., 1999]. Rates as measured in the field (cm/yr) are converted to g cm^{-2}/yr using a site-averaged bulk density.

2.1.6. Precipitation and discharge data. Monthly precipitation data (1895–2002) for the Virginia Coastal Plain were obtained from National Climatic Data Center (NCDC) [NOAA, 2003].

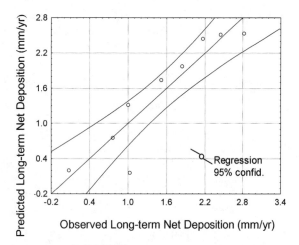

LTDep=0.23+1.06(Elevation)-0.35(Flood Frequency)
+0.83(Distance Channel) R^2=0.85, p=0.02

Figure 6. Predicted plotted against measured long-term net deposition rates (mm/year) (dendrogeomorphic measurements) for Upham Brook on the Chickahominy River (R2=0.85, F=9.55, and p=0.02) with 1:1 line shown for reference. Variables: Flood frequency, the number of inundated days per year; Distance Channel, measurement distance from channel margin (m); Elevation: elevation relative to site minimum (cm); and LTDep, long-term net deposition.

Water-table and floodplain water levels were determined from U.S.G.S. streamflow-gaging station information where available and floodplain stage-recorder wells. USGS streamflow-gaging stations are located 9 km upstream of WX on the Pocomoke River (Station 01485000) (1950 to present); at PF on the Chickahominy River (1942 to present) (Station 02042500); and at DR on Dragon Run (1981 to present) (Station 01669520). A USGS streamflow-gaging station was operational at Upham Brook on the Chickahominy River from 1990 until 1994 (Station 02042426). The streamflow-gaging station on the Pocomoke River is within the channelized portion of the river. The Chickahominy River streamflow-gaging station is located approximately 109 km from the headwaters, approximately 80 km downstream of the Richmond area. The ground-water/surface-water interface was investigated at three sites (PX on the Pocomoke River, BB on the Chickahominy River, and at DR on Dragon Run, Figure 1). AQUAPOD water-level measurement systems (DACOM Technologies) were installed to measure water levels to approximately 1 m depth and at least 2.5 m above the floodplain at 15-minute intervals.

A suite of hydroperiod characteristics was determined for each site and the elevation was determined of each deposition location measurement. The elevations of the deposition measurement locations were surveyed along with the site topographic survey. Elevations within the floodplain reaches near the USGS streamflow-gaging stations were tied to the streamflow-gaging station. At the other sites, the elevations within the floodplains were surveyed and related to the channel water level on the day of the reach survey. The relative elevation of each sampling point was calculated as the difference between the point elevation and the minimum elevation at the reach. The total number of inundated days/yr (flood frequency), average duration of an inundation event (duration), and the average number of inundation events/yr (event frequency) were determined for each deposition measurement location from the last 25 years of hydrologic data record.

2.1.7. Vegetation. Woody vegetation was assessed within the floodplain sites by use of point-centered quarter techniques with at least 20 points per site [*Mueller-Dombois and Ellenberg*, 1974]. Vegetation was identified to species and diameter at breast height was measured. Density and diversity measures were calculated to describe the community structure and composition within each site. Diversity for each location was characterized by the Shannon Diversity Index (H'),

$$H'=-\Sigma p_i \log p_i \qquad (2)$$

where p_i is the proportion of the total number of individuals of species i [*Shannon and Weaver*, 1949].

2.2. Data Analysis

One goal of this study is to determine whether net sedimentation rates and patterns within Coastal Plain floodplains can be predicted from various independent variables, including elevation, distance from the channel, distance from a slough (if present), flooding duration, flooding frequency, and potential for deposition. First, Pearson correlation matrices were produced to investigate the direct relations among the variables. Relations that were statistically significant above the 95% confidence interval (p<0.05) and 90% confidence intervals (p<0.10) are noted in the results. Stepwise backwards multiple regressions were conducted to predict annual and long-term net deposition rates from the independent variables. All analyses were conducted with Statistica, Version 5. Each independent variable was checked for statistical significance with respect to the dependent variable, net annual or long-term deposition rates. Redundancy among the independent variables was checked between the strongly intercorrelated independent variables. All measurements were standardized prior to statistical analyses. The best-fit model for each analysis was determined from the resulting F-value and P-value.

3. RESULTS AND DISCUSSION

3.1. Floodplain Site Scale

Sediment deposition patterns within the studied floodplains were affected strongly by topography and floodplain water levels. The floodplain reaches vary in fluvial geomorphic complexity from sites with relatively uniform relief (BL, Pocomoke River and BI, Dragon Run) and hydroperiod to those dissected by multiple sloughs with a relatively wide range in elevations and, thus, hydroperiod characteristics (WX and PX, Pocomoke River; DR, Dragon Run; and UB, BB, and PF, Chickahominy River). Multiple regressions of empirical data suggest that variability in measured deposition rates can be explained well by the connectivity between the floodplain and a sediment source (channel or slough) and the flow path through the floodplain, the hydroperiod (elevated water table and flood event duration), and the potential for deposition of suspended sediment. Results are provided from Blades, Pocomoke River (uniform topography); and Mascot, Dragon Run and Upham Brook, Chickahominy River (complex) that exemplify the patterns and trends observed within the other study sites.

3.1.1. Uniform relief and hydroperiod. Annual net deposition rates were measured along two transects extending from 5 to 255 m from the channel margin at the Blades site on the Pocomoke River. Accumulation of material was dominated by organic matter (Figure 2), with total net deposition greatest farthest from the channel where velocities were typically very low (3 mm/yr). Net mineral deposition along the two transects quickly dropped to very little within a short distance of the channel (1.83 to 0.24 mm/yr). Although these trends were fairly strong, neither net mineral nor net organic deposition rates were significant with distance from the channel margin. Net mineral deposition increased slightly in the middle of Transect One past the low-lying area at the base of a low mound, which may reflect sediment transported within the floodplain or material transported as overland flow that accumulated at the base of the slope. In locations with very low mineral deposition rates, greater than 90% of the sediment was coarse silt or finer (4 phi).

Near the channel, the organic matter was allochthonous, mostly small woody material, with a minor component of autochthonous leaf material (Figure 2B). In the backswamp, nearly all the organic material was autochthonous because velocities are rarely high enough to remobilize and export this material from the backswamp. Grain-size analyses support this; the substrate mineral fraction nearest the channel was significantly coarser and became better sorted, finer, and more uniform with distance from the channel (Figure 2C). There-

fore, the coarser-grained mineral sediment was deposited as the water entered the floodplain over the banks, similar to the diffusive processes described by *Pizzuto* [1987]. Total suspended solid concentrations of fine material (finer than 4 phi) (TSS) from an overbank flow event in the spring 2001 also followed this pattern; TSS concentration decreased with distance from the channel margin, with the majority of the mineral material deposited nearest the channel (Figure 3).

3.1.2. Complex floodplain topography. The spatial patterns on the floodplain study sites with more complex floodplain topography, usually with slough networks, are more complex than those observed within the floodplain sites with uniform topography. Concentrations and grain sizes did not decrease systematically with distance from the channel, as was observed in sedimentation patterns at Upham Brook, Chickahominy River (Figure 4). At this site, two main sloughs flowed laterally into the floodplain. Peaks in deposition occurred further into the floodplain than expected by diffusion alone. Sedimentation patterns, however, clearly were coincident with the dominant flow paths through the site. Comparisons of temporal and spatial sediment delivery trends can explain these differences. Total suspended solid concentrations during an overbank flow event in August 2001 declined with distance from the channel margin, indicative of diffusion (Figure 5, upper). If this were the dominant process, deposition rates would be expected to be greatest near the channel. The observed sedimentation patterns, however, did not correspond with the TSS patterns during overbank flow events.

Nonoverbank flow events that delivered sediment through sloughs occurred 80 to 90% of the time and transported 40 to 60 percent of the mineral sediment onto the floodplain during the study period. Sediment was conveyed further into areas far from the channel (50 to 100s meters) where the sloughs dissipate or within slough channels during the waning of the hydrograph, to be later mobilized and transported (Figure 5, lower). In general, it appears that sediment was delivered by two dominant mechanisms in the complex sites, by overbank flows and conveyance in the sloughs. The observed deposition patterns reflected the more complex topography and delivery mechanisms. However, the largest magnitude events sampled had recurrence intervals of only two to four years. Larger, infrequent floods may deliver the majority of sediment to the floodplains.

Annual net deposition rates at Upham Brook were not significantly associated with any fluvial geomorphic properties of the site, which probably is reflective of the topographic complexity of the site and the flashy flood regime (addressed in further detail in the next section) (Table 2). Long-term net deposition rates were greater in areas that were inundated for shorter periods during fewer events. As expected, hydrope-

Table 4: Pearson correlation matrix for Mascot (DR) at Dragon Run, a site with complex topography. Grain-size parameters (size and sorting) are in phi units.

		Hydrology			Geomorphology			Substrate			
		Flood Frequency	Duration	Event Frequency	Distance Channel	Distance Slough	Elevation	OM	Sand: Clay	Mean Grain Size	Sort
Deposition	ST Dep	**-0.72**	**-0.73**	**0.69**	**-0.72**	*-0.54*	**0.76**	0.18	0.51	*-0.57*	*0.54*
	LT Dep	**-0.93**	**-0.91**	**0.88**	**-0.86**	-0.46	**0.93**	-0.24	**0.62**	**-0.86**	**0.79**
Substrate	OM	0.27	0.09	-0.08	0.07	-0.23	-0.35				
	Sand:Clay	**-0.69**	*-0.54*	0.36	-0.41	-0.36	**0.66**				
	Mean Grain Size	**0.82**	**0.73**	**-0.66**	**0.64**	0.35	**-0.81**				
	Sort	**-0.79**	**-0.68**	*0.58*	*-0.57*	-0.29	**0.75**				
Hydrology	Duration	**0.97**									
	Event Frequency	**-0.92**	**-0.95**								
	Distance Channel	**0.83**	**0.87**	**-0.85**							
	Distance Slough	0.46	0.50	-0.37	0.45						
	Elevation	**-0.94**	**-0.91**	**0.88**	**-0.87**	-0.42					

Significance levels   **p<0.05**
                      p<0.10

Abbreviations

OM: Percent Organic
Flood Frequency: Total number of inundated days per year
Event Frequency: Number of events per year
STDep: Annual net deposition (mm/yr)
LTDep: Long-term net deposition (mm/yr)

riod and elevation variables were strongly correlated. Lower elevations with an elevated water table had lower net deposition rates. Organic content of the substrate was not related to any of the fluvial geomorphic variables included in the analyses.

Despite the lack of explanation of net deposition rates by the individual fluvial geomorphic variables, grain-size characteristics appeared to be fairly well explained. Grain size (in phi units) decreased (became finer) with the total number of inundated days per year, inundation duration, lower elevations and with distance from the channel. Thus, grain sizes were smaller further from the channel in the lower elevations, although deposition rates were not as great in these locations. Deposited sediments were better sorted (in phi units) in the lower elevations farther from the channel with longer hydroperiods (longer inundation durations, total annual inundation, and inundation event frequency). These observations suggest that at UB, the coarsest material was deposited as the flow entered the site during short-duration flows.

Long-term net deposition rates from dendrogeomorphic measurements were related to the fluvial geomorphic variables by stepwise backwards multiple regression analyses at Upham Brook. No significant linear regression models were possible for predicting annual net deposition rates from the clay pad measurements because the fluvial geomorphic and topographic variables were not able to predict the high recent deposition rates in the middle of the floodplain site. Net long-term deposition rates are best explained by a combination of the distance from a sediment source:

(Distance Channel), elevation, and total inundation (Flood Frequency): LTDep 0.23+1.06(Elevation)-0.35(Flood Frequency)+0.83(Distance Channel) (Table 3, Figure 6).

Flood Frequency was not significant individually, but the best-fit model was produced when it was included. Thus, net deposition rates were greater in the higher elevation areas at distances from the channel that were not inundated for prolonged periods. The flow regime at UB was flashy and the high velocities that had been observed during flows and in the sloughs may have prohibited deposition of significant fines within most of the site. These findings indicate a balance between extensively ponded low elevations with minimal deposition rates in which the waters do not readily mix with incoming sediment-laden channel flows and/or are scoured during high velocity flows, and low elevations that are frequently inundated by channel flows for extended durations such that suspended fines are able to settle.

The Mascot site at Dragon Run also is an example of a floodplain with complex topography. In comparison to the Upham Brook site, the Mascot floodplain has a relatively high natural sand levee that is overtopped during overbank events approximately every 2 years. A prominent crevasse splay indicated that advection is an important delivery process for the

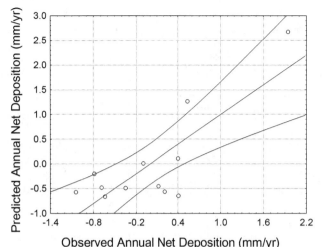

STDep=0.59(Elevation)-0.46(Distance Slough)+0.89(Event Frequency)
R^2=0.65, p=0.03

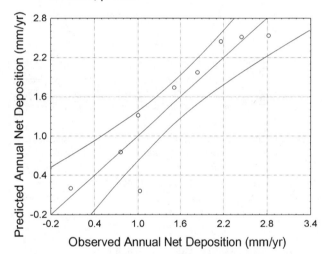

LTDep=-0.03-0.60(Flood Frequency)-0.23(Distance Channel)+0.18(Event Frequency)
R^2=0.85, p=0.02

Figure 7. Predicted plotted against measured annual net deposition rates (mm/year) (from clay pad measurements) for Mascot site on Dragon Run (R2=0.65, F=4.98, and p=0.03) with 1:1 line shown for reference (upper). Predicted plotted against measured long-term net deposition rates (mm/year) (dendrogeomorphic measurements) for Mascot site on Dragon Run (R2=0.85, F=15.22, and p=0.004) (lower). Variables: Flood frequency, the number of inundated days per year; Event Frequency, mean number of inundation events per year; Distance Channel or Slough, measurement distance from channel margin (m); Elevation, elevation relative to site minimum (cm); STDep, annual net deposition; and LTDep, long-term net deposition.

site. A slough is located about 30 m from the channel margin that is oriented in the down-valley direction of the channel.

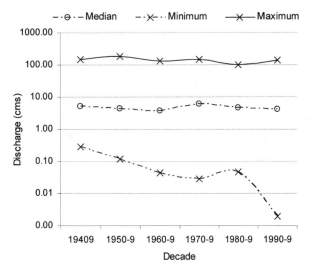

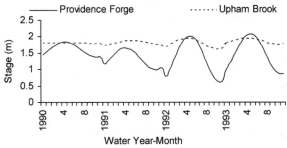

Figure 8. Harmonic analyses of decadal trends in minimum, maximum, and median discharges at Providence Forge (PF) on the Chickahominy River, 1940–1999 (upper). Harmonic analyses of the mean monthly stages between water years 1990 and 1994 at Upham Brook (UB) (dotted line, upstream site) and Providence Forge (PF) (solid line, downstream site) (lower).

Based on Pearson correlation analyses, annual net deposition was greater near the channel on the banks (higher elevations) and near the slough (sediment sources) where surfaces were inundated more frequently by new events (Table 4). Smaller grain sizes and greater sorting are correlated with lower deposition rates. Annual deposition rates tended to be lower in areas with prolonged periods of inundation (elevated water table) and inundation duration per event. Long-term net deposition rates followed similar patterns. Rates were greatest nearest the channel where the floodplain surface is inundated by a large number of events per year and vegetation roughness increases and velocities decreases. Hydroperiod and elevation variables were strongly correlated (Table 3), as expected. Locations with an elevated water table and long-duration events had lower net long-term deposition rates. The percentage of organic matter within the surface sediments was not correlated with any of the apriori geomorphic properties of the floodplain reach.

Mean grain-size and sorting decreased with reduced long-term deposition rates. However, the sand:clay (S:C) ratio, which reflects the capacity of the flow to winnow away fines and to transport sands, increased with floodplain elevation. The high sand content near the banks was controlled by the close proximity to the sediment source and larger net flows, rather than the relative elevation. The sand:clay ratio decreased with the number of inundated days per year and the average inundation duration per event. Deposited sediments became better sorted with increased inundated days per year, event duration, and distance from the channel. Sediments were less sorted at the higher elevations on the banks. The mean grain size increased with total annual inundation, event duration, and distance from the channel. Grain-size distributions were also coarser with increased event frequency.

Annual net deposition rates determined from clay pad measurements were related to the fluvial geomorphic variables by stepwise backwards multiple regression analyses. Annual net deposition rates are best explained by a combination of the distance to a sediment source (e.g. slough or channel), elevation, and event frequency, as:

STDep=-0.59 (Elevation) -0.46 (Distance Slough) +0.89 (Event) (Table 3, Figure 7).

Relative elevation and distance to the nearest slough were not significant individually, but their combined inclusion provided the best-fit model. Proximity to the main channel was considered, but was not significant. Net deposition rates increased with closer proximity to a sediment source (sloughs). The number of inundation events, as the source of the sediment, rather than inundation duration per event or the total annual inundation, was the most important hydroperiod predictor of floodplain deposition-rate trends. Rates were lower at higher elevations and farther from sloughs. This suggests that the lower lying areas of the floodplain that are within close proximity to the sloughs and that receive influxes of channel water during moderate flow events have the highest deposition rates. In addition, these findings suggest that areas within the floodplain that may be at the lowest elevation and inundated for the longest duration per event may not necessarily have the highest deposition rates in Coastal Plain floodplains.

Long-term net deposition rates determined by the dendrogeomorphic technique were similarly related to the fluvial geomorphic variables by stepwise backwards multiple regression analyses (Figure 7),

LTDep=-0.03-0.60(Flood Frequency)-0.23(Distance Channel)+0.18(Event Frequency).

Net deposition rates were best predicted by the proximity to the channel (Distance Channel), total annual inundation (Flood Frequency), and frequency of flood events (Event Frequency). Rates were greater in areas near the channel at relatively high elevations that were flooded fairly frequently.

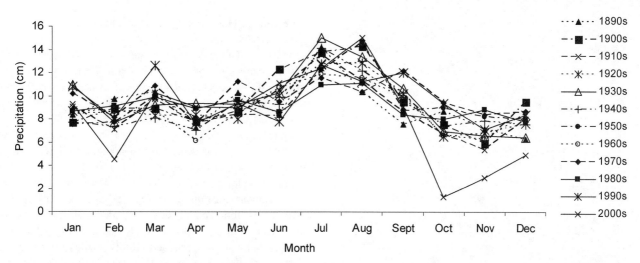

Figure 9. Monthly variation in precipitation patterns (cm) within the Virginia Coastal Plain from the 1890s through 2002 [NOAA, 2003]. High rainfall amounts during the late summer and early fall reflect localized intense thunderstorms.

Although the role of sloughs was diminished in predicted long-term deposition rates, the findings were very similar to those determined from the clay pad analyses; deposition rates were low in the lowest elevations that were flooded for extended durations.

3.2 Watershed Scale and the Influence of Land Use

The mechanisms by which sediment is transported into the floodplain and stored are also evident at the watershed scale. Sediment storage has been altered within the study floodplain sites in which hydroperiods and sediment generation have changed due to urbanization and channelization within the watersheds. Altering the hydrologic regime, in particular by increasing the frequency of flashy flows, may limit the capacity of floodplains to trap elevated sediment loads generated during the initial stages of urbanization. Likewise, in channelized reaches, low connectivity, higher velocities, and shorter hydroperiods caused by confined flows likely would limit sediment deposition and storage within the watershed.

3.2.1. Urbanization Effects on the Chickahominy River. Urbanization within the upper watershed has affected the discharge trends and annual hydroperiod along the floodplains of the Chickahominy River. Water use to meet urban needs has increased progressively in the Richmond area the past 15 years, which has resulted in a drop in water-table elevations of 6 to 10 m since the early 1970s, particularly in the late fall (data available in the USGS Virginia District Office in Richmond). Similar ground-water trends were not observed in the Providence Forge area. Decadal median annual discharges at Providence Forge demonstrate interdecadal variation, with no

trends. However, minimum discharges have decreased progressively since the 1940s (Figure 8). This trend does not appear to be attributed to decadal precipitation trends during the past 60 years within the Coastal Plain (Figure 9).

The effects of urbanization on the peak flows in the upper portions of the watershed appear to attenuate with downstream distance. A comparison between the monthly mean stage at the upstream site and downstream site between the 1990 and 1994 water years (the upstream gage was only operational for this short duration), illustrates the differences in recent hydrologic regimes along the Chickahominy River (Figure 8). The downstream site at PF has the typical annual hydroperiod of Coastal

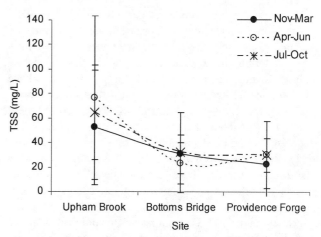

Figure 10. Mean total suspended solid concentration in overbank areas with downstream distance along the Chickahominy River (2000–02). Total suspended solid concentrations (TSS) are separated by seasons, Nov–Mar, Apr–Jun, and Jul–Oct, 2000–2002. Error bars note standard deviations.

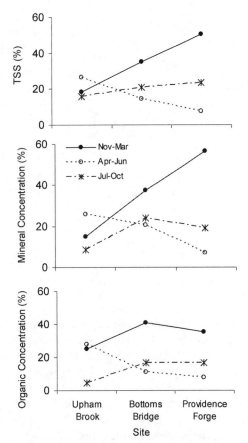

Figure 11. Seasonal variations in the proportion of total suspended solids (TSS), mineral concentration, and organic concentrations that were delivered by sloughs along the Chickahominy River (2000–02), during Nov–Mar, Apr–Jun, and Jul–Oct.

Plain rivers, which are prolonged periods of high stages during the spring and winter with drier periods in the summer and fall. Conversely, the upstream site appears to have little inter- and intra-annual variation; events are flashy—lasting only a few hours or a day at most. These differences have implications for sediment deposition patterns and storage.

During the 2000–02 monitoring of suspended sediment within the floodplain, TSS concentrations in overbank areas decreased in the downstream direction (Figure 10). Similar to deposition patterns within a site, TSS concentrations were also highly variable. These data include all measurements made within the site and consequently reflect concentrations near the channel, in the sloughs, and in the backswamp areas. Seasonal trends in TSS concentrations are not evident at this scale, excepting a slight decrease in TSS transport during the winter months. Sloughs at the three Chickahominy River sites were substantial conveyors of sediment into the floodplain (Figure 11). Transport within sloughs is less important during the spring when large portions of the lower sites are ponded,

reflecting a lack of mixing of the ponded water with the channel-derived sediments.

Mean annual net deposition rates (2000–02) increase in the downstream direction, from 3 to 4.6 to 7.2 mm/yr, with significant differences between the upstream site at Upham Brook and the two downstream sites. High intra-site variability in net deposition patterns is reflected in the high standard deviations (Figure 12). Therefore, a greater proportion of transported sediment was deposited and stored in the lower reaches than in the upper reaches of the river during the period of study. Mean long-term net deposition rates showed little trend in the downstream direction. Decadal analyses of the long-term deposition data, however, indicated a distinct transition in downstream net deposition rates occurred around 1940, when urbanization proceeded rapidly in the upper portions of the watershed. The upper study site is within the outskirts of the current urban area and the land surrounding the two downstream sites still is predominately forested and agricultural. The pre- and post-1940 net deposition rates at UB are not significantly different (1.04 ± 0.79 to 1.51 ± 1.28 mm/yr); however, rates have significantly increased from 0.46 ± 0.44 to 1.32 ± 1.13 mm/yr (p<0.01) at BB and from 0.44 ± 0.59 to 4.20 ± 4.29 mm/yr (p<0.001) at PF.

When the sedimentation rates are converted to storage, two interesting trends are evident at the Chickahominy River sites. First, sediment storage has increased 4-fold within the watershed since the 1940s (from 3,925 to 12,875 metric tons of sediment /year). Prior to 1940, storage appears to have been more evenly distributed within the watershed, with a greater proportion retained within the upstream reaches (2,620 metric t-sediment/yr between UB and BB and 1,305 metric t-sediment/yr between BB and PF). Since the 1940s, sediment storage has increased 6-fold in the downstream reaches, from

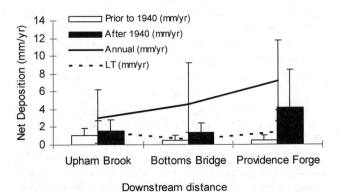

Figure 12. Net deposition rates (annual and long-term (LT)) with downstream distance along the Chickahominy River. Long-term rates have been separated into before and after 1940, when urban growth accelerated in the upper portions of the watershed (bars). Whiskers note the standard deviation.

1,300 to 8,100 metric t-sediment per year, with the storage in the upstream reaches approximately doubling (from 2,620 to 4,775 metric t-sediment/yr). Recent shifts in sediment deposition along the river are supported by field observations at the sites. Many of the trees have exposed root systems within

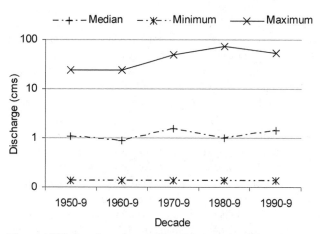

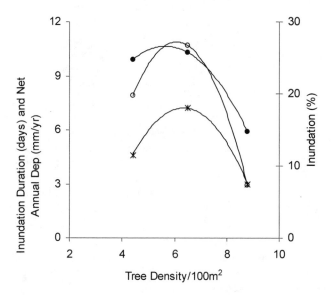

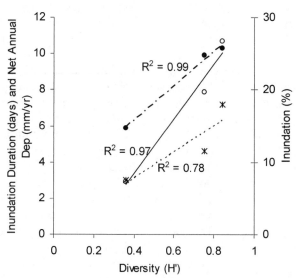

Figure 14. Harmonic analyses of decadal trends in minimum, maximum, and median discharges at the USGS stream-flow gaging station at Willards on the Pocomoke River, 1950–1999.

the floodplain at UB. This indicates recent net erosion because roots had once been below the ground surface. Flood powers within the Upham Brook floodplain likely have increased in the past 60 years in response to greater impervious areas upstream. Exposed roots were not observed at the other two floodplain sites on the Chickahominy River.

Changes in the fluvial geomorphic processes along the Chickahominy River since the 1940s appear to be affecting the vegetation community structure patterns (Figure 13). The forested wetlands are an environment in which sediment retention and vegetation patterns are closely connected to one another. The geomorphic processes control both. The density of woody vegetation was greater in the least frequently flooded sites, but the individual trees were smaller. Mean basal area increased from 390 at UB to 930 at BB, and 935 cm²/individual at PF. Vegetation structure has implications in sediment processes as a roughness agent, which influences flow velocity and depth. Similarly, community diversity, given as H' on the Shannon-Weaver Index, is directly related to deposition and hydroperiod, with $R^2=0.78$ for annual net deposition rates, $R^2=0.97$ with the percent of inundation within a year, and $R^2=0.99$ for the length of inundation per event. Thus, greater diversity was observed in the floodplains in which water levels had strong seasonal trends, and overbank flows, which carry sediment and facilitate topographic heterogeneity through localized deposition and erosion, were frequent. Changes in the hydrology and geomorphic character of the floodplain in response to continuing urban growth within the watershed would be expected to impact future diversity patterns along the Chickahominy River.

3.2.2. Channelization and Agriculture in the Pocomoke River. Channelization of the upper Pocomoke River in the

• Inundation Duration (days/event)
✕ Net annual deposition (mm/yr)
○ Inundation (%)

Figure 13. Vegetation community structure (tree density/100 m2) (upper) and diversity (H') (lower) patterns in relation to inundation duration per event (days), the annual proportion of time that the mean elevation is inundated (percent), and the annual net deposition rates (Net Annual Dep) (mm/yr) at the study sites along the Chickahominy River.

early 1940s has affected the ground-water and discharge trends and annual hydroperiod (Figure 14). Ground-water records are scarce prior to the 1940s; however, the extensive network of ditches and early studies, such as *Beaven and Oosting*, [1939] indicate that the upper portion of the basin was flooded for prolonged periods. Ground-water levels since the 1940s show a gradual decline (about 5 cm per year) within the upper portions of the basin. Maximum flows have greatly increased since the 1950s, whereas, temporal trends have not been indicated in other parameters. The loss of overbank flows in the upstream channelized portion of the Pocomoke watershed resulted in the measured increases in peak flows and flow velocities. Little urban development had occurred within the upper portions of the watershed (1% in 1990, [*Chesapeake Bay Program,* 2002]), and consequently, has had limited effects on the observed hydrologic trends. The channelized sections of the Pocomoke River end downstream of the upper study site at Whiton's Crossing (Figure 1).

Long-term mineral net deposition trends demonstrate significant temporal and spatial trends, and clearly illustrate the effects on deposition of the loss of connectivity between the

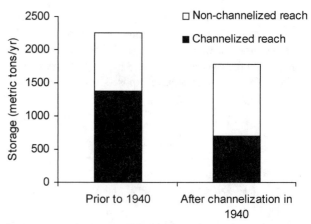

Figure 16. Comparison of floodplain storage (metric tons/yr) in channelized and non-channelized reaches before and after channelization along the Pocomoke River. Channelization of the main Pocomoke channel was conducted during the 1940s. Storage was calculated from long-term net deposition measurements using dendrogeomorphic techniques.

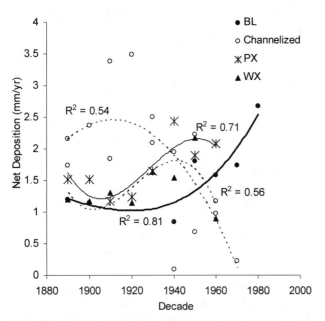

Figure 15. Trends in long-term net mineral deposition rates (mm/yr) with downstream distance along the Pocomoke River from the channelized sections to Blades (BL) between 1990 and pre-1890 (the oldest trees date to the 1820s, calculated from dendrogeomorphic measurements). Net deposition rates within the two upstream channelized sites (o, —), have declined, particularly since the 1930s. WX is also channelized, although a prominent breach in the levee permits channel water to flow into the swamp (triangle, —). PX is 5 km farther downstream is below the extent of channelization (*, -). Lastly, BL is 11.5 km downstream (closed o, -).

channel and floodplain (Figure 15). Prior to the 1930s, net deposition slightly decreased in the downstream direction. During the period between the 1920s and the 1940s, a profound transition in deposition rates is observed. Rates in the upstream reaches have dropped to almost zero as few flows have been able to overtop the high manmade levees (p=0.09). Similarly, rates within WX also have dropped within the past 60 years, although trends are not significant. However, past the downstream extent of channelization, deposition rates have increased significantly, first observed at PX (p=0.04) and later at BL (p=0.09). Deposition rates within the sites are also significantly different temporally (by decade, excluding BL) (p=0.007). Deposition rates would be expected to increase downstream of channelization because this site is the first location for the flow to enter the floodplains. Furthermore, the agricultural fields within the watershed have been ditched extensively, which provides a direct conduit between the fields and the channel. This additional sediment would be expected to be deposited within these floodplains.

Sediment storage was distinguished between pre- and post-1940s for the analyses, when channelization of the main channel of the Pocomoke River was completed. The sediment storage capacities of the study reach along the Pocomoke River appear to have decreased; decreasing from 2,255 to 1,785 metric t-sediment/yr between 1940 and the present (Figure 16). Prior to channelization, greater amounts of sediment and organic matter were stored within the upstream floodplain reaches. Since that time, storage has decreased by 50%. Although current trends in storage are greater in the nonchannelized reaches, sediment storage only has increased by 20%.

4. CONCLUSIONS

Fluvial geomorphic processes associated with sediment erosion, deposition and transport are the primary agents of floodplain landscape development. A two-year field study was conducted between 2000 and 2002 along three Coastal Plain rivers within the Chesapeake Bay watershed; the Chickahominy River, Pocomoke River, and Dragon Run to determine the main controls on sediment deposition and to investigate the effects of urbanization and channelization on floodplain geomorphic processes and deposition patterns.

Sediment is delivered and transported throughout the floodplains during overbank flows, with concentrations generally decreasing with distance from the channel. In addition, sediment enters the floodplains during low and moderate events within favored flow paths (e.g. sloughs). Observed annual and long-term depositional patterns reflect both mechanisms of sediment delivery onto the floodplains and are strongly affected by local topography, elevation, and distance from the channel. In topographically uniform sites, net deposition rates decrease with distance from the channel. Patterns of deposition and erosion are affected also by water elevations on the floodplain prior to a flood event. Ponded water appears to restrict and/or substantially slow incoming channel water. As a result, deposition rates are more variable in the more topographically complex sites, often with high rates near flow paths. Sediment properties, including grain size and sorting, are also strongly related to hydrologic and geomorphic characteristics of the floodplain. Thus, from an ecological perspective, sediment processes that are affected by local topography appear to be extremely important in maintaining high surface heterogeneity (variations in elevations) and substrate grain-size and organic characteristics. These processes affect vegetation communities by controlling the spatial and temporal environmental conditions within floodplains, including the hydroperiod, ground elevations, and the delivery and exchange of nutrients and organic matter.

At the watershed scale, long-term analyses indicate that spatial and temporal differences in sediment deposition/storage within the studied watersheds can be related to urbanization, water use, and channel changes. Increased urbanization within the Chickahominy River watershed has led to a lower water table, which has altered the annual hydroperiod within the upstream sections of the river. Deposition rates have increased over time, with a proportionally greater volume stored within the lower reaches. Channelization along the Pocomoke River has isolated the floodplains in the upper section of the watershed from incoming channel water except during the highest flows. Storage has increased substantially over time in the downstream nonchannelized sections; however, total storage within the watershed has decreased over

time. The Coastal Plain floodplains have an important role in the water-quality trends of the Chesapeake Bay. It is important to consider how human activity within the tributary watersheds has impacted and will continue to affect the geomorphic conditions of the floodplains, thereby determining the evolution of these floodplains, the state of the vegetation community, and fluxes of sediments and related contaminants.

5. ABBREVIATIONS

Study Sites

DR	Mascot, Dragon Run (USGS streamflow-gaging station)
BI	Big Island, Dragon Run
UB	Upham Brook, Chickahominy River
BB	Bottom's Bridge, Chickahominy River
PF	Providence Forge, Chickahominy River (USGS streamflow-gaging station)
WX	Whiton's Crossing, Pocomoke River
PX	Porter's Crossing, Pocomoke River
BL	Blades, Pocomoke River

Acknowledgements. Funding for this work was provided by the USGS Chesapeake Bay Science Program, University of Virginia Dissertation Year Fellowship (2002–03), and a Moore Research Grant from the University of Virginia, Department of Environmental Sciences (2000–01). We gratefully acknowledge the assistance of Lauren Alexander-Augustine, Daniel Kroes, and Michael Schening of the U.S. Geological Survey in our field efforts. This manuscript benefited greatly from comments and suggestions by Faith Fitzpatrick and Allen Gellis of the U.S. Geological Survey; and an anonymous reviewer. The authors thank them for their time and efforts.

REFERENCES

Beaven, G. F., and H. J. Oosting, Pocomoke Swamp: A study of a cypress swamp on the Eastern Shore of Maryland, *Bulletin of the Torrey Botanical Club*, *66* (6), 367–389, 1939.

Brower, J., J. Zar, and C. vonEnde, *Field and Laboratory Methods for General Ecology*, William C. Brown Publishers, 1989.

Brush, G. S., Rates and patterns of estuarine sediment accumulation, *Limnology and Oceanography*, *34*, 1235–1246, 1989.

Chesapeake Bay Program, URL: www.chesapeakebay.net, 2002.

Cooper, S. R., and G. S. Brush, A 2,500-year history of anoxia and eutrophication in Chesapeake Bay, *Estuaries*, *16* (3B), 617–626, 1993.

Costa, J. E., Effects of agriculture on erosion and sedimentation in the Piedmont Province, Maryland, *Geological Society of America Bulletin*, *86*, 1281–1286, 1975.

Dauer, D. M., S. B. Weisberg, and J. A. Ranasinghe, Relationships between benthic community condition, water quality, sediment quality, nutrient loads, and land use patterns in Chesapeake Bay, *Estuaries*, *23* (1), 80–96, 2000.

Dunne, T., L. A. K. Mertes, R. H. Meade, J. E. Richey, and B. R. Forsberg, Exchanges of sediment between the flood plain and channel of the Amazon River in Brazil, *Geological Society of America Bulletin*, *110*, 450–467, 1998.

Edwards, T. K., and G. D. Glysson, Field methods for measurement of fluvial sediment: Techniques of Water-Resources Investigations of the U.S. Geological Survey, Book 3, Applications of Hydraulics, pp. 89, U.S. Geological Survey, 1999.

Happ, S. C., Sedimentation in South Carolina Piedmont Valleys, *American Journal of Science*, *243*, 113–126, 1945.

Hollis, G. E., The effect of urbanization on floods of different recurrence intervals, *Water Resources Research*, *11* (3), 431–435, 1975.

Hupp, C. R., Relations among riparian vegetation, channel incision processes and forms, and large woody debris, in *Incised River Channels*, edited by S. E. Darby, and A. Simon, pp. 219–245, John Wiley and Sons, Ltd., Chichester, U.K., 1999.

Hupp, C. R., Hydrology, geomorphology, and vegetation of Coastal Plain Rivers in the southeastern United States, *Hydrological Processes Special Issue*, *14*, 2991–3010, 2000.

Hupp, C. R., and D. E. Bazemore, Temporal and spatial patterns of wetland sedimentation, West Tennessee, *Journal of Hydrology*, *141*, 179–196, 1993.

Hupp, C. R., and E. E. Morris, A dendrogeomorphic approach to measurement of sedimentation in a forested wetland, Black Swamp, Arkansas, *Wetlands*, *10*, 107–124, 1990.

Hupp, C. R., and W. R. Osterkamp, Riparian vegetation and fluvial geomorphic processes, *Geomorphology*, *14*, 277–295, 1996.

Hupp, C. R., M. D. Woodside, and T. M. Yanosky, Sediment and trace element trapping in a forested wetland, Chickahominy River, Virginia, *Wetlands*, *13*, 95–104, 1993.

Jones, K. B., A. C. Neale, M. S. Nash, R. D. VanRemortel, J. D. Wickman, K. H. Ritters, and R. V. O'Neill, Predicting nutrient and sediment loadings to streams from landscape metrics: A multiple watershed study from the United States Mid-Atlantic Region, *Landscape Ecology*, *16*, 301–312, 2001.

Langland, M. J., J. D. Blomquist, L. A. Sprague, and R. E. Edwards, Trends and status of flow, nutrients, and sediments for selected nontidal sites in the Chesapeake Bay Watershed, 1985–1998., pp. 46, U.S. Geological Survey Open-File Report 99–451, 2000.

Meade, R. H., Sources, sinks, and storage of river sediments in the Atlantic drainage of the United States, *Journal of Geology*, *90* (3), 235–252, 1982.

Mertes, L. A. K., Documentation and significance of the perirheic zone on inundated floodplains, *Water Resources Research*, *33*, 1749–1762, 1997.

Messina, M. G., and W. H. Connor, *Southern Forested Wetlands*, 341 pp., Lewis Publishers, New York, 1998.

Middelkoop, H., and M. VanDerPerk, Modelling spatial patterns of overbank sedimentation on embanked floodplains, *Geografiska Annaler*, *80A*, 95–109, 1998.

Mitsch, W. J., and J. G. Gosselink, *Wetlands*, John Wiley & Sons, Inc., New York, 1993.

Mueller-Dombois, D., and H. Ellenberg, *Aims and Methods of Vegetation Ecology*, 547 pp., John Wiley and Sons, New York, 1974.

Nanson, G. C., and J. C. Croke, A genetic classification of floodplains, *Geomorphology*, *4*, 459–486, 1992.

The Nature Conservancy, in *Virginia News*, pp. 11, The Nature Conservancy, 2002.

NOAA, National Climatic Data, URL: lwf.ncdc.noaa.gov/oa/climate/research, 2003.

Owens, P. N., D. E. Walling, and G. J. L. Leeks, Deposition and storage of fine-grained sediment within the main channel system of the River Tweed, Scotland, *Earth Surface Processes and Landforms*, *24*, 1061–1076, 1999.

Patterson, G. G., G. K. Speiran, and B. H. Whetstone, Hydrology and its effects on distribution of vegetation in Congaree National Monument, South Carolina., U.S. Geological Survey, Water-Resources Investigations Report, 1985.

Phillips, J. D., Fluvial sediment storage in wetlands, *Water Resources Bulletin*, *25* (4), 867–873, 1989.

Phillips, J. D., Fluvial sediment delivery to a Coastal Plain estuary in the Atlantic drainage of the United States, *Marine Geology*, *98*, 121–134, 1991.

Pizzuto, J. E., Sediment diffusion during overbank flows, *Sedimentology*, *34*, 301–317, 1987.

Raymond, P. A., and J. E. Bauer, Riverine export of aged terrestrial organic matter to the North Atlantic Ocean, *Nature*, *409*, 497–500, 2001.

Ross, K. M., Geomorphic and vegetation responses to flooding disturbance on the Rapidan River, Virginia, M. S. Thesis thesis, University of Virginia, Charlottesville, 1999.

Shannon, C. E., and W. Weaver, *The Mathematical Theory of Communication*, University of Illinois Press, Urbana, IL., 1949.

Sheridan, J. M., and R. K. Hubbard, Transport of solids in streamflow from Coastal Plain watersheds, *Journal of Environmental Quality*, *16* (2), 131–137, 1987.

Simon, A., Adjustment and recovery of unstable alluvial channels: identification and approaches for engineering management, *Earth Surface Processes and Landforms*, *20*, 611–628, 1995.

Syphard, A. D., and M. W. Garcia, Human- and beaver-induced wetland changes in the Chickahominy River watershed from 1953 to 1994, *Wetlands*, *21* (3), 342–353, 2001.

Trimble, S. W., Denundation studies: can we assume stream steady state?, *Science*, *188*, 1207–1208, 1975.

Walling, D. E., and Q. He, The spatial variability of overbank sedimentation on river floodplains, *Geomorphology*, *24*, 209–223, 1998.

Walling, D. E., Q. He, and A. P. Nicholas, Floodplains as suspended sediment sinks, in *Floodplain Processes*, edited by M.G. Anderson, D.E. Walling, and P.D. Bates, pp. 399–439, John Wiley & Sons, Ltd., 1996.

Wharton, C. H., W. M. Kitchens, E. C. Pedleton, and T. W. Sipe, The Ecology of Bottomland Hardwood Swamps of the Southeast: A Community Profile, pp. 1–126, U.S. Fish and Wildlife Service, U.S. Dept of the Interior, Washington, D.C., 1982.

Wolman, M. G., A cycle of sedimentation and erosion in urban river channels, *Geografiska Annaler*, *49A*, 385–395, 1967.

A.D. Howard, University of Virginia, Department of Environmental Sciences, 291 McCormick Road, P.O. Box 400123, Charlottesville, VA.

C. R. Hupp, U.S. Geological Survey, Water Resources Division, 12201 Sunrise Valley Drive MS 430, Reston, VA.

K. M. Ross, University of Virginia, Department of Environmental Sciences, 291 McCormick Road, P.O. Box 400123, Charlottesville, VA.

Vegetation Propagule Dynamics and Fluvial Geomorphology

A. M. Gurnell and J. M. Goodson

Department of Geography, King's College London, London, UK.

P. G. Angold, I. P. Morrissey, G. E. Petts

School of Geography, Earth and Environmental Sciences, University of Birmingham, Birmingham, UK.

J. Steiger

Departement de Géographie, Université d'Angers, Angers, France.

The character of river landscapes is primarily a product of fluvial processes. However, this paper synthesises published research by the authors to illustrate that riparian vegetation can also have important direct influences on these landscapes. The hydrodynamic properties of vegetation propagules influence the way in which they are transported by river systems. Riparian zone form and roughness and flood properties constrain both propagule and sediment deposition patterns. Thus analogies can be drawn between sediment and propagule dynamics in fluvial systems. Riparian trees are particularly important in influencing rates of riparian zone aggradation and degradation. The reproductive strategies adopted by riparian trees, particularly the timing of propagule release in relation to the river flow regime, and the rate at which propagules can develop into an established vegetation cover, are both extremely important in affecting the level of control imposed by tree species on the fluvial system. Sexual and vegetative propagules may establish at different rates under the same environmental conditions. Sheltering effects of large wood that is deposited with the propagules can modify the microenvironment to enhance rates of vegetation establishment by sheltering young plants and encouraging the deposition of fine sediment and additional vegetation propagules. Thus variations in riparian tree species and their management may have a significant influence on the character of a river and on the interplay between fluvial processes and vegetation in both time and space.

1. INTRODUCTION

The character of river landscapes is largely a function of fluvial processes. Temporal and spatial variability in disturbance by river flows, and the erosion and deposition of sediment, structure riparian vegetation patterns; resulting in close associations between vegetation composition and age, and geomorphological features [*Bendix and Hupp*, 2000; *Hupp*, 2000]. For example, meander migration represents a natural distur-

Riparian Vegetation and Fluvial Geomorphology
Water Science and Application 8
Copyright 2004 by the American Geophysical Union
10.1029/008WSA15

bance that is driven by fluvial processes and creates multi-successional stages and a dynamic mosaic of habitat types across the river's margin and floodplain [e.g. *Salo et al.*, 1986]. Along meandering rivers, plants establish and are replaced on point bar surfaces as the bars aggrade, become incorporated into the floodplain, and experience changes in sediment, moisture, and flow disturbance. At the same time competition between species also influences the developing plant communities. As a result, predictable vegetation patterns emerge that reflect the local geomorphology of point bars and, more generally, the pattern of river planform change [*Robertson and Augspurger*, 1999]. Braided rivers can support island development, represented by a dynamic pattern of islands of different age and vegetation species composition [e.g. *Kollmann et al*, 1999]. Although individual islands may be created and destroyed, the area of the river's active zone occupied by islands and the age structure of these islands may maintain a dynamic equilibrium in relation to contemporary flow and sediment transport processes, or may adjust in response to changes in the frequency-magnitude characteristics of flow and sediment transfer events [e.g. *Gurnell and Petts*, 2002]. Even on more established floodplains, vegetation patterns are also determined to a significant degree by fluvial processes. In their study of a reach of the River Ain, France, *Piégay et al.* [2000] showed how floodplain wetlands varied in their size, planform, elevation, and degree of connectivity with the main channel, and thus their flood inundation frequency and flow velocities, rate and calibre of sedimentation, and mix of river water with local surface and groundwater. Wetlands of different type were the product of different geomorphological histories and supported distinct plant communities [*Bornette et al.*, 1998].

Recently increasing emphasis has been placed on the active role of riparian vegetation in influencing the character of river landscapes [e.g. *Abernethy and Rutherfurd*, 1998; *Friedman et al.*, 1996; *Hupp and Simon*, 1991; *Johnson*, 1997, 2000; *McKenney et al.*, 1995; *Nanson and Knighton*, 1996; *Rowntree and Dollar*, 1999; *Tooth and Nanson*, 2000]. *Tabacchi et al.* [2000] reviewed three distinct influences of riparian vegetation on hydrological processes: (i) the physical impact of living and dead plants on flow patterns (hydraulics); (ii) the impact of plant physiology on water cycling; and (iii) the impact of riparian vegetation on water quality. A key element in the physical impact of riparian vegetation on fluvial processes is the first of these influences. Vegetation stands vary greatly in their hydraulic resistance and root strength, and thus the degree to which they can resist erosion and induce sedimentation. Variations in such characteristics occur between species and between stands of different age and density, so that the mosaic of vegetation patches that characterises natural river corridors is also a mosaic of different potential vegetation-fluvial process inter-

actions. This theme was developed by *Gurnell and Petts* [2002], where they noted four types of mechanisms by which riparian vegetation actively influences fluvial processes. Two of these were abiotic mechanisms: the flow resistance of the vegetation canopy; and the impact of root systems on the erodibility of sediments. In addition two groups of biotic mechanisms were highlighted: the reproductive strategies adopted by the plant species, and the nature, magnitude and timing of propagule dispersal. These biotic mechanisms are extremely important in influencing the potential distribution and establishment rates of riparian vegetation, although many other abiotic and biotic controls influence where and at what rate vegetation actually establishes. Moreover these two groups of mechanism have received relatively little research attention and so provide the focus for this paper.

2. VEGETATION PROPAGULES: DISPERSAL, DEPOSITION, GROWTH

Plant species vary widely in their ability to reproduce from seeds or plant fragments, and the longevity and germination requirements of seeds are also highly variable between species. Some species produce large numbers of short-lived seeds [e.g. *Salix* spp., *Karrenberg et al*, 2002], whereas other species produce very few seeds, reproducing largely by vegetative propagation. In the following discussion, the term propagule applies to both seeds and plant fragments from which new plants can develop. Plant fragments may consist of entire uprooted plants, or any smaller above or below ground part of the plant, according to the vegetation species.

Propagules may be dispersed by many mechanisms. They may be deposited close to the parent plant or be dispersed over considerable distances by wind, water or animals. Of particular relevance to riparian species is the process of propagule transport by water: here termed 'hydrochory' [*Danvind and Nilsson*, 1997]. Hydrochory forms an important mechanism for initial propagule dispersal and also for the mobilisation of propagules that have been deposited from local vegetation or have been initially dispersed by other mechanisms into the riparian zone.

The timing of initial propagule dispersal in relation to the river flow regime can be important in governing deposition sites. For example, *Van Splunder* [1998] observed Salicaceae seedlings of early dispersing species at higher levels within river margins than later dispersing species along the River Rhine, reflecting the water level at the time of dispersal. Whilst the initial rise in water levels at the commencement of the high flow season can remobilise seeds and vegetative propagules to achieve significant propagule dispersal [*Cellot et al.*, 1998], many species are seasonally specific in their time of propagule release [*Stainforth and Cavers*, 1976; *Kubitzki and Ziburski*,

1994] and so may occur synchronously with particular river flow sequences.

The quantity, hydrodynamic (flotation) characteristics, type (seeds, vegetative fragments) and longevity of propagules provide other important ways in which vegetation can actively influence the probability of successful growth and establishment after transport and deposition by water [*Goodson et al.*, 2001]. In particular, flotation characteristics may influence not only the deposition site but also reflect the character of the materials deposited at the same time and site.

Given appropriate environmental conditions, the rate of growth of riparian plants is often considerably greater as a result of vegetative propagation rather than from seedlings. For example, *Barsoum and Hughes* [1998] grew seedlings and cuttings of *Populus nigra* in sediments of differing calibre and under different water table regimes. They showed that cuttings developed above- and below-ground biomass more rapidly than seedlings, regardless of the experimental treatment. Thus reproductive strategies of individual species influence not only the number of propagules and the area over which they are dispersed (particularly when dispersed by other media in addition to water), but also the rate at which the young plants are likely to grow under different environmental conditions. The ability of some species (particularly the Salicaceae) to regrow from large vegetative fragments, including whole uprooted trees, can support particularly rapid rates of vegetation establishment [e.g. *Edwards et al.*, 1999].

In this paper, recent research by the authors is reviewed and combined to provide a perspective upon propagule dynamics and fluvial processes, particularly within the riparian zone of river systems. In the context of this paper, the riparian zone is defined as 'that part of the biosphere supported by, and including, recent fluvial landforms and is inundated or saturated by the bankfull discharge. This includes many floodplains, riparian wetlands, banks and all fluvial landforms below the bankfull elevation' [Hupp and Osterkamp, 1996].

3. MINERAL AND ORGANIC SEDIMENT DEPOSITION WITHIN THE RIPARIAN ZONE DURING INDIVIDUAL FLOOD EVENTS

Observations of the distribution, quantity and particle size of mineral and organic sediment deposited during individual flood events, illustrates clear associations with flood magnitude and river margin character.

Steiger et al. [2001] used artificial turf mats to investigate variations in the weight, particle size and organic matter content of flood-deposited sediment around a meander bend on the River Severn, UK, during four flood events (flood peak discharges = 197, 305, 325, 346 m^3 s^{-1} Q$_{2.33}$ = 303 m^3 s^{-1}). Whilst, planform location was found to influence some of the varia-

tion in sediment properties, the primary influence was site elevation within the riparian zone. In each flood, the heaviest sedimentation occurred within a 1 m elevation band. Sand content of the sediment was approximately 35 per cent in a band extending from the low flow channel to the upper limit of the 1 m band of heaviest sedimentation. Sand content fell to approximately 10 per cent above this level. Organic matter content was low (less than 5 per cent) in sediment deposited up to the upper limit of the 1m band of heaviest sedimentation. Above this level, there was a gradual increase in organic matter content. These observations illustrate a close link between flood level, a narrow band of relatively heavy sedimentation, and clear vertical zonation in the calibre and organic content of the deposited sediment.

In another study of sediment deposited by three floods along a 71 km reach of the Garonne River, France [*Steiger and Gurnell*, 2002], the morphology and vegetation distribution within the riparian margins of the river were more complex than those of the River Severn study site. Therefore, the sampling design used to investigate flood-deposited sediment was stratified to reflect landforms constructed by fluvial processes (e.g. floodplain, lateral benches, side-channels and point bars), that had become exposed to varying degrees as a result of incision of the low-flow channel, and had been colonised by riparian vegetation including grasses and herbs, riparian shrubs, and more mature riparian woodland, or on the highest benches and floodplain, had been commercially planted with poplar. The quantity of sediment deposited varied significantly with flood event, planform context, landform type and vegetation cover and in some cases with sample location within the landform. The quantity of sediment deposited increased with flood peak discharge. Sites under natural riparian vegetation experienced higher sedimentation than poplar plantations. Sites on concave (outer) banks received less sedimentation than those on convex (inner) banks. Sedimentation on floodplain sites and higher benches was less than on low benches, point bars and side channels. Sediment particle size was generally coarser at locations with higher amounts of sedimentation. The quantity of total organic carbon increased with the quantity of deposited sediment and the concentration of total organic carbon increased significantly with an increase in the percentage of silt and clay in the deposited sediments.

These observations from the Rivers Severn and Garonne led to the development of a conceptual model (Figure 1) to depict how sedimentation within riparian zones may vary according to flow dynamics. During a small flood event (Figure 1a), low connectivity between the river and the riparian zone produces highest sediment accumulations close to the river channel. During a larger flood event two major sedimentation patterns may occur according to the geomorphological setting. Within a constrained setting (Figure 1b) connectivity does not increase, but

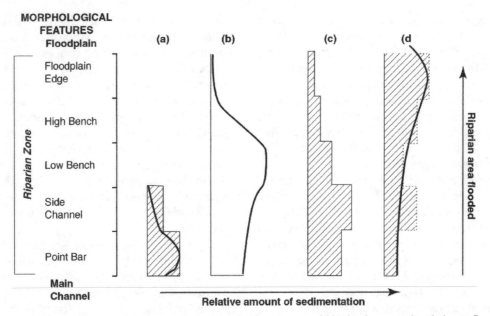

Figure 1. A conceptual model of the pattern in deposition of sediment mass within riparian zones in relation to flood magnitude. (a) Sedimentation pattern generated by a small flood within a geomorphologically simple and constrained (thick line) or complex and less constrained (shaded) riparian zone. (b) Sedimentation pattern generated by a larger flood within a geomorphologically simple and constrained setting. (c) Sedimentation pattern generated by a larger flood within a geomorphologically complex and less-constrained setting. (d) Sedimentation pattern generated by a very large flood within a geomorphologically simple and constrained (thick dashed line) or complex and less constrained (shaded and dashed borders) riparian zone [Steiger et al., 2003].

the zone of maximum sediment accumulation occurs at a higher elevation to reflect flood inundation patterns, as was observed on the River Severn. Within an unconstrained heterogenous riparian zone (Figure 1c) the zone of maximum sedimentation is dictated by morphological complexity as well as flood magnitude. For even larger floods, the zone of maximum sedimentation might be expected to occur at the floodplain margins (e.g. Figure 1d).

Although none of this research was concerned with propagule dispersal, propagules are included within the organic fraction, and research described in the following section illustrates the link between the quantity, calibre, and organic content of deposited sediment and the viable seeds deposited with the sediment.

4. VIABLE SEEDS IN WINTER FLOOD-DEPOSITED SEDIMENTS

Research at three sites within a 30 km reach of the River Dove, UK, by *Goodson et al.* [2003], has illustrated how viable seeds were mobilised after the peak season of seed release, and deposited on river margins by winter floods in association with mineral and organic sediment. In this research samples were obtained for the aggregate deposition from all floods that

occurred between October 1999 and March 2000. Paired artificial turf mats were installed at three elevations within the three sites: 'Top' (the top of the bank on the grassland floodplain surface within 2 m of the bank edge), 'Mid' (at least 0.3 m below the bank top edge within the central section of the bank profile and above the toe), and 'Low' (within the toe area of the bank), in order to sample vertical variations in viable seed and sediment deposition along different bank profiles. One mat from each pair was used to analyse sediment quantity, particle size, and organic matter content, and the other mat was used in a 10-week germination trial to determine the number and species of viable seeds that had been deposited.

Analysis of the sample data revealed significant power relationships between the number of viable seeds per m^2 and both the dry weight of sediment and the weight of organic matter deposited at the same site. The samples were analysed using CANOCO v4 [*ter Braak and Smilauer*, 1998]. Detrended Correspondence Analysis [DCA, *Hill and Gauch*, 1980] was applied to the seed species information collected from 90 samples. Figure 2a shows a biplot of these samples in relation to the first two DCA axes. Since the position of each sample within the biplot reflects its floristic composition, it is apparent that there are differences in floristic composition between samples collected at different elevations across the river-banks (the three colours

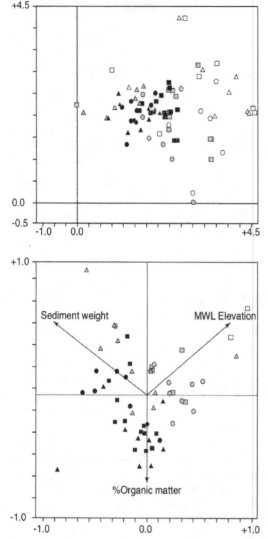

Figure 2. (a). Biplot showing the distribution of 90 samples in relation to the first two DCA axes. (b). Biplot showing the distri-bution of 56 samples and 3 environmental variables in relation to the first two RDA axes (Site 1 = circles, Site 2 = triangles, Site 3 = squares; Top = white, Mid = grey, Low = black) [Goodson et al., 2003].

species, and environmental variables. Figure 2b shows the distribution of sample scores from the analysis. Despite only 56 samples being used for the analysis, drawn mainly from Mid and Low mat locations, there is a clear separation between samples from different elevation groups, reflecting differences in their floristic composition. Three significant environmental variables (P<0.05) reflect differences between the samples. Elevation relative to local mean water level (MWL) provides a far more precise measure of elevation and thus inundation characteristics than the Top, Middle, Low classification, and differences between samples are also reflected in their percent organic content and the quantity of sediment deposited.

Although seeds may have been mobilised from their original depositional sites and redeposited on the mats by many mechanisms over the winter sampling period, the importance of fluvial processes in supporting the patterns shown in Figure 2 is demonstrated by:

(i)   the small quantities of viable seeds and sediment that were found deposited on mats that were not inundated;

(ii)   the changes in the floristic composition of the viable seeds with topographic position, and particularly the smaller differences in floristic composition between samples at Low sites, where inundation duration was greatest;

(iii)   the association between the floristic composition of the deposited seeds and the fluvially-related environmental variables (MWL, deposited sediment quantity and organic content).

Thus the quantity and floristic composition of the deposited seeds were found to reflect elevation of the depositional site in relation to mean water level and the quantity, calibre and organic matter content of sediment deposited at the same location. Although the viable-seed studies were based on aggregate samples from winter floods, whereas the studies of sediment quantity, calibre and organic content described in the previous section were based on samples taken during individual flood events, it is clear that if viable seed quantities and species composition are related to flood-deposited sediment characteristics, then patterns of flood inundation coupled with form and vegetation roughness properties of the riparian zone are important determinants of patterns in the deposition of viable seeds.

5. DEPOSITION OF LARGE VEGETATIVE PROPAGULES (LIVING DRIFTWOOD)

All rivers transport sediment and vegetation propagules but rivers draining natural corridors also transport significant quantities of driftwood. Along the braided, gravel-bed Tagliamento River, Italy, the margins of islands and the floodplain forest release large and small pieces of dead and living driftwood. Deposition of both dead and living driftwood contributes to the development of new patches of vegetation on exposed gravel

of symbol, although overlapping are displaced from the right to the left of the plot in the order Top, Mid, Low). In addition, the Low samples are distributed within a more restricted zone of the plot than the Mid samples, which in turn are distributed within a more restricted zone of the plot than the Top samples. This suggests greater similarity in the floristic composition of the Low samples in comparison with the Mid or High samples.

Fifty-six samples supported viable seeds and retained sufficient sediment (>50g) for an analysis of sediment characteristics. Redundancy Analysis [RDA, *Rao*, 1964, cited in *Jongman et al.*, 1995] was used to explore relationships between samples, surfaces [*Gurnell et al.*, 2001]. However, in the context of the present discussion of propagules, it is the living driftwood that has particular relevance.

Gurnell et al. [2000] noted that wood mobilised by large flood events on the Tagliamento was preferentially retained during the falling stages of the floods in quite narrow elevation bands within the active zone of the river and also in association with three main types of geomorphological feature. The largest quantities of wood were stranded on bar crests, and the other main retention sites were secondary channels and the margins of established islands. More recent detailed surveys within two reaches of the Tagliamento have illustrated that over 75% of the large wood (sample sizes 445, 353) was retained, respectively, within a 1 and 1.4m elevation range, although the active zones within the two reaches possessed a typical total elevation range of 2 and 3 m. Within the two reaches, the proportion of living driftwood, which was potentially capable of regeneration, was 9% and 37% of the total driftwood, respectively. In general, the living driftwood was retained at lower elevations (perhaps reflecting greater moisture availability) and the piece sizes were larger than the dead driftwood. Thus the fate of these large propagules is closely associated with fluvial processes and river morphology as is the fate of the viable seeds and organic matter previously discussed.

6. INTERACTIONS BETWEEN PROPAGULES AND FLUVIAL PROCESSES ALONG THE TAGLIAMENTO RVER, ITALY

Research on the Tagliamento River has focussed upon island development, where islands are defined as discrete areas of woodland vegetation surrounded by either water-filled channels or exposed gravel [*Ward et al.*, 1999]. The research has explored the complex relationships between propagules and fluvial processes, including the various trajectories that support vegetation development [*Gurnell et al.*, 2001]; important interactions between vegetation growth, sedimentation and the hydrology of the alluvial aquifer [*Gurnell et al.*, 2001, *Gurnell and Petts*, 2002]; and the role of floods in the development and destruction of islands [*Gurnell and Petts*, 2002].

Gurnell et al. [2001] proposed a model of island development based upon three trajectories of vegetation establishment on gravel bars along the Tagliamento River (Figure 3). Assuming controlled environmental conditions (particularly climate, sediment calibre, subsurface hydrological regime, and nutrient availability), vegetation growth is slowest when it commences as a dispersed cover of seedlings and small vegetative fragments because of the poor retention of fine sediment to support plant growth by the relatively sparse vegetation (Figure 3, trajectory A). Dead driftwood deposited on gravel bar surfaces provides a focus for fine sediment and

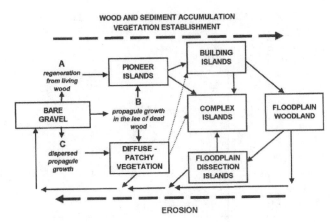

Figure 3. A conceptual model of island development for the Tagliamento River, Italy [Gurnell et al., 2002].

organic matter accumulation, which can support more rapid seedling and sapling growth than trajectory A (Figure 3, trajectory B). When large pieces of living driftwood including uprooted trees are deposited, certain species can send out adventitious roots and shoots, increasing their biomass so that they attract even more rapid accumulation of fine, moisture retentive sediment than trajectory B. Regrowth from these large living wood pieces coupled with the growth of sheltered seedlings and vegetation fragments results in potentially the fastest overall rate of vegetation development (Figure 3, trajectory C). Trajectory C is the first stage in the development of pioneer islands (fine sediment accumulated around a living driftwood core and supporting growth of a range of seedlings and saplings), which may then enlarge and/or coalesce to form building islands (Figure 3). Thus the rate of vegetation establishment on gravel bar surfaces is actively controlled by the interaction between large wood and propagules in relation to the three trajectories. In the case of each trajectory, the accumulation of fine sediment around the growing vegetation can enhance vegetation growth by retaining moisture, but excessive sediment deposition may smother and kill the growing plants. As the vegetation grows, it becomes more resistant to fluvial erosion. *Gurnell and Petts* [2002] suggested that the predominant trajectory of vegetation establishment influences how quickly the vegetation grows, the nature of its interaction with sedimentation (aggradation thresholds for increasing or retarding growth), and the degree to which it is capable of resisting erosion by floods of a particular magnitude and frequency. Vegetation growth is associated with increasingly extensive root anchorage of the vegetation to the gravel surfaces of the active zone; the aggradation of fine sediment around the vegetation to raise the surface of the active zone towards that of the surrounding floodplain; and the development of adventitious roots within the accumulating fine sediment, which help to prevent undermining of the aggrading vegetated surface by fluvial processes. Thus vegetation growth is associated with increasing elevation of the vegetated surface, so reducing inundation frequencies, and is also associated with reinforcement of pre-existing and accumulating sediment, so increasing the resistance of the sediment to erosion. Where all three vegetation growth trajectories can be supported by local riparian tree species, the rate of increase in both surface elevation and root reinforcement of sediment is hypothesised to be highest under trajectory C and lowest under trajectory A (Figure 3).

Gurnell and Petts [2002] conceptualised the dynamics of islands within a braided river reach over ca. 60 years in response to the series of annual discharge maxima shown in the top graph of Figure 4. The remaining three graphs in Figure 4 illustrate the hypothetical impact of this flood series on the number and area of islands and on aquatic habitat diversity according to the three vegetation development trajectories (Figure 3). The predominant form of island activity is indicated in relation to trajectory C (a, islands predominantly building; b, islands merging; c, islands being eroded / destroyed). The contrasting significance of the vegetation trajectories on aquatic habitat diversity is indicated in the bottom graph by comparing the situation where living driftwood dominates vegetation development (trajectory C) with that where diffuse seedlings and saplings predominate (trajectory A). The structural control upon the reach as islands develop through wood accretion, sedimentation processes, and vegetation growth and succession, has a fundamental impact upon the seasonally-exposed terrestrial habitat mosaic and upon physical habitat diversity.

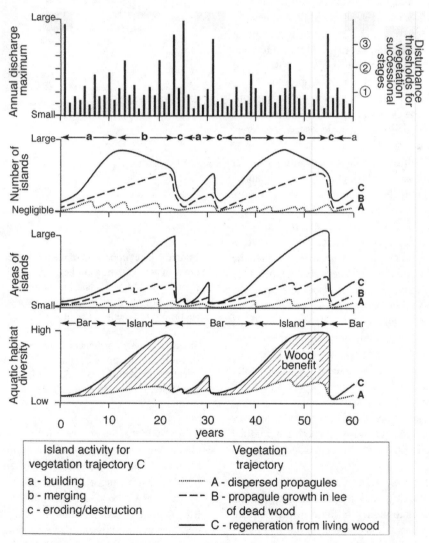

Figure 4. Temporal significance of three vegetation recruitment and growth trajectories for the number and areal extent of islands and for aquatic habitat diversity [Gurnell et al., 2002].

7. DOWNSTREAM VARIATIONS IN THE INFLUENCE OF VEGETATION PROPAGULES ON ISLAND-BUILDING

Research on the Tagliamento River has illustrated important interactions between propagules and fluvial processes, which may affect the spatial as well as the temporal characteristics of the river's island-braided pattern.

The relative importance of the three vegetation establishment trajectories (Figure 3) may vary from reach to reach, and some of the controls on this variation relate to the character of the vegetation [*Gurnell et al.*, 2001]. For example, the timing, quantity, and longevity of seeds will influence the degree to which trajectories A and B (Figure 3) can lead to significant vegetation growth. The degree to which the vegetation species present along the river can reproduce vegetatively will impact significantly on all of the trajectories, but particularly trajectory C. Thus changes in riparian tree species may favour one or more of the trajectories, and the most rapid rate of vegetation establishment and island development is hypothesised to be associated with species that can regrow once uprooted, transported, and deposited by the river. Changes in the proportions of wood stored on open gravel bars along the Tagliamento River that is dead or living (sprouting) and that is deposited as different types of accumulation (whole shrubs/trees, jams, logs), or is already accumulating sediment and evolving into a pioneer island [Figure 5, Gurnell et al., 2000], in part reflects changes in riparian tree species along the river. Whilst a variety of Salix species are found along the river (*Salix alba, S. daphnoides, S. eleagnos, S. purpurea, and S. triandra*), *Alnus incana* is the most important riparian tree species in the headwaters, whereas *Populus nigra* is the most important riparian tree species in the lower reaches. Uprooted *Populus nigra* trees sprout vigorously and, coupled with resprouting of the *Salix* species, support the dominance of trajectory C in the lower reaches of the river. *Alnus incana* that has been uprooted and deposited by flood waters, does not resprout very readily. As a result, trajectory C is of less importance in the upper reaches of the river. The impact of differences in tree species is reinforced by the relative lack of fine sediment available to support resprouting wood in the headwaters. Thus the way in which vegetation actively influences island dynamics varies along the river with changes in the main riparian tree species and reflects both longitudinal variations in climate [*Gurnell et al.*, 2001] and sediment calibre [*Petts et al.*, 2000].

Whilst vegetation propagule dynamics can result in important influences on the character of river reaches, fluvial processes have the power to moderate these influences and indeed to erode or bury vegetation completely during powerful flood events. Subtle interactions between vegetation propagules and fluvial processes can be illustrated by observations

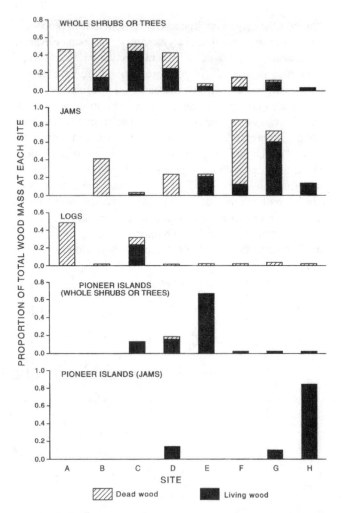

Figure 5. The proportion of the total wood mass stored in different types of accumulation on the open gravel areas of the active zones of eight sites (A to H) distributed in a downstream sequence along the Tagliamento River, Italy. The wood is subdivided into five classes according to whether it is in the form of individual logs, jams or whole shrubs/trees. In the latter two cases, the wood can be subdivided according to whether it is exposed or is partly covered by sediment to form a pioneer island. The proportions of the different wood accumulation types add up to 1.0 for each individual survey site (A to H) [Gurnell et al., 2000].

along river reaches where physical processes and properties change but vegetation species remain constant. In the lower reaches of the Tagliamento River there is a transition from a braided to a meandering planform, which is largely dictated by a change in the local sediments across an old marine shoreline. A distinct reduction in active zone width, reflecting the decrease in erodibility of marginal sediments, leads to increasing unit stream power during floods through most of the reach. Living and dead driftwood is common throughout the reach,

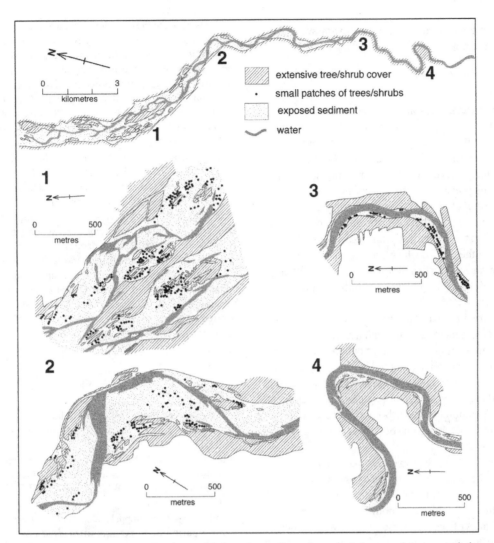

Figure 6. Distribution of riparian woodland, woodland patches and wood accumulations, and open gravel along a reach of the Tagliamento River, Italy. The maps are interpretations from air photographs taken in 1985/6 (whole reach), 1997 (locations 1 and 2) and 1990 (locations 3 and 4).

and pioneer islands, frequently cored by whole uprooted *Populus nigra,* are an important component of island development. Figure 6 illustrates that wood accumulations and pioneer islands (small patches of riparian shrubs) are present throughout the reach (Figure 6, black dots), but that their distribution and alignment change along the reach. In general, the upstream section of the reach (e.g. Figure 6, location 1) exhibits freely-formed, irregular islands on the highest bar surfaces, surrounded by small, irregularly distributed vegetated patches. As the upper braided section narrows and the main channel pattern changes from multi- to single-thread (e.g. Figure 6, location 2), islands adopt a more elongated, streamlined shape, become confined to high bar surfaces close to the edge of the active zone, and are surrounded by more widely distributed

patches of wood and vegetation. Within the meandering section (e.g. Figure 6, location 3), islands have become attached to the active zone margin and the small, vegetated patches are now aligned in ribbons parallel to river flow along the highest parts of point bars. By location 4 (Figure 6), the continuing reduction in channel slope and negligible channel narrowing result in a reduction in unit stream power. In this reach driftwood and other vegetation propagules are incorporated into the aggrading channel margin to form scroll bars through which the propagules sprout to define streamlined islands along the concave bank, floodplain forest margin.

This example illustrates the active role of vegetation propagules, particularly resprouting trees, in driving island development and the way in which the changing power of the fluvial processes moderates the expression of the developing riparian vegetation in the number, distribution, and shape of islands.

7. CONCLUSIONS

This paper has presented research examples drawn from four different rivers, subject to different levels of management intervention in three different European countries (France, Italy, United Kingdom) where environmental conditions and plant species differ widely. Nevertheless, common themes emerge.

(i)   The results presented in this paper develop the ideas presented by *Gurnell et al.* [2002] by illustrating that propagules can be viewed as part of the total mineral and organic sediment load of rivers. The hydrodynamic properties of the propagules dictate the way in which they are transported and deposited. Riparian zone form and roughness and flood properties constrain deposition patterns for both sediment and propagules.

(ii)   Riparian trees are particularly important in enhancing flow resistance and sediment cohesion within riparian zones, and thus in actively influencing rates of aggradation and degradation. The reproductive strategies adopted by riparian trees, particularly the timing of propagule release in relation to the river flow regime, and the rate at which propagules can develop into an established vegetation cover, are both extremely important in affecting the level of active control imposed by tree species on the fluvial system.

(iii)   Sexual and vegetative propagules may establish at different rates under the same environmental conditions. Moreover, the sheltering effects of other materials, particularly large wood, deposited with the propagules can modify the micro-environment to enhance rates of vegetation establishment by sheltering young plants and encouraging the deposition of fine sediment and additional propagules.

(iv) Although the character of riparian zones may be primarily influenced by fluvial processes, there is an increasing body of evidence to support the view that vegetation can also actively influence riparian zone character. Thus variations in riparian tree species and their management may have a significant influence on the character of a river, particularly its riparian margins and islands, and on the interplay between fluvial processes and vegetation in both time and space. These ideas are well-illustrated along the Tagliamento River, Italy, where there is an intimate relationship between propagules delivered from the river's relatively unmanaged margins and islands, downstream gradients in climate and sediment calibre, reach scale changes in stream power, and temporal patterns in the magnitude and frequency of floods.

Acknowledgements: On the River Severn we thank Mr J. Wingfield, Mr P. Roberts, Mr J. Cooke, and Ms S. Townsend and her colleagues from the Preston Montford Field Centre for their support and permission for access. On the River Dove, we thank Mr Davies, Mr Langridge, Mr Palmer, Mr Sessions and Mr Weston, for their permission for access. We also gratefully acknowledge the help of the Environment Agency in providing river stage and discharge data for both rivers and in collaborating on the River Dove research programme. We particularly thank Mr R. Stokes of the Environment Agency for his continuing interest and support. Research on the UK rivers was funded by a TMR Marie Curie post-doctoral research training grant to Dr J Steiger from the European Commission (ERBFMBICT983285), by a UK Natural Environment Research Council (NERC) studentship (GT16/98/CONN/8) to Ms. J.M. Goodson, and by NERC research grant GR3/CO038. Research on the Garonne River was undertaken while J. Steiger was in receipt of a NaFöG (Nachwuchsförderungsgesetz) post-graduate research fellowship from the German Federal State of Berlin. The research was carried out at the Centre d'Ecologie des Systèmes Aquatiques Continentaux, CNRS-UPS, Toulouse, France and the U.F.R. Géographie - Aménagement, Institut Daniel Faucher at the University of Toulouse le Mirail. Research on the Tagliamento River was supported by grants GR9/3249, GR3/11909 and NER/B/S/2000/00298 from the UK NERC.

REFERENCES

Abernethy, B., and I. Rutherfurd, Where along a river's length will vegetation most effectively stabilise stream banks?, *Geomorphology*, 23, 55–75, 1998.

Barsoum N., and F. M. R. Hughes, Regeneration response of black poplar to changing river levels, in *Hydrology in a Changing Environment*, edited by H. Wheater, and C. Kirkby, pp. 397–412, Wiley, Chichester, 1998.

Bendix J., and C. R. Hupp, Hydrological and geomorphological impacts on riparian plant communities, *Hydrological Processes*, 14, 2977–2990, 2000.

Bornette G., C. Amoros, and N. Lamoroux, Aquatic plant diversity in riverine wetlands: the role of connectivity, *Freshwater Biology*, 39, 267–283, 1998.

Cellot B., F. Mouillot, and C. P. Henry, Flood drift and propagule bank of aquatic macrophytes in a riverine wetland, *Journal of Vegetation Science*, 5, 631–640, 1998.

Danvind M., and C. Nilsson, Seed floating ability and distribution of alpine plants along a Swedish river, *Journal of Vegetation Science*, 8, 271–276, 1997.

Edwards P. J., J. Kollmann, A. M. Gurnell, G. E. Petts, K. Tockner, and J. V. Ward, A conceptual model of vegetation dynamics on gravel bars of a large Alpine river, *Wetlands Ecology and Management*, 7, 141–153, 1999.

Friedman J. M., W. R. Osterkamp and W. R. Lewis, The role of vegetation and bed-level fluctuations in the process of channel narrowing, *Geomorphology*, 14, 341–351, 1996.

Goodson J. M., A. M. Gurnell, P. G. Angold, and I. P. Morrissey, Riparian seed banks: structure, process and implications for riparian management, *Progress in Physical Geography*, 25, 301–325, 2001.

Goodson J. M., A. M. Gurnell, P. G. Angold, and I. P. Morrissey, Evidence for hydrochory and the deposition of viable seeds within winter flow-deposited sediments: the River Dove, Derbyshire, UK. *River Research and Applications*, in press, 2003.

Gurnell A. M., G. E. Petts, D. M. Hannah, B. P. G. Smith, P. J. Edwards, J. Kollmann, J. V. Ward, and K. Tockner, Wood storage within the active zone of a large European gravel-bed river, *Geomorphology*, 34, 55–72, 2000.

Gurnell A. M., G. E. Petts, D. M. Hannah, B. P. G. Smith, P. J. Edwards, J. Kollmann, J. V. Ward, and K. Tockner, Riparian vegetation and island formation along the gravel-bed Fiume Tagliamento, *Earth Surface Processes and Landforms*, 26, 31–62, 2001.

Gurnell, A. M., H. Piégay, F. J. Swanson, and S. V. Gregory, Large wood and fluvial processes, *Freshwater Biology*, 47, 601–619, 2002.

Gurnell, A. M., and G. E. Petts, Island-dominated landscapes of large floodplain rivers, a European perspective, *Freshwater Biology*, 47, 581–600, 2002.

Hill, M. O., and H. G. Gauch, Detrended correspondence analysis, an improved ordination technique, *Vegetatio*, 42, 47–58, 1980.

Hupp C. R., Hydrology, geomorphology and vegetation of coastal plain rivers in the south-eastern USA, *Hydrological Processes*, 14, 2991–3010, 2000.

Hupp, C. R., and W. R. Osterkamp, Riparian vegetation and fluvial geomorphic processes, *Geomorphology*, 14, 277-295, 1996.

Hupp C. R., and A. Simon, Bank accretion and development of vegetated depositional surfaces along modified alluvial channels, *Geomorphology*, 4, 111–124, 1991.

Johnson W. C., Equilibrium response of riparian vegetation to flow regulation in the Platte River, Nebraska. *Regulated Rivers*, 13, 403–415, 1997.

Johnson W. C., Tree recruitment and survival in rivers: influence of hydrological processes. *Hydrological Processes,* 14, 3051–3074, 2000.

Jongman, R. H. G., C. J. F. Ter Braak, and O. F. R. Van Tongeren, (Eds.), *Data analysis in community and landscape ecology*, 299 pp., Cambridge University Press, Cambridge, 1995.

Karrenberg, S., P. J. Edwards, and J. Kollmann, The life history of Saliceae living in the active zone of flood plains, *Freshwater Biology*, 47, 733–748, 2002.

Kollmann, J., M. Vieli, P. J. Edwards, K. Tockner, K., and J. V. Ward, Interactions between vegetation development and island formation in the Alpine river Tagliamento, *Applied Vegetation Science*, 2, 25–36, 1999.

Kubitzki, K., and A. Ziburski, Seed dispersal in flood-plain forests of Amazonia, *Biotropica*, 26, 30–43, 1994.

McKenney, R., R. B. Jacobsen, and R. C. Wertheimer, Woody vegetation and channel morphogeneisis in low-gradient, gravel-bed streams in the Ozark Plateaus, Missouri and Arkansas, *Geomorphology*, 13, 175–198, 1995.

Nanson, G. C., and A. D. Knighton, Anabranching rivers: their cause, character and classification, *Earth Surface Processes and Landforms*, 21, 217–239, 1996.

Petts G. E., A. M. Gurnell, A. J. Gerrard, D. M. Hannah, B. Hansford, I. P. Morrisey, P. J. Edwards, J. Kollmann, J. V. Ward, K. Tockner, and B. P. G. Smith, Longitudinal variations in exposed riverine sediments: a context for the ecology of the Fiume Tagliamento, Italy, *Aquatic Conservation: Marine and Freshwater Ecosystems*, 10, 249–266, 2000.

Piégay H., G. Bornette, A. Citterio, E. Herouin, B. Moulin, and C. Statiotis, Channel instability as a control on silting dynamics and vegetation patterns within perifluvial aquatic zones, *Hydrological Processes*, *14*, 3011–3029, 2000.

Rao, C. R., The use and interpretation of principal components analysis in applied research, *Sankhya A 26*: 329–358, 1964.

Robertson, K. M., and C. K. Augspurger, Geomorphic processes and spatial patterns of primary forest succession on the Bogue Chitto River, USA, *Journal of Ecology*, 87, 1052–1063, 1999.

Rowntree, K. M., and E. S. J. Dollar, Vegetation controls on channel stability in the Bell River, Eastern Cape, South Africa, *Earth Surface Processes and Landforms*, 24, 127–134, 1999.

Salo J., R. Kalliola, I. Hakkinen., Y. Makinen, P. Niemela, M. Puhakka, and P. D. Coley, River dynamics and the diversity of Amazon lowland forest, *Nature*, 322, 254–258, 1986.

Stainforth R. J., and P. B. Cavers, An experimental study of water dispersal on *Polygonum* spp, *Canadian Journal of Botany*, *54*, 2587–2597, 1976.

Steiger, J., and A. M. Gurnell, Spatial hydrogeomorphological influences on sediment and nutrient deposition in riparian zones: observations from the Garonne River, France, *Geomorphology*, in press, 2002.

Steiger J., A. M. Gurnell, and G. E. Petts, Sediment deposition within the margins of a reach of the River Severn, UK, *Regulated Rivers: Research and Management* 17, 441–458, 2001.

Tabacchi E., L. Lambs, H. Guilloy, A.-M. Planty-Tabacchi, E. Muller, and H. Décamps, Impacts of riparian vegetation on hydrological processes, *Hydrological Processes*, 14, 2959–2976, 2000.

Ter Braak, C. J. F., and P. Smilauer, *CANOCO Reference Manual and User's Guide to Canoco for Windows: Software for Canonical Community Ordination (version 4)*. Microcomputer Power, Ithaca, NY, 352pp, 1998.

Tooth, S., and G. C. Nanson, The role of vegetation in the formation of anabranching channels in an ephemeral river, Northern plains, arid central Australia, *Hydrological Processes*, 14, 3099–3117, 2000.

Van Splunder, I.., *Floodplain forest recovery: softwood forest development in relation to hydrology, riverbank morphology and management* PhD thesis, Catholic University of Nijmegen, Nijmegen, The Netherlands, RIZA report 98.001, 1998.

Ward J. V., K. Tockner, P. J. Edwards, J. Kollmann, G. Bretschko, A. M. Gurnell, G. E. Petts, and B. Rossaro, A reference river system for the Alps: the Fiume Tagliamento, *Regulated Rivers: Research and Management*, 15, 63-75, 1999.

P. G. Angold, I. P. Morrissey, and G. E. Petts, School of Geography, Earth and Environmental Sciences, University of Birmingham, Birmingham, UK.

A. M. Gurnell, and J. M. Goodson, Department of Geography, King's College London, London, UK.

J. Steiger, Departement de Géographie, Université d'Angers, Angers, France.

Floodplain Stabilization by Woody Riparian Vegetation During an Extreme Flood

Eleanor R. Griffin and J. Dungan Smith

U.S. Geological Survey, Boulder, Colorado

Dense woody riparian vegetation acts to reduce flow velocities and boundary shear stresses on floodplain surfaces during deep overbank flows. Where woody vegetation is sparse and the slope sufficiently steep, the floodplain surface is vulnerable to high rates of erosion during floods. Once erosion of a floodplain surface begins, it proceeds rapidly and can lead to floodplain unraveling (transformation of a narrow, single-threaded stream and its floodplain to a much wider braided or partially-braided stream). On June 16, 1965, an extreme flood occurred along East Plum Creek and its tributaries, resulting in the unraveling of East Plum Creek and then Plum Creek to its confluence with the South Platte River. Large-scale aerial photographs taken two days after the flood show that Carpenter Creek, a headwater tributary of East Plum Creek, also unraveled over much of its length. A tributary of Carpenter Creek that was also subjected to a deep overbank flow, but where the floodplain was protected by dense shrub willows, did not unravel. Boundary shear stresses on the floodplain surfaces were calculated for each of four sites using a process-based model that includes drag on shrub willows. The four sites cover the transition from locations where the floodplain surface and vegetation remained intact to locations where erosion resulted in a much wider channel and almost all vegetation within the flood channel was removed. The results indicate that the critical shear stress for erosion of a grass and sparse shrub-covered floodplain is between 0.4 and 2.0 N/m^2, and that dense shrubs reduced the boundary shear stresses on floodplain surfaces by up to three orders of magnitude.

1. INTRODUCTION

Erosion of floodplains can be caused by both lateral bank cutting during near-bankfull or slightly overbank flows and by direct erosion of the floodplain surface during deep overbank flows. Our work addresses the second process and quantifies the effects of woody riparian vegetation in reducing floodplain surface erosion by imposing considerable form drag on the flood flow. Although floodplain surface erosion only occurs during overbank flows, once it begins, it can proceed rapidly as flow depths increase in channels that form down-valley on the floodplain. As flow depths in floodplain channels increase, the boundary shear stresses on their beds rise, increasing the sediment transport rate and leading to increasing rates of channel widening and deepening [*Smith and Griffin*, 2002]. This process leads to undercutting and removal of vegetation on the floodplain surface. Sediment removed from the floodplain surface is deposited in the channel immediately downstream from the site of erosion, elevating the bed and forcing even more water overbank. If overbank flow is of sufficient duration or magnitude, rapid removal of the floodplain surface can occur, with sediment redistributed downstream creating a wider, shallower, braided or partially-braided stream. This process is referred to here as floodplain unrav-

Riparian Vegetation and Fluvial Geomorphology
Water Science and Application 8
This paper is not subject to U.S. copyright. Published in 2004 by the American Geophysical Union
10.1029/008WSA16

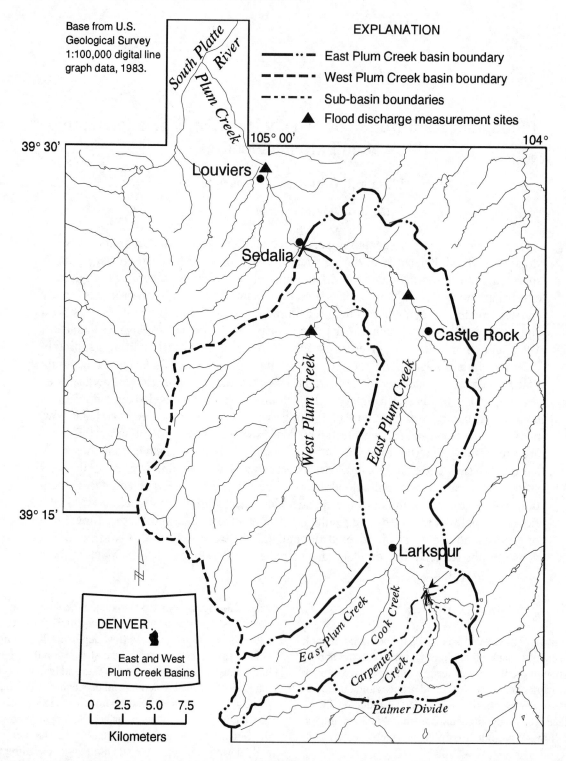

Figure 1. Map showing the location of the study area. Model calculations were made for flow over the floodplain in the vicinity of the confluence of three headwater tributaries of East Plum Creek at the location indicated by the arrow. Basin area for the unnamed tributary immediately east of Carpenter Creek is similar to that for Carpenter Creek (22.9 and 24.4 km^2, respectively). The smaller basin to the northeast has an area of 8.6 km^2, and that unnamed tributary flows into the first about 400 m upstream from its confluence with Carpenter Creek.

eling.

On June 16, 1965, several intense convective rainstorms occurred between Denver, Colorado, and the Palmer Divide to the south, directly over the East Plum Creek basin. The Palmer Divide extends east from the foothills of the Rocky Mountains to form the southern boundary of the Plum Creek basin (Figure 1). The intense, sustained rainstorms caused extensive flooding along East Plum Creek from Larkspur to Sedalia, along Plum Creek from Sedalia to its mouth at the South Platte River, and along the South Platte River through Denver. Bridges were destroyed from Larkspur through Sedalia, and substantial structural damage occurred along the floodplain of the South Platte River through Denver. Peak discharge of the South Platte River was almost twice the previous maximum in the 67-year record [*Matthai*, 1969], and most of the flow came from the Plum Creek basin. The floods caused a geomorphic transformation from single-thread, meandering streams to much wider, shallower braided streams along both East Plum Creek and Plum Creek. Effects of the flood on Plum Creek near Louviers were studied in detail by *Osterkamp and Costa* [1987], and *Griffin and Smith* [2001] examined the effects of the flood along East Plum Creek, from Larkspur to Sedalia. In this paper, predictions based on boundary shear stresses calculated using a model for flow over shrub-covered floodplains [*Smith*, 2001; this volume] are compared to documented effects of the flood along two headwater tributaries of East Plum Creek – Carpenter Creek and the first of two unnamed tributaries of Carpenter Creek (Figure 1). Both are perennial streams with low base flows, and drainage areas upstream from their confluence are similar (24.4 and 22.9 km^2).

Along East Plum Creek, downstream from Larkspur, the magnitude of the 1965 flood was so great that resistance from woody vegetation along the channel was inconsequential in protecting the floodplain surface (Figure 2). Along Carpenter Creek, East Plum Creek, and Plum Creek, the unraveling process led to the development of a hydraulic geometry in equilibrium with the peak flood discharge in less than three hours [*Griffin and Smith*, 2001]. Within that short time, the channels of these streams widened by an order of magnitude. For example, channel width of East Plum Creek in a 4.9 km reach between Larkspur and Castle Rock increased from an average of 8.3 m before the flood to an average of 105 m two days after the flood. Along the unnamed tributary of Carpenter Creek, upstream from its confluence with a second tributary, the shrub-protected floodplain surface remained intact and shrubs remained rooted on the floodplain.

Previous studies have examined the influence of hydrogeomorphic processes on bottomland plant distributions [e.g., *Sigafoos*, 1964; *Hupp and Osterkamp*, 1985, 1996; *Friedman et al.*, 1996] and recognized that distributions of bottomland

plant species are controlled in part by their susceptibility to destructive flooding. These authors observed that dense woody vegetation reduces mean flow velocities and increases floodplain roughness, but did not quantify the effects of woody vegetation on overbank flows.

Burkham [1976] estimated effective roughness produced by flow through dense woody vegetation by comparing hydraulic properties for floods of similar magnitude that occurred before and after large-scale removal of floodplain shrubs along the Gila River, Arizona. In that case, other factors affecting flood flow properties, such as changes in the channel and floodplain topography after the first flood and shrub density prior to removal, were not known, so these results cannot be used to predict the effects of woody vegetation on deep overbank flows elsewhere. Other researchers have combined numerical methods with experimental data to determine the flow resistance of vertical cylinders extending through the flow depth, representing flow over vegetatively-roughened floodplains [e.g., *Pasche and Rouvé*, 1985]. However, methods that rely on empirically-derived constants to determine the roughness coefficient [*Pasche and Rouvé*, 1985] or that are applicable for specific limited distributions of shrubs or trees [e.g., *Li and Shen*, 1973] are limited in their predictive capabilities.

The purpose of our work is to quantify the effects of woody vegetation on overbank flows using the physical characteristics and distribution of shrubs to calculate the reduction in mean flow velocity and boundary shear stress resulting from form drag on shrub stems, branches and leaves. Our objective is to be able to predict under what flow conditions unraveling will occur, given the physical characteristics and density of woody vegetation on the floodplain. In this paper, we apply a predictive, process-based model for flow over shrub-covered floodplains developed by *Smith* [2001; this volume] to a test case in which effects of a large flood were documented in aerial photographs taken two days after the flood. The study reach includes areas where the floodplain and woody vegetation remained intact after the flood, areas where some, but not all, shrubs were removed, and an area where nearly all floodplain vegetation and pre-flood topography were removed. Calculated values of actual boundary shear stresses on the floodplain surfaces, which can be used to predict whether or not the floodplain surface will erode under given flow conditions, are compared to the documented effects of the flood. Results from application of the overbank flow model to this case, where the effects of the flood on the floodplain and its vegetation are known, serve as a test of the model.

1.1. Geomorphic Setting

East Plum Creek is a sand-bed stream that flows north from the Palmer Divide to its confluence with West Plum Creek at

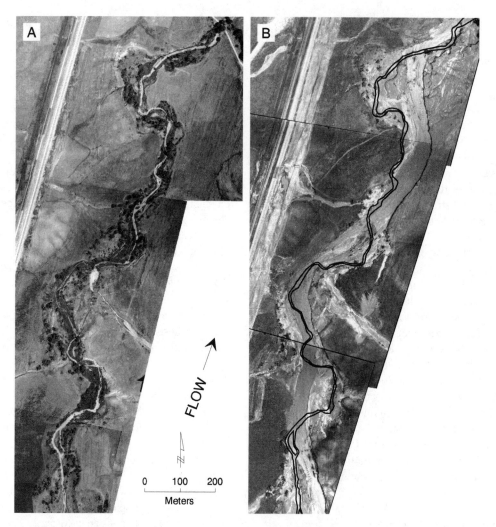

Figure 2. Aerial photographs showing the effects of the 1965 flood in a 2-km reach of East Plum Creek south of Castle Rock, demonstrating the concept of unraveling. Comparison of the 1956 (A) and 1965 (B) photographs shows that the flood cut a new channel straight through the floodplain surface on the inside of the former meander bends and redistributed the sediment, filling in the pre-flood channel and creating a much wider braided system. As a result of the flood, channel width in this 2-km reach increased from an average of 7.4 m in May 1956 to an average of 96 m on June 18, 1965. The edges of the 1956 channel (solid lines) are drawn over the flood photographs. Sinuosity in this reach decreased from 1.25 in 1956 to 1.05 in 1965, after the flood. Interstate Highway 25, which was under construction at the time of the flood, is visible along the upper left edge of the 1965 photographs.

Sedalia, south of Denver, Colorado (Figure 1). The stream cuts through Upper Cretaceous and Paleocene sedimentary rocks (sandstone, siltstone, claystone and conglomerates of the Dawson Formation) throughout most of its basin area. Channels in a small segment of the basin in the southwest corner (about 15% of the total basin area), where terrain is steep (Figure 3), cut Pikes Peak Granite. Upper Pleistocene alluvial terraces extend along both banks of East Plum Creek east of the foothills and along much of Carpenter Creek [*Trimble and*

Machette, 1979a,b]. The terraces are primarily made up of Broadway Alluvium (7.6 to 12.2 m above the present stream level) and Louviers Alluvium (up to 21 m above the present stream level). Both deposits include bouldery, cobbly, poorly-sorted gravel as well as well-stratified gravel, sand, silt, and clay. Terrain is relatively steep east of the foothills, both in the stream channels and on adjacent hillslopes. Much of the basin area is non-irrigated rangeland covered by sparse grasses

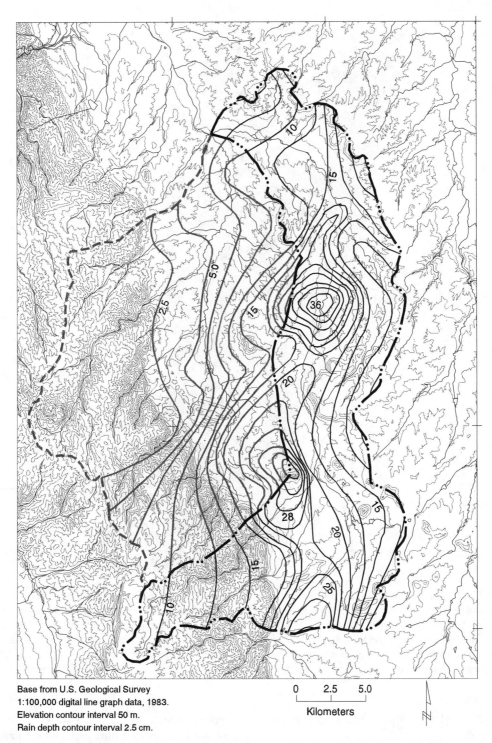

Base from U.S. Geological Survey
1:100,000 digital line graph data, 1983.
Elevation contour interval 50 m.
Rain depth contour interval 2.5 cm.

0   2.5   5.0
Kilometers

Figure 3. Topographic contours (thin gray lines), streams (thin black lines) and rain depth contours (heavy lines) in the area of the Plum Creek basin. Total rain-depth contours are from an isohyetal map of the storms of June 16, 1965 [Matthai, 1969], with the black lines showing rain depths over the East Plum Creek basin and gray lines showing rain depths over the West Plum Creek basin. Total rain depth at the centers of the three intense cells were about 360 mm, and the contours are in 25-mm increments (several are labeled). The southern-most intense rain cell was centered over the Palmer Divide at the upper end of the Carpenter Creek basin.

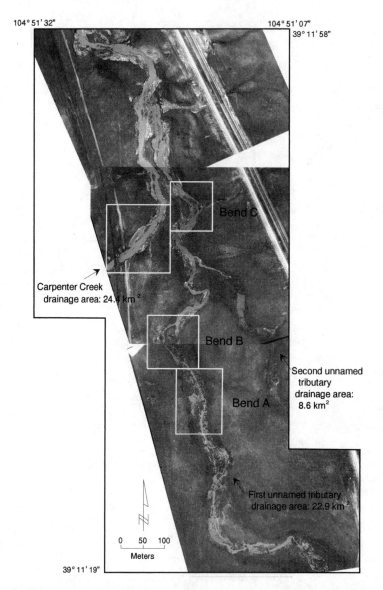

Figure 4. Series of aerial photographs taken on June 18, 1965, two days after the flood, showing the confluence of Carpenter Creek and the first and second unnamed tributaries. The overbank flow model was applied to conditions at bends A, B, and C on the first unnamed tributary and to Carpenter Creek upstream from its confluence with the unnamed tributary. Although the flood flow was estimated to be about 3 m deep over the first unnamed tributary upstream from its confluence with the second, smaller tributary, this stream did not unravel until downstream from the confluence with the smaller tributary. Carpenter Creek unraveled at a point upstream from the railroad bridge on the left side of the middle photograph.

common in this semi-arid environment, and, during intense storms, runoff is high.

1.2. Description of the Storms

The storms that caused the floods of June 16, 1965 were the result of instability caused by warm, moist air at low levels that moved into the area from the south and encountered a cooler, dryer air mass at high levels to the west. This instability was enhanced by terrain lifting and daytime heating [*Reidel et. al.*, 1969]. The combination of high moisture and high instability resulted in "an extreme release of convective rainfall" [*Reidel et. al.*, 1969]. The resulting storms were nearly stationary bursts that lasted about 6 hours, between about 1200 and 1800 hours MDT. Total rain depth at the centers of three intense cells was about 360 mm, compared to an average total

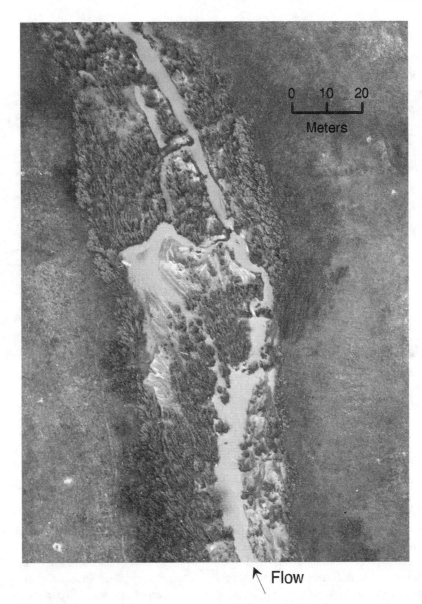

↑ Flow

Figure 5. Aerial photograph taken June 18, 1965 showing details of bend A on the first unnamed tributary two days after the flood. The floodplain remained intact and dense shrubs remained rooted despite being subjected to about a 3-m deep flood flow. Lines of shrubs nearly perpendicular to the flow indicate the presence of beaver dams. Several of these also remained intact, with only short breaches resulting from the flood flow.

annual precipitation at Castle Rock (Figure 1) of 430 mm, and the mean total rain depth over the East Plum Creek basin was 170 mm. Two of the intense cells formed almost directly over the East Plum Creek channel (Figure 3), and the storms moved slowly from south to north, following the channel downstream. The third cell formed over the Palmer Divide, at the upper end of the Carpenter Creek basin (Figure 1). Based on the isohyetal map of the storms [*Matthai*, 1969], average rain depths over the basins of Carpenter Creek and the

unnamed tributary were 230 mm and 170 mm, respectively. An average of 150 mm of rain fell over the basin of a smaller, second unnamed tributary that flows into the first tributary about 400 m upstream from its confluence with Carpenter Creek.

Only one official weather station was located in the East Plum Creek basin (at Castle Rock; Figure 1), where rain depths of 11 and 13 mm were recorded from noon to 1300 and from 1300 to 1400 hours on June 16, 1965. Hourly rain depths

Figure 6. Photograph of the first unnamed tributary, taken in February 2000, looking downstream at a point near the southern end of the bottom photograph in Figure 4. Dense sandbar willows and thick grasses cover the floodplain. The willows in this area were 2 to 3 m high. (A small pack and water bottle are shown at the lower right for scale.)

measured at this station between 1400 and 2100 hours were lumped (154 mm for the 7-hour period). Total rainfall measured at Castle Rock on that date (nearly 180 mm) is more than twice the depth of any other daily total recorded between 1948 and 2001 at that station [*Western Regional Climate Center*, 2003]. The flood destroyed the only operating streamflow-gaging station in the Plum Creek basin, on Plum Creek near Louviers. As a result, precise timing of rain and flood events is not known. Eyewitness accounts of the events provide the limited available information on their timing [*Matthai*, 1969]. Flood peaks were reported to have arrived at West Plum Creek near Sedalia about 1500 hours and at East Plum Creek, just north of Castle Rock, about 1600 hours on June 16, 1965. The flood peak along East Plum Creek near Castle Rock was estimated to be 44 times a 50-year flood [*Matthai*, 1969], yet the valley floor along Plum Creek near Louviers was inundated for only about 2.5 hours [*Osterkamp and Costa*, 1987].

1.3. Available Data

East Plum Creek and some of its tributaries flow along major highways from the Palmer Divide to Sedalia, and the Colorado Department of Transportation has obtained large-scale aerial photographs covering East Plum Creek and sections of its trib-

utaries at various times over the past several decades. Large-scale (about 1: 6,000) aerial photographs taken in May 1956, showing East Plum Creek from Larkspur to Sedalia, were examined to determine the condition of the East Plum Creek channel and floodplain before the flood. Aerial photographs with scales ranging from about 1:2,500 to 1:4,000 taken on June 18, 1965, two days after the flood, showed the effects of the flood along East Plum Creek and some of its tributaries. The flood photographs cover East Plum Creek from Larkspur to Sedalia as well as Carpenter Creek and parts of its tributaries (Figure 4) beginning in the vicinity of the arrow in Figure 1. The photographs were scanned at 600 dots per inch, then rectified to map coordinates using a geographic information system (GIS).

Field observations in February 2000 provided information on the physical characteristics of shrubs along both Carpenter Creek and its main tributary as well as bankfull channel dimensions of both streams. Because much of the floodplain and woody vegetation along the first and second unnamed tributaries remained intact after the 1965 flood, some of the typical shrub dimensions could be estimated from the aerial photographs. In the area of deep overbank flow along the first unnamed tributary, the shrubs still lay flat two days after the flood, and shrub height, canopy diameter, and shrub clump spacing were estimated from the rectified photographs (Figure 5). The average height, shrub canopy width, and shrub

Figure 7. Photograph of Carpenter Creek, taken in February 2000, looking upstream toward the railroad bridge seen on the left side of the middle photograph in Figure 4. Sparse, small willows are distributed over the floodplain, which is covered by grazed grasses. Shrub canopy was estimated to cover 10 to 15 percent of the floodplain surface. Terraces about 6 m high are present on both sides of the floodplain, and the bankfull channel is about 2.5-m wide and 0.6-m deep. On June 18, 1965, the flood-channel bed covered the entire width of the floodplain, creating a smooth bed about 5 times wider than the channel seen here.

clump spacing in the 1965 photographs appear to be consistent with the shrub dimensions observed in 2000. Therefore, additional shrub characteristics observed in 2000, including typical stem diameter and number of stems per shrub clump, were applied in the model of the 1965 flood.

Rainfall data in the form of an isohyetal map for the storms of June 16, 1965 were obtained from *Matthai* [1969]. Indirect flood peak discharge measurements were made on East Plum Creek near Castle Rock, on West Plum Creek near Sedalia, and on Plum Creek near Louviers (USGS, Colorado District, written communication, 2000). U.S. Geological Survey digital map data (1:100,000 and 1:24,000-scale) were used to measure drainage areas, estimate local valley cross-sections, and determine reach-averaged stream and floodplain slopes. Flood peak discharges along Carpenter Creek and at bends A, B, and C (Figure 4) along the first unnamed tributary were estimated from the available data. These estimates were based on: 1) total rain depths over upstream drainage areas; 2) the indirect measurement of flood peak discharge on East Plum Creek near Castle Rock; and 3) an infiltration loss calculated from a comparison of total rain volumes and flood peak discharges for East and West Plum Creeks [*Griffin and Smith*, 2001].

2. METHODS

2.1. Flow Model

The *Smith* floodplain flow model [2001; this volume] is a generalized model designed to accept as inputs physical parameters for particular locations including floodplain topographic roughness, floodplain slope, and physical parameters of woody vegetation on the floodplain. Density and physical characteristics of the woody vegetation are used to calculate the net drag and resulting boundary shear stress on a large area of floodplain. As applied here, the model treats flow on the floodplain as steady and horizontally uniform. Flow accelerations can be added to the model, but this was not done for these applications because of a lack of sufficient data on floodplain topography and shrub distribution at the time of the flood. Instead, estimates of average floodplain roughness were used along with the assumption of uniformly distributed shrubs. These assumptions enable application of the model to provide a good first estimate of the fluid mechanical effects of woody vegetation on deep overbank flows.

The floodplain flow model is applicable to situations where the shrubs have rigid stems extending all the way through the

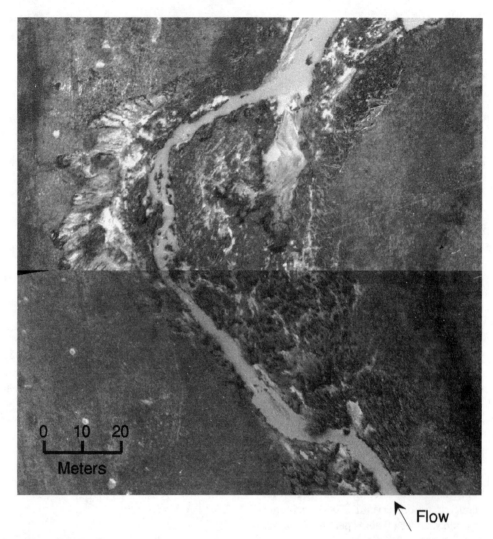

Figure 8. Aerial photograph of bend B on the first unnamed tributary, taken two days after the flood, where the floodplain also remained intact, even though the shrub canopy was somewhat less dense here than at bend A. The unvegetated area in the downstream half of the bend was likely a former beaver pond, with the dam at the downstream end of the bend removed by the flood flow. This bend is upstream from the confluence with the second, smaller tributary.

flow, or where shrubs with flexible stems bend over so that much of the stem length is generally parallel to the ground in a deep overbank flow. The first situation can be modeled using a single flow layer, with the cross-sectional area of the shrubs affecting the flow over the entire depth [*Smith and Griffin*, 2002; Smith, this volume]. The shrubs along Carpenter Creek and its tributaries fall into the second category, in which form drag produced by shrubs bent over during an overbank flow affects the flow in three distinct layers [*Smith*, 2001]. A low-velocity zone is produced beneath the layer of bent-over stems and branches, where flow is through a field of vertical stems that penetrate the entire layer. The middle layer consists of flow through leaves, branches, and nearly horizontal stems. In the bottom and middle layers,

spatially varying skin friction and form drag forces are subtracted from the shear stress per unit volume. The top layer is normal channel flow over the top of the layer of stems, branches and leaves, with no drag forces. A substantial shear is supported across the middle layer, and for the middle and upper layers, shear velocities are calculated from the shear stresses at the bases of the layers.

The three-layer model [*Smith*, 2001] was applied to calculate the reduction in the boundary shear stress resulting from form drag on shrubs in varying densities across the floodplains of Carpenter Creek and its tributaries. Form drag on dense shrubs also increases the depth of flow. Shear stress resulting from the depth-slope product, without shrubs present, was calculated

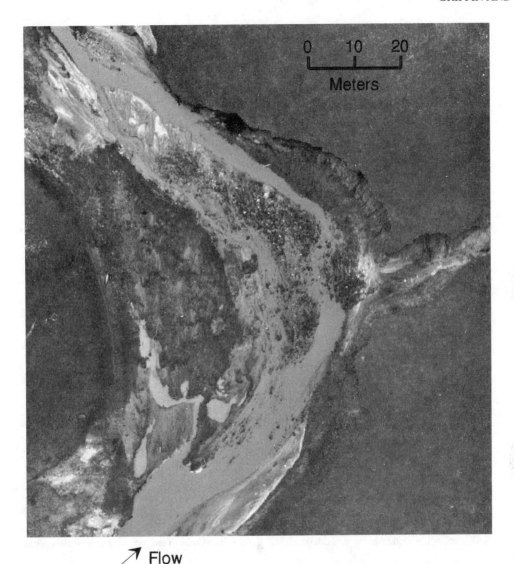

↗ Flow

Figure 9. Aerial photograph of bend C on the first unnamed tributary downstream from the confluence with the second unnamed tributary and about 400 m upstream from its confluence with Carpenter Creek, taken two days after the flood. At this location, the flood eroded a much wider channel through the bend, but did not remove all shrubs lying on the former floodplain surface.

using an estimated flow depth, which was obtained by assuming flow was about critical and iterating depth until the Froude number was 1 and discharge matched the flood peak (253 m³/s at bend A, for example).

We examined three sites along the first unnamed tributary of Carpenter Creek, beginning at bend A (Figure 4), where the floodplain remained intact after the flood, and extending downstream to a point where the floodplain unraveled. We also examined a site along Carpenter Creek, upstream from its confluence with the tributary (Figure 4). As stated above, the shrubs were modeled as if their distribution was uniform in

the horizontal. At each site, the estimated flood peak discharge [*Griffin and Smith*, 2001], average flood flow depth, average floodplain gradient, and typical floodplain roughness and shrub parameters were used to calculate the resulting boundary shear stress on the floodplain surface.

2.2. Floodplain Shrubs

Sandbar willow (*Salix exigua*), the dominant shrub, is present in varying densities along the streams in the East Plum Creek basin, including the headwater tributaries. Although it

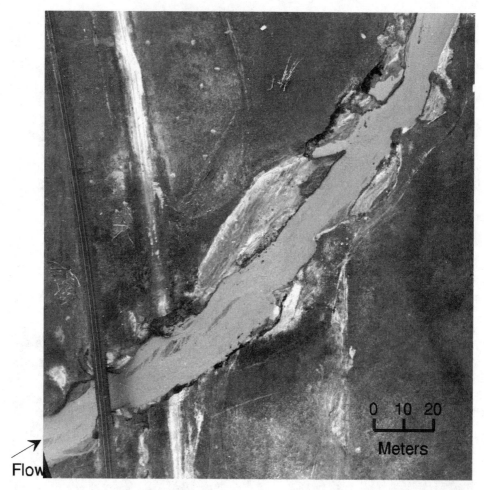

Figure 10. Aerial photograph showing the 200-m reach of Carpenter Creek upstream from its confluence with the tributary, two days after the flood. The water surface appears to have been nearly flat on July 18, 1965, indicating that the flood-channel bed was also nearly flat with an average width of 12.7 m. Only a few shrubs in the path of the flood flow remain rooted.

spreads by rhizomes, stems tend to be clustered in groups with characteristic stem numbers and spacing. Scaling relations were developed for the willows so that important fluid-mechanical properties could be estimated from mean stem diameters, stem spacing, stem group diameters, and stem group spacing. Stem group spacing and diameter are related to shrub-canopy spacing and diameter, which were determined from the aerial photographs. Two major age groups of sandbar willows were present on the floodplain during a field visit in February 2000. Average stem diameter in the older group, consisting of apparently dead shrubs, was about 0.04 m, and the shrubs had grown to a height of about 4.5 m. The dead stems were surrounded by smaller healthy stems with an average stem diameter of about 0.03 m near the ground and a height of about 100 times the average stem diameter (3 m; Figure 6). Where the shrubs are dense, they grow in clumps with about 36 stems per clump. The average clump diameter at the ground is about 0.5 m, and the average canopy size is 1 to 2 m, depending on the density of the shrubs. Average canopy size for a clump estimated from the post-flood aerial photographs was essentially the same as that for the younger shrubs in 2000, so the characteristics of that group were used in the overbank flow model.

Canopy diameter is typically three times the clump diameter near the ground (0.5 m) when the shrubs are spaced far enough apart so that their canopies do not touch. Stems were distributed uniformly under a canopy 1.5 m in diameter for the model application. Where the shrubs are dense, there are enough stems over a large area that they are modeled as if uniformly distributed, because flow cannot accelerate significantly between stem groups. Where the shrubs are sparse (Figure 7), they do not grow in clumps and are modeled as if uniformly spaced over the floodplain because their actual distribution at the time of the flood is unknown. Using these values, canopy cover was translated to an average stem spacing on the floodplain surface.

Shrub densities at each site at the start of the 1965 flood were estimated from the characteristics observed in the aerial photographs and during the field visit of 2000. At bend A along the first unnamed tributary (Figure 5), the floodplain and vegetation remained fully intact. Dense shrubs on the floodplain were still bent over parallel to the ground two days after the flood, and pre-flood canopy coverage was estimated to be about 64 percent of the floodplain area. This translates to an average stem spacing on the floodplain of 0.27 m. Calculations of boundary shear stress on the floodplain surface were made using this spacing as an input to the flow model. In the flood photograph, lines of shrubs generally perpendicular to the flow with open water upstream indicate the presence of beaver dams. The presence of beaver dams is important because the dams produce wetlands that support the rapid growth of willows and herbaceous understory vegetation. During overbank flows, the dense shrubs and grasses provide high resistance to erosion.

The extent of overbank flow at bend A is indicated by the flood-trained shrubs with their stems lying in the down-valley direction (Figure 5). The shrubs extend from the edge of the terrace on the left side of the stream to the edge of the terrace on the right side, and the longest beaver dam extends across most of that length. In order to examine the hydraulic effects of the beaver dams, they were modeled as large bedforms (dunes) using the method of *Smith and McLean* [1977]. The bedform height used in the calculation was the height of the beaver dam, and the bedform length was the length of the pond in the up-valley direction.

Most of the floodplain surface at bend B (Figure 8) also remained intact, even though the shrubs were somewhat less dense than at bend A. Shrubs are visible on the upstream side of the floodplain surface within the bend as well as lying in the flood channel, still rooted. Pre-flood canopy coverage was estimated to have been about 45 percent over the entire floodplain surface at bend B. This includes the area of a former beaver pond located on the downstream side of the bend and upstream from a dam that washed out during the flood. Canopy cover affecting the flow on the upstream half of the floodplain at this bend is about 55 percent, which leads to an average stem spacing of 0.30 m. Canopy density on the upstream half of the bend was used in the model application at this location.

At bend C (Figure 9), downstream from the confluence of the first and second unnamed tributaries, much of the floodplain surface along the main path of the flood flow eroded. Only a few shrubs remained rooted on the former floodplain

surface. Based on the density of shrubs still visible on the higher surface on the inside of bend C, pre-flood canopy coverage was estimated to have been 21 percent at that site. This translates to an average stem spacing of 0.48 m. The vegetation appears to have been similar to that observed along Carpenter Creek in February 2000 (Figure 7), with grazed rangeland grasses and sparse shrubs.

Nearly all pre-flood vegetation along the floodplain of Carpenter Creek near its confluence with the unnamed tributary was removed by the flood flow (Figure 10). Therefore, pre-flood shrub density along this reach was estimated from the post-flood conditions nearby on the unnamed tributary at bend C (Figure 9) and the conditions observed along Carpenter Creek in February 2000 (Figure 7). Shrub canopy density was estimated to have been about 13 percent, with average stem spacing calculated to be 0.60 m. Stems were assumed to be distributed uniformly over the floodplain.

2.3. Conversion of Percent Canopy Cover to Average Stem Spacing

The physical characteristics of the floodplain shrubs were used to convert the percent of the floodplain area covered by shrub canopy, as seen in the aerial photographs, to an average stem spacing on the floodplain surface using the method developed by Smith, described below. The average stem spacing and stem dimensions then become inputs to the floodplain flow model. Because shrub stems are assumed to be distributed uniformly over the floodplain surface, the average stem spacing can be determined by dividing the canopy area per shrub by the number of stems per shrub (n_s) and the fraction of floodplain area covered by shrub canopy (F_{cc}). For the model application, this calculation is simplified by converting the area of shrub canopy (generally a circle) to a square with an equivalent area. A square with the same area as a circle with diameter D has a side of length L, where $L = D\sqrt{\pi/4}$. The equation for average stem spacing, λ_{ss}, is then:

$$\lambda_{ss} = \frac{\sqrt{\pi/4} \cdot D_c}{\sqrt{n_s} \cdot \sqrt{F_{cc}}}$$

where D_c is the average diameter of the shrub canopy.

When the shrub clumps are distributed so that their canopies are just touching, floodplain canopy coverage is about 79 percent (the difference between the area of a circle and the area of a square with D = L). When the canopy diameter is 1.5 m, the number of stems per shrub is 36, and the fraction of floodplain area covered by shrub canopies is 1, the average stem spacing is 0.22 m. An average stem spacing of about 7.3 times the typical stem diameter (0.03 m) is then the maximum spacing that leads to 100 percent canopy coverage of the floodplain.

As the shrubs get closer, their canopies intermingle and average stem spacing decreases while the canopy coverage remains 100 percent. Although average stem spacing can be less than 7.3 times the average stem diameter, the decreased spacing cannot be detected from the canopy cover visible in the aerial photographs. Therefore, the minimum average stem spacing that can be estimated from the aerial photographs is about 0.22 m. The most dense shrubs observed during the site visit were spaced about 1.0 m apart (center-to-center) and had an average stem spacing of only 0.18 m, or 6 times the average stem diameter.

3. RESULTS

Results from the application of the *Smith* [2001] floodplain flow model are presented in Table 1. At each location, the average floodplain gradient (0.0103 at all sites) determined from the topographic data was used. Flood peak discharges estimated by *Griffin and Smith* [2001] and flood-flow widths and average stem spacing derived from the aerial photographs were also model inputs. Flood-flow velocities were reduced by form drag on the vertical stems and branches in the bottom flow layer and by flow through the nearly horizontal stems, branches, and leaves in the middle layer. Shear stress and velocity were matched at the boundaries of each flow layer through an iterative process, and the resulting boundary shear stress on the floodplain surface was calculated.

3.1. Results for Reaches that Remained Intact

In the middle of the floodplain at bend A (Figure 5), upstream from where the floodplain began to unravel, a large beaver pond and dam are visible in the post-flood photograph. Most of the dam remained intact, with only a narrow breach in the main channel about the width of one shrub clump diameter. In the main flow channel, three other beaver dams are visible with

Table 1. Model input parameters and calculated flow velocities and boundary shear stresses along Carpenter Creek and the first unnamed tributary

	Bend A	Bend B	Bend C	Carpenter Creek
Canopy Coverage (%)	64	55	21	13
Average stem spacing (m)	0.27	0.30	0.48	0.60
Discharge (m³/s)	253	253	309	571
Flow depth (m)	3.1	3.0	2.9	3.8
Flow velocity (m/s)	2.6	2.7	3.3	4.8
Calculated boundary shear stress (N/m²)	0.3	0.4	2.0	6.3

similar-sized breaches (Figure 5). Shrubs growing along the tops of the dams remained rooted after the flood and anchored the dams during the flood flow. Shear stresses calculated for flow over the dam sequence, neglecting the shrubs, are much higher than those resulting from the presence of high-density shrubs alone. Therefore, the beaver dams and ponds probably would not have provided sufficient flow resistance during the flood to protect the floodplain surface. Instead, the dams produced wetlands that encouraged the dense growth of shrubs and herbaceous vegetation that did protect the floodplain surface.

The resulting boundary shear stress on the actual floodplain surface was reduced from about 185 N/m² (shear stress resulting from the depth-slope product with no drag-producing elements in the flow) to about 0.3 N/m² (Table 1). The average flood-flow depth without shrubs present would have been about 1.8 m (versus 3.1 m), and the mean flow velocity would have increased from 2.6 m/s to about 4.2 m/s. Although the model calculations indicate form drag on the dense shrubs increased the flood-flow depth by more than a meter, they also indicate the boundary shear stress on the floodplain surface was reduced by up to three orders of magnitude. These results are similar to those of *Smith* [this volume] and *Smith and Griffin* [2002], who demonstrate that, for overbank flows in areas with high stem densities, the boundary shear stress depends as much on stem density as on flow depth. Manning's n determined using the values of depth and velocity for flow through the dense shrubs along with the floodplain slope is 0.036, a typical value for flow over floodplains covered by high grasses [*Chow*, 1959].

The floodplain surface at bend B (Figure 8), with about 55 percent canopy cover, was also protected from erosion. A beaver dam on the downstream side of the bend washed out, leaving a wide gap at the downstream end of its pond. However, most of the floodplain surface remained intact, with many shrubs still rooted on the surface, particularly within the upstream half of the bend. Several shrubs remained rooted in the flood channel, and another beaver dam at the upstream end of the bend was breached but not entirely removed. The extent of the flood flow was again confined to an area between the two adjacent terraces, which are about 40 m apart. The calculated boundary shear stress on the floodplain surface at bend B was 0.4 N/m². This low value is consistent with the intact condition of the floodplain as seen in the aerial photograph taken two days after the flood.

3.2. Results for an Unraveled Reach

At bend C (Figure 9), the flood transformed the narrow, single-thread channel into a much wider and shallower multi-thread channel. Much of the floodplain surface was removed, and only a few shrubs remained rooted on the former floodplain within the flood channel. Bend C is downstream from the confluence of the two unnamed tributaries. The second, smaller tributary contributed an estimated 56 m³/s to the peak flow discharge, increasing the peak flow on the first unnamed tributary by about 22 percent. No evidence remains of beaver dams or ponds in the post-flood 1965 photograph, although both were in this area in February 2000. Based on the density of shrubs remaining on the upper floodplain surface in the inside of the bend, the pre-flood canopy density was estimated to be about 21 percent, with the remaining area likely covered by rangeland grasses rather than by dense wetland grasses.

The calculated boundary shear stress at bend C (2.0 N/m²) is greater than that required for erosion of the mineral soil, but is probably less than would be required to erode a surface covered by dense herbaceous vegetation. This is consistent with the observed effects of the flood, which indicate that there was significant floodplain surface erosion through this bend. Existing bed roughness, including cobbles at the mouth of a smaller channel, and the lack of finer material being transported into the reach from upstream, where the floodplain surface did not erode, prevented smoothing of the flood channel bed and transition to a channel in equilibrium with the flood peak discharge.

The boundary shear stresses on the floodplain surfaces calculated for the three tributary bends indicate that the threshold for erosion of the floodplain surface is between 0.4 N/m², the critical shear stress for erosion of the mineral soil determined using the method of Shields [Graf, 1971], and 2.0 N/m², the boundary shear stress at bend C, where much of the floodplain surface eroded. Bend C appears to have had similar vegetation prior to the flood as existed along Carpenter Creek in February 2000.

3.3. Results for Carpenter Creek Upstream from its Confluence with the Tributary

The flood flow along Carpenter Creek upstream from its confluence with the unnamed tributary created a new channel about 5 times wider than the estimated 2.5 m wide pre-flood bankfull channel. The calculated boundary shear stress on the floodplain surface, 6.3 N/m² (Table 1), is well above the calculated shear stress at bend C along the tributary (2.0 N/m²), where the floodplain surface eroded and where pre-flood vegetation coverage was likely similar to that of Carpenter Creek. In this reach of Carpenter Creek, the flood eroded the floodplain and redistributed the sediment, filling in the original channel, to create a nearly flat bed with an average width of 12.7 m, as seen in the flood photograph (Figure 10). Anti-dunes are visible on the surface of the water still flowing two days after the flood. A smooth bed covered with sand-sized sediment is necessary to form anti-dunes. In addition, the pres-

ence of anti-dunes indicates that the Froude number was near the critical value (1), and the channel was transformed to the point of developing a hydraulic geometry in equilibrium with the peak flood discharge.

4. DISCUSSION

The results of the process-based model for flow over floodplains are shown herein to be consistent with the conditions of the floodplain surfaces seen in the post-flood photographs of Carpenter Creek and the first unnamed tributary, confirming the value of the modeling approach. Using the observed physical characteristics and densities of the floodplain shrubs, the model calculated low boundary shear stresses at bends A and B on the first unnamed tributary, predicting that they were protected from unraveling by dense willow and wetland flora. Downstream from the confluence of the first and second unnamed tributaries, where discharge was higher and shrub density lower, the model-calculated boundary shear stress associated with the observed unraveling of bend C provides an upper estimate of the erosion threshold (0.4 to 2.0 N/m²) for a floodplain surface covered by sparse shrubs and rangeland grasses. This range of values is similar to that found for the floodplain within the meander belt of the Clark Fork of the Columbia River in the Deer Lodge Valley, Montana [Smith and Griffin, 2002].

Floodplain flow model results show that form drag on the dense shrub willows caused flow depths during the flood to increase more than a meter over depths for the same discharge without shrubs present. Where stem densities are low, the boundary shear stresses increase with flow depth similar to flows with only bed roughness. Model calculations also indicate that form drag on dense shrubs reduced mean flow velocities by about 40 percent and reduced actual boundary shear stresses on the floodplain surfaces by up to three orders of magnitude. With flow depths on the order of 3.0 m and mean flow velocities of about 3.3 m/s, the average stem spacing needed to reduce the boundary shear stress on the actual floodplain surface below the threshold for erosion is between 0.30 and 0.48 m, or between 55 and 21 percent canopy coverage for shrubs with the physical characteristics described here.

The verified, predictive, process-based model permits transfer of the information obtained from empirical analysis of this unraveling event to other geomorphic and riparian situations.

5. CONCLUSIONS

The results of the floodplain flow-model applications presented here demonstrate that dense shrubs on the floodplain reduced boundary shear stresses below critical values during a 3-m deep overbank flow, thus protecting the floodplain surface from erosion. Reduction of flow velocities and boundary shear stresses were quantified using form drag calculated from characteristic shrub dimensions, the valley cross-section and slope, and flood peak discharge estimated from rainfall and other data. Shrub densities with an average stem spacing of about 10 times the typical stem diameter (0.03 m) were sufficient to protect the floodplain surface during this flood, whereas canopy densities leading to an average stem spacing of about 16 times the typical stem diameter did not provide sufficient floodplain protection to prevent unraveling. Predictions of floodplain vulnerability to surface erosion made based on the model results are consistent with observed effects of the 1965 flood along two headwater tributaries of East Plum Creek. Therefore, this process-based model provides a means of calculating potential effects of woody vegetation on deep overbank flows in other cases where shrub and floodplain topographic characteristics are known or can be estimated.

REFERENCES

Burkham, D. E., Hydraulic effects of changes in bottom-land vegetation on three major floods, Gila River in southeastern Arizona, *U.S. Geol. Surv. Prof. Paper,* 655-J, 14 pp., 1976.

Chow, V. T., *Open Channel Hydraulics*, 680 pp., McGraw-Hill Book Company, New York, 1959.

Friedman, J. M., W. R. Osterkamp, and W. M. Lewis Jr., The role of vegetation and bed-level fluctuations in the process of channel narrowing, *Geomorphology*, 14, 341–351, 1996.

Graf, W. H., *Hydraulics of Sediment Transport*, 513 pp., McGraw-Hill Book Company, New York, 1971.

Griffin, E. R., and J. D. Smith, Computation of bankfull and flood-generated hydraulic geometries in East Plum Creek, Colorado, in *Proceedings of the Seventh Federal Interagency Sedimentation Conference*, Reno, Nevada, vol. 1, section II., pp. 50–56, 2001.

Hupp, C. R., and W. R. Osterkamp, Riparian vegetation and fluvial geomorphic processes, *Geomorphology*, 14, 277–295, 1996.

Hupp, C. R., and W. R. Osterkamp, Bottomland vegetation distribution along Passage Creek, Virginia, in relation to fluvial landforms, *Ecology*, 66(3), 670–681, 1985.

Li, R-M., and H. W. Shen, Effect of tall vegetations on flow and sediment, *Journal of the Hydraulics Division, Proceedings of the American Society of Civil Engineers*, vol. 99, No. HY5, 793–814, 1973.

Matthai, H. F., Floods of June 1965 in South Platte River Basin, Colorado, *U.S. Geol. Surv. Water-Supply Paper*, 1850-B, 64 pp., 1969.

Osterkamp, W. R., and J. E. Costa, Changes accompanying an extraordinary flood on a sand-bed stream, in *Catastrophic Flooding*, edited by L. Mayer and D. Nash, pp. 201–223, Allen and Unwin, Boston, 1987.

Pasche, E., and G. Rouvé, Overbank flow with vegetatively roughened flood plains, *Journal of Hydraulic Engineering*, 11(9), 1262–1278, 1985.

Riedel, J. T., F. K. Schwarz, R. L. Weaver, *Probable Maximum Precipitation Over South Platte River, Colorado and Minnesota River, Minnesota*, Hydrometeorological Report No. 44, Weather Bureau, U.S. Department of Commerce, Washington, D.C, 122 pp., 1969.

Sigafoos, R. S., Botanical evidence of floods and flood-plain deposition, *U.S. Geol. Surv. Prof. Paper,* 485-A, 35 pp., 1964.

Smith, J. Dungan, and E. R. Griffin, Relation between geomorphic stability and the density of large shrubs on the flood plain of the Clark Fork of the Columbia River in the Deer Lodge Valley, Montana, *U.S. Geol. Surv. Water-Resources Investigations Report*, 02-4070, 25 pp., 2002.

Smith, J. Dungan, On quantifying the effects of riparian vegetation in stabilizing single threaded streams, in *Proceedings of the Seventh Federal Interagency Sedimentation Conference,* Reno, Nevada, vol. 1, section IV, pp. 22–29, 2001.

Smith, J. Dungan, and S. R. McLean, Spatially averaged flow over a wavy surface, *J. Geophys. Res.*, 82(12), 1735–1746, 1977.

Trimble, D. E., and M. N. Machette, Geologic map of the greater Denver area, Front Range urban corridor, Colorado, *U.S. Geol. Surv. Misc. Inv. Series Map,* I-856-H, 1979a.

Trimble, D. E., and M. N. Machette, Geologic map of the Colorado Springs-Castle Rock area, Front Range urban corridor, Colorado, *U.S. Geol. Surv. Misc. Inv. Series Map*, I-857-F, 1979b.

Western Regional Climate Center, *Climatological Data Summary for Colorado*, Desert Research Institute, 2003, accessed through URL http://www.wrcc.dri.edu/climsum.html.

Eleanor R. Griffin, U.S. Geological Survey, 3215 Marine Street, Suite E-127, Boulder, Colorado, 80303; email: egriffin@usgs.gov.

J. Dungan Smith, U.S. Geological Survey, 3215 Marine Street, Suite E-127, Boulder, Colorado, 80303; email: jdsmith@usgs.gov.

Flow and Boundary Shear Stress in Channels with Woody Bank Vegetation

Jason W. Kean and J. Dungan Smith

U.S. Geological Survey, Boulder, Colorado.

Determination of patterns of erosion and deposition in rivers with tree-covered banks and floodplains requires a detailed knowledge of the near-boundary flow and boundary shear stress fields. Drag on the stems, branches, and exposed roots of woody vegetation on channel banks, reduces the flow within the vegetation and on near-bank portions of the channel bed. The near-bank flow is also reduced by friction on the adjacent bank. Both of these effects must be taken into account in calculating the boundary shear stress field appropriate for erosion and deposition calculations. A model is presented that explicitly calculates the drag force on rigid vegetation and includes this force as a term in the equations of motion. The velocity and boundary shear stress fields in straight channels with steady flow are found by solving these equations numerically using a ray-isovel turbulence closure that accommodates lateral boundaries and includes the effects of vegetation on the turbulence. The model is compared to laboratory data, and then used to examine the effects of vegetation density and channel geometry on the distribution of velocity and boundary shear stress in the channel. The model shows that in channels with sparse bank vegetation, both drag on the vegetation and stress on the banks contribute significantly to reducing the near-bank flow and boundary shear stress, while in channels with dense bank vegetation, drag on the vegetation is the dominant effect. The model also shows that relatively sparse vegetation by itself can have effects comparable to a sloping bank alone.

1. INTRODUCTION

The banks and floodplains of many natural rivers and streams are vegetated with trees and shrubs. This vegetation significantly affects the flow and boundary shear stress, and, consequently, influences the patterns of erosion and deposition. Woody plants affect the flow and boundary shear stress both through drag on the topographic features associated with the vegetation, such as hummocks and root balls, and through drag on the stems, branches, and roots themselves. Quantifying the effect of the latter is the focus of this study. By developing a model to calculate the drag, flow, and boundary shear stress in channels with vegetation on the banks and floodplains, this work provides a framework for determining the role of vegetation in shaping channel morphology.

Early studies of the effects of vegetation have used methods based on Manning's equation. The roughness for the entire channel is expressed in terms of a coefficient, which is either determined empirically, estimated based on various guidelines [e.g. *Chow*, 1959], or estimated by visual comparison to similar channels with known roughness [e.g. *Barnes*, 1967]. A limitation of this approach is that the roughness coefficient cannot be related in a simple manner to important characteristics of the vegetation such as stem spacing. Moreover, these approaches cannot resolve the spatial distribution of boundary shear stress needed for calculating patterns of erosion and deposition.

Riparian Vegetation and Fluvial Geomorphology
Water Science and Application 8
10.1029/008WSA17

Recent work has focused on explicitly calculating the drag force on the stems and branches and including it as a body force in the equations of motion [*Simões*, 2001; *Smith*, 2001; *Smith and Griffin*, 2002 a, b; *Smith and Griffin* this volume; *Griffin and Smith*, this volume]. The drag on the stems is determined using the drag coefficient of a single stem and the mean velocity within the vegetation, which is calculated as part of the solution. This approach allows the flow solution to be more easily related to the characteristics of the vegetation. Furthermore, the effects of vegetation throughout the flow domain can be taken into account. *Smith* [2001] used this approach to obtain analytic expressions for the velocity and boundary shear stress in a vegetated floodplain using an algebraic eddy viscosity model. Algebraic eddy viscosity models were also used by *Simões* [2001] to calculate flow and boundary shear stress numerically in a compound channel with sparse vegetation on the floodplain. Although the algebraic eddy viscosity used in the *Simões* study simplified the computation, it did not include the lateral effects of the banks or the effects of vegetation on the turbulence, thus restricting the application of the method to gradually sloping banks and sparse floodplain vegetation. Steep banks are common in natural rivers, and, like vegetation, they have a significant effect on the near-bank flow and boundary shear stress.

Other studies of this type have sought to include the effects of vegetation on the turbulence using k-ε turbulence models. *Shimizu and Tsujimoto* [1994] and *López and García* [1997] included drag terms in the momentum equation as well as in the equations for k and ε. These two studies used standard values for the five k-ε turbulence constants and used laboratory experiments to determine two additional drag-related turbulence constants. The experimental setups in both studies consisted of a rectangular flume covered with submerged vertical rods. The turbulence constants and boundary conditions for the k-ε turbulence models used in these studies, however, do not apply to channels of interest here, which have emergent vegetation on the banks and floodplains.

The goals of this study are to develop a process-based model for predicting the flow and boundary shear stress fields in channels with banks and floodplains protected by woody vegetation, and then to use this model to explore the relative effect vegetation density and channel geometry have on these fields. Drag on the vegetation is calculated explicitly, and the turbulence closure employed is based on the ray-isovel scheme of *Houjou et al.* [1990], which is modified herein to include the effects of the vegetation on the turbulence. This scheme is well suited for channels with shrub-protected and steeply sloping banks. Our model is tested using the laboratory measurements made by *Pasche* [1984]. Recent field work by the authors has focused on the Rio Puerco,

Figure 1. Downstream view into a gradual bend of the inset channel of the Rio Puerco arroyo near Belen, NM. The banks are vegetated with sandbar willow and tamarisk. The channel is approximately 2.7 m deep and the unvegetated bed is approximately 5 m wide.

NM, which has a narrow vegetated channel inset into the arroyo floor as shown in Figure 1. Using a channel of the approximate geometry shown in the figure, comparisons of the velocity and boundary shear stress fields are made for flows with different vegetation density and stage. These flows are also compared to flow in a vegetated channel without lateral boundaries.

2. MODEL DEVELOPMENT

Houjou et al. [1990] developed a theoretical model that calculates the velocity and shear stress fields for uniform flow and applied this model numerically to rectangular channels with known bed and bank roughness. This study adapts that model to include the drag effects of vegetation. In addition, the new model uses a different structure for the eddy viscosity in the interior of the flow and is applied to study more natural channel shapes. Because this study is focused on natural streams and rivers, secondary circulations, such as those that have been observed to occur in straight, prismatic channels [e.g. *Knight et al.*, 1994], are not included. The natural variability of the boundary, planform, and vegetation will likely disrupt such secondary circulations. Moreover, to determine whether or not they can exist in channels with natural geometries, a model with a zero-order flow field is an essential first step.

Many of the principles governing flow in channels with shrub-protected banks are derived from the simpler case of flow in an infinitely wide channel without vegetation. For this reason, we will present this simpler case first, extend it to two dimensions, and finally add the vegetation.

2.1. Infinitely Wide Channel

The momentum equation for unaccelerated, horizontally uniform flow in an infinitely wide channel without vegetation is

$$0 = \rho g S + \frac{\partial \tau_{zx}}{\partial z} \tag{1}$$

where, ρ is the fluid density, g is the gravitational acceleration, S is the water surface slope, and τ_{zx} is the zx component of the deviatoric stress tensor. Integrating (1) with respect to z and applying the boundary condition that $\tau_{zx} = 0$ at the water surface, gives

$$\tau_{zx} = \rho g S H (1 - z/H) \tag{2}$$

where H is the flow depth. This makes the boundary shear stress, τ_b, equal to $\rho g S H$. The shear stress is related to the mean velocity gradient by

$$\tau_{zx} = \tau_b (1 - z/H) = \rho K \frac{\partial u}{\partial z} \tag{3}$$

where, u is the streamwise velocity and K is the kinematic eddy viscosity. A generalization of the algebraic expression for K of *Rattray and Mitsuda* [1974] has been shown by *Shimizu* [1989] to give good agreement with the experimental open-channel velocity measurements of *Einstein and Chien* [1955]. This generalization has the form

$$K = \kappa u_* z \left(\tau_{zx}/\tau_b\right) = \kappa u_* z (1 - z/H) \text{ for } z/H \leq 0.2 \tag{4a}$$

$$K = \kappa u_* H / \beta \text{ for } z/H > 0.2 \tag{4b}$$

where κ is von Karman's constant, which equals 0.408 [*Long et al.*, 1993], u_* is the shear velocity, which equals $\sqrt{\tau_b/\rho}$, and β is a constant, which equals 6.24. Substituting this expression into (3), equating it with (2), and integrating with respect to z using the boundary condition that $u = 0$ at the roughness

height, $z = z_o$, gives a two part expression for the velocity profile, which is given by

$$u = \frac{u_*}{\kappa} \ln\left(\frac{z}{z_o}\right) \text{ for } z/H \leq 0.2 \tag{5a}$$

$$u = \frac{u_*}{\kappa}\left[\ln\left(\frac{0.2H}{z_o}\right) + \beta\left(\frac{z}{H} - \frac{1}{2}\frac{z^2}{H^2} - 0.18\right)\right]$$

$$\text{for } z/H > 0.2 \tag{5b}$$

2.2. Channel with Lateral Boundaries

The momentum equation for steady, streamwise uniform flow in an unvegetated channel with lateral boundaries is

$$0 = \rho g S + \frac{\partial \tau_{yx}}{\partial y} + \frac{\partial \tau_{zx}}{\partial z} \tag{6}$$

where τ_{yx} is the yx component of the deviatoric stress tensor. The additional term in the momentum equation complicates the fluid-mechanical problem considerably. Fluid stresses are transmitted to both the bed and the banks, making analytic solutions for the shear stress, such as (3) for the infinitely-wide channel, only possible for specific two-dimensional geometries such as pipes. Nevertheless, the force balance for uniform flow, which is a balance between the downstream component of the weight of water in the channel and the friction along the wetted perimeter, provides a strong constraint on the boundary shear stress for all channel geometries. This balance requires that the average boundary shear stress for the entire channel be equal to

$$\overline{\tau}_b = \rho g S \, A / p_b \tag{7}$$

where A is the cross-sectional area of the channel and p_b is the wetted perimeter. A similar balance exists for the friction along an incremental length of wetted perimeter, δp_b,

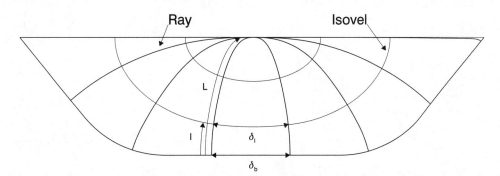

Figure 2. Diagram of orthogonal ray-isovel grid.

and the downstream component of the weight of a fraction of the water in the channel. This balance gives an expression for the boundary shear stress on a section of the boundary that is analogous to (7), where δp_b takes the place of p_b, and A is replaced by the fractional area that is responsible for producing the shear stress on that portion of the boundary. This area is defined using the natural, curvilinear coordinate system of the flow, which consists of lines (rays) that are perpendicular to lines of constant velocity (isovels) as shown in Figure 2. The rays, which are perpendicular to the boundary and extend to the water surface, define streamwise surfaces of zero shear. The area responsible for producing the shear on the portion of the boundary, δp_b, is simply the total area between two rays separated by δp_b at the boundary. In this framework, the boundary shear stress is expressed as

$$\tau_b = \rho g S \frac{\int_0^L dA}{\delta p_b} \tag{8}$$

where L is the total length of the ray. The shear stress along each ray in the interior is given by the expression

$$\tau_{lx} = \rho g S \frac{\int_l^L dA}{\delta p_l} \tag{9}$$

where, l is the distance along the rays from the boundary, and δp_l is the length along an isovel between the two adjacent rays.

Like τ_{yx} and τ_{zx}, the kinematic eddy viscosity for flow near lateral boundaries varies in both the y and z directions. This variation can be simplified using the ray-isovel coordinate system. If it is assumed that the rays define the direction of shear along which the mixing occurs, the spatial variation in K can be reduced to a one-dimensional variation along each ray. Thus, in the ray-isovel coordinate system, a scalar eddy viscosity can be defined by the expression

$$\tau_{lx} = \rho K \frac{\partial u}{\partial l} \tag{10}$$

The eddy viscosity near the boundary depends on the distance from the boundary, l, the shear velocity at the base of each ray, $\sqrt{\tau_b / \rho}$, and a non-dimensional function of τ_{lx} and τ_b. This function is $\kappa \tau_{lx} / \tau_b$. For infinitely wide channels $\tau_{lx}/\tau_b = 1 - z/H$, and for circular pipes $\tau_{lx}/\tau_b = 1 - r/R_p$. In the latter expression, R_p is the radius of the pipe and r is the radial distance from the wall of the pipe to the interior. Note that the former expression is used in (4a). If it is assumed that this function is valid for all channel geometries between these two limiting cases, then K near the boundary may be expressed as

$$K = \kappa u_* l \frac{\tau_{lx}}{\tau_b} \tag{11}$$

In the *Houjou et al.* model [1990], K is given by (11) from $l = 0$ to l_m, which is the ray distance that is within 20% of the area between the two rays. From $l = l_m$ to L, K is held constant at its value at $l = l_m$. This definition of K does not adequately

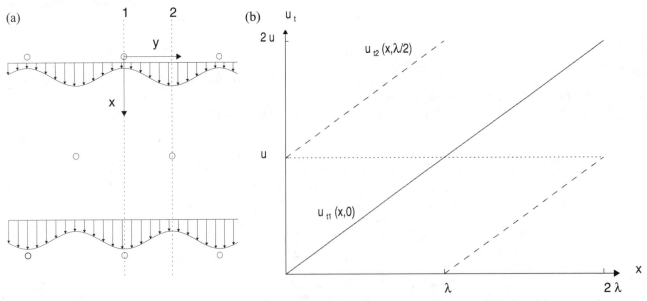

Figure 3. (a) Local velocity variations, $u_t(x,y)$, through a regular staggered array of stems, and (b) plot of the streamwise velocity $u_t(x,0)$ and $u_t(x,\lambda/2)$ along paths 1 and 2. The average of $u_t(x,y)$ in both the x and y directions equals u.

account for the mixing in the interior of the channel, which occurs at a length scale associated with the flow depth. Although the *Houjou et al.* formulation matches (3) for the limit of the infinitely wide channel, it incorrectly assigns near boundary mixing to the interior in channels with lateral boundaries. In this investigation, K along each ray is defined to increase according to (3) until it reaches the channel scale eddy viscosity, K_o, which is given by the equation

$$K_o = \kappa \sqrt{gSH}\,\frac{H}{\beta} \qquad (12)$$

Having defined K for the entire channel, the momentum equation for steady flow is given by

$$0 = \rho gS + \frac{\partial}{\partial y}\left(\rho K \frac{\partial u}{\partial y}\right) + \frac{\partial}{\partial z}\left(\rho K \frac{\partial u}{\partial z}\right) \qquad (13)$$

with the boundary conditions that $\partial u/\partial z = 0$ at the water surface, $z = H$, and $u = 0$ at the roughness height $l = l_o$. Given an initial guess of the shear stress on the boundary and the velocity in the interior, the computation alternatively solves the momentum equation for u and the equations for τ_b, τ_{lx}, and K until the flow solution converges.

2.3. Channels with Vegetation

In steady, uniform flow in channels with vegetated banks, the downstream component of the weight of water is balanced by both the friction on the wetted perimeter and the drag on the array of stems and branches. The drag force per unit volume due to the stems may be expressed as

$$F = \tfrac{1}{2}\rho C_D \alpha\, u_{ref}{}^2 \qquad (14)$$

where C_D is the drag coefficient of a single stem and u_{ref} is the reference velocity. The vegetation density parameter, α, which has dimensions of length^{-1}, is the cross-sectional area of the stems oriented perpendicular to the flow direction per unit volume. For a random array of vertical stems extending from bottom to top of the unit volume, the vegetation density parameter may be expressed as

$$\alpha = nD_s = \frac{D_s}{\lambda^2} \qquad (15)$$

where n is the number of stems per unit area, D_s is the average stem diameter, and λ is the average stem spacing. The vegetation density parameter may be a function of y and z due to spatial changes in shrub architecture, such as an increase in the number of branches towards the surface or an increase in size of stems away from the center of the channel.

The reference velocity for drag on a segment of a stem is defined as the velocity that would be present if the stem segment were removed from the flow, averaged over the space the stem occupied. Defining u_{ref} in this manner allows us to use the value of C_D for a single roughness element, which, in this case, is an individual stem or branch. This approach has been shown to be valid for determining the drag on a Gaussian shaped roughness element embedded in a series of identical elements protruding from a flat surface [*Kean*, 2003]. The stems and branches are modeled as circular cylinders, and the Reynolds numbers for the flows of interest in this study are in the range where C_D for a cylinder is nearly constant and equal to about 1.2. Although the cross-sections of the stems and branches may deviate slightly from circular, the error associated with the cylinder approximation is negligible relative to uncertainties in parameters such as average stem diameter and spacing. When the drag force is expressed on a per volume basis rather than for an entire stem, u_{ref} can be approximated using the local mean velocity $u(y,z)$. Including the drag per unit volume in the momentum balance gives the equation

$$0 = \rho gS + \frac{\partial}{\partial y}\left(\rho K \frac{\partial u}{\partial y}\right) + \frac{\partial}{\partial z}\left(\rho K \frac{\partial u}{\partial z}\right) - \frac{1}{2}\rho C_D \alpha u^2 \qquad (16)$$

The boundary shear stress in the unvegetated region of the channel is given by (8). In vegetated regions of the channel, however, τ_b is reduced relative to its unvegetated state as a result of the drag on the shrubs. The boundary shear stress in these regions is determined by subtracting the total drag force within the area between two adjacent rays from the downstream component of the weight water in that area. This may be expressed as

$$\tau_b = \frac{\int_0^L (\rho gS - F)\,dA}{\delta p_b} \qquad (17)$$

where F is given by (14).

2.3.1 Turbulence within the vegetation. Although the dominant effect of vegetation on the flow is through drag, vegetation can also affect the flow by changing the turbulence. For sparse vegetation, where the mean stem spacing is greater than the dominant mixing length scale for the channel ($\lambda > H$), the channel-scale eddies persist through the vegetation. However, for denser vegetation ($\lambda < H$) the stems and the small-scale wake turbulence that they produce break up the channel-scale eddies, changing the turbulence and eddy viscosity. For the situation where bed effects are negligible, the eddy viscosity is estimated based on the production of turbulent kinetic energy by the flow through the stems.

Following the work of *Nepf* [1999], it is assumed that all of the energy extracted from the mean flow by the drag on the stems goes into producing turbulent kinetic energy. The production of turbulence due to shrub stems or tree trunks, P_t, is then the work done by drag per unit time and mass, which may be expressed as

$$P_t = \frac{1}{2} C_D \frac{D_s}{\lambda^2} u^3 \tag{18}$$

Assuming the stems are vertical, the production of turbulent kinetic energy due to vegetation may also be expressed as

$$P_t = K_t \left(\frac{\partial u_t}{\partial y} \right)^2 \tag{19}$$

where, K_t is the eddy viscosity within the stems, and $u_t(x,y)$ is the local velocity within the stems. The spatial average of $u_t(x,y)$ in the x direction is equal to u, but the exact distribution throughout the stems is unknown. The eddy viscosity is assumed to have the form

$$K_t = A \lambda \, \Delta u_t \tag{20}$$

where A is an undetermined constant and Δu_t is the streamwise average difference between the maximum and minimum values of u_t in the y direction. In order to obtain an estimate of Δu_t, the natural, random arrangement of the vegetation is modeled as a staggered array as shown in Figure 3a. In this arrangement the maximum change in u_t in the y-direction occurs between $y = 0$ and $y = \lambda/2$. Thus, $\partial u_t / \partial y$ is approximately equal to $\Delta u_t / (\lambda/2)$. Using this approximation and the further assumption that Δu_t is related to u by a constant of proportionality, $C_t = \Delta u_t / u$, (19) may be rewritten as

$$P_t = 4A \frac{(C_t u)^3}{\lambda} \tag{21}$$

Justification for relating the mean velocity, u, to Δu_t is made by comparing the velocity profile within the stems along two streamwise paths as shown in Figure 3b. The velocity along each path accelerates from zero velocity just behind each stem to a maximum value somewhere before it encounters the next stem. In the staggered arrangement, the two streamwise velocity profiles have the same shape but differ in phase by half of a wavelength (λ). The streamwise average difference in velocity between the two profiles, Δu_t, is proportional to u because the average of the profiles must equal u. The value of C_t depends on the shape of the streamwise velocity profile. If it is assumed that u_t along each path varies linearly from zero to a maximum just before it encounters the next stem, then $C_t = 1$. The value of A is found by equating (21) with (18). The expression for K_t can be rewritten as

$$K_t = \frac{1}{8} C_t^{-2} C_D D_s u \tag{22}$$

2.3.2 Combination of bed and tree turbulence. In channels such as the one shown in Figure 1, turbulence is produced by both the boundary and the vegetation. Up to this point, the turbulence has been described separately for each source. Close to the boundary, turbulence is produced by the shear on the boundary and the eddy viscosity, K_b, is given by (11) and (12). Within the trees, turbulence is produced by drag on the stems, and the eddy viscosity is K_t. In regions of the chan-

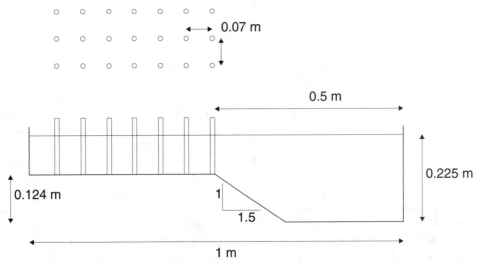

Figure 4. Geometry and dimensions of *Pasche's* [1984] laboratory channel. The plan view arrangement of the simulated floodplain vegetation is shown above the channel. The parameters of the experimental run to which comparisons are made are $H = 0.225$ m, $\lambda = 0.07$ m, $D_s = 0.012$ m, $\alpha = 2.4$ m^{-1}, $S = 0.001$, and $Q = 0.061$ m^3/s.

nel where the turbulence is produced by both the boundary and the vegetation, a simple definition of the eddy viscosity is not possible because the velocity and length scales are not well defined. In lieu, of a general turbulence closure, the eddy viscosity in areas affected by both bed and tree turbulence is taken to be a weighted average of K_t and K_b based on the distance along the ray, l, and the mean stem spacing, λ. It is given by the expression

$$K = \frac{1}{1+(\lambda/l)^2} K_b + \frac{1}{1+(l/\lambda)^2} K_t \qquad (23)$$

Though not strictly correct, (23) provides a means to close the problem and is valid in the limit of small λ/l and l/λ. Sensitivity tests have shown the general patterns of velocity and boundary shear stress are not significantly affected by the value of K as adjusted by (23). Given the uncertainty in other parameters such as bed roughness and vegetation density, this approximation is adequate to provide the desired insight into these problems.

3. COMPARISON TO LABORATORY MEASUREMENTS

Confidence in the model is built by comparing model results to laboratory measurements made by *Pasche* [1984]. The experiments were conducted using a tilting recirculating flume 25.5 m long and 1 m wide. Velocity measurements were made using a laser Doppler current meter. The channel geometry and dimensions are shown in Figure 4. The walls of the main chan-

nel are smooth, and the wall of the flume is located where the centerline of a symmetric compound channel would be. Floodplain vegetation was simulated using vertical wooden rods that extend the full depth of the flow. The rods were arranged in a rectangular array oriented such that rows of rods were parallel to the flow direction. The experimental case to which model results are compared is also reported in *Pasche and Rouvé* [1985].

A comparison of depth-averaged measured and calculated velocities across the channel is shown in Figure 5. The results of the model generally match the shape of the measured velocity profile. The model routes more of the flow through the main channel than was the case in the experiment, and as a result, slightly over predicts the depth-averaged velocity in the main portion of the channel, and under-predicts it on the vegetated floodplain. The good agreement near the right wall of the flume demonstrates that the model is correctly accounting for the lateral effects of the wall. One reason for the discrepancy between the measured and calculated velocity on the vegetated floodplain is likely due to the spatial arrangement of the vertical cylinders. Our model is formulated with the assumption that the cylinders are randomly spaced throughout the floodplain with a mean spacing λ. Although this assumption is reasonable for modeling flow through stem arrangements that typically occur in nature, it is not well suited for modeling flow through the rectangular array of the experiment. The experimental cylinder arrangement is more similar to a planted orchard on the floodplain than to a natural arrangement of trees. The rectangular array allows for preferential flow paths between the rows of cylinders, which results in increased flow

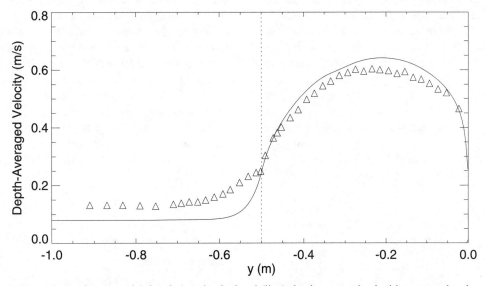

Figure 5. Comparison of measured (triangles) and calculated (line) depth-averaged velocities across the channel. The vertical dotted line marks the interface between the vegetated floodplain and the main channel. The decrease in measured and calculated velocity approaching $y = 0$ is due to the flume wall. Measured velocities are from *Pasche* [1984].

Table 1. Bankfull channel parameters for the modeled Rio Puerco channel.

Depth	Top Width	Bank Angle	Area	Wetted Perimeter	Slope	z_o
2.7 m	11 m	50°	23.2 m^2	13.2 m	0.001	0.001 m

through the array relative to a random arrangement. This interpretation is supported by the fact that the flow visualization performed during the experiment revealed large coherent eddies at the interface between the channel and the floodplain. These structures are produced by a shear instability between fast flow in the main channel and the slower jet between the first two rows of cylinders. A random arrangement of cylinders would tend to disrupt such structures.

Given the difference in vegetation arrangements between the experiment and the model, the 30% difference between measured and calculated velocity on the floodplain seems reasonable. Future experiments aimed at simulating natural vegetation will provide a better test of this model. If conducted, these experiments should use random vegetation arrangements, rather than rectangular ones.

4. APPLICATION OF THE MODEL

4.1. Channel Cross-section

The channel geometry used for most of the calculations is based on a cross-section of the inset channel of the Rio Puerco arroyo near Belen, NM (Figure 1). The cross-sectional geometry is about the same for 65 km and the banks of the inner channel are vegetated nearly continuously with tamarisk and willows for this distance. The cross-section was chosen for two reasons. First, the combined characteristics of rough banks and a narrow channel highlight the need for models that correctly incorporate lateral effects. Secondly, this model is ultimately intended to be used to help determine the role vegetation has played in recent channel changes along the Rio Puerco.

Significant aggradation has occurred in the Rio Puerco over the past several decades [Elliot et al., 1999]. This aggradation has coincided with the establishment of the exotic shrub tamarisk on the arroyo floor. Prior to the introduction of

tamarisk in the early 1920's the most common shrub species in the arroyo was sandbar willow. Field measurements have shown that the vegetation density of present day stands of sandbar willow is less than that of tamarisk. In addition, the zone where sandbar willow grows is limited to the banks and floodplain immediately beside the inner channel, while tamarisk is found throughout the arroyo floor. A process-based, coupled channel/floodplain model, such as the one presented here, allows for a comparison of the flow and boundary shear stress in the present and pre-tamarisk channel.

The main properties of the modeled channel are listed in Table 1. It is straight and roughly trapezoidal in cross section with rounded corners near the bed. The bed is extremely smooth and composed of clayey silt. A small roughness height, z_o, was chosen (0.001 m) to conform to the bed of the prototype channel. The roughness height is taken to be the same for the bed and banks of the channel so that the dominant roughness of the channel is from the vegetation. The parameters varied in the model calculations, such as stem diameter and spacing, span the range found along the 65 km reach of the Rio Puerco.

Despite the fact that the medium to large diameter branches found on the banks of the Rio Puerco are fairly rigid, the stiffness of the Rio Puerco bank vegetation does not entirely meet the rigid vegetation assumption of the model. During bankfull flow events in the Rio Puerco, some bending and low velocity movement of the stems and branches will occur in response to vorticity shedding by the stems and branches. Provided the vegetation extends above the water surface, the effects of the bending and movement of the stems on the drag will be small. The bending and movement of the stems will add a small positive or negative velocity perturbation to the mean velocity. Although the total velocity is squared in the drag equation, this added effect appears from preliminary estimates and observations on other streams to be very small relative to the square of the mean value.

Table 2. Vegetation and flow parameters for the bankfull channel cases.

Vegetation density	D_s (m)	λ (m)	D_s/λ^2 (m^{-1})	Q (m^3/s)	$\overline{\tau}_b / (\rho g S R)$	Fr	n (s/m$^{1/3}$)
None	-	-	0	50.7	1	0.48	0.021
Sparse	0.03	1.73	0.01	42.7	0.68	0.44	0.025
Medium	0.03	0.54	0.1	33.6	0.43	0.41	0.032
Dense	0.03	0.17	1.0	29.2	0.36	0.40	0.037

4.2. Variation of Bankfull Flow Properties with Vegetation Density

Calculations are presented for bankfull flow in four channels with different densities of rigid vegetation on the banks. The four cases correspond to dense, medium, sparse, and no

vegetation, as determined by measurements of stem density along the Rio Puerco in April 2002. The values of $\alpha = D_s/\lambda^2$ for the four cases are: 1, 0.1, 0.01, and 0 m^{-1}. The same model cross-section is used for each case. Table 2 lists the vegetation and flow parameters for each of the cases. The flow parameters are: calculated bankfull discharge, Q, ratio of perimeter-

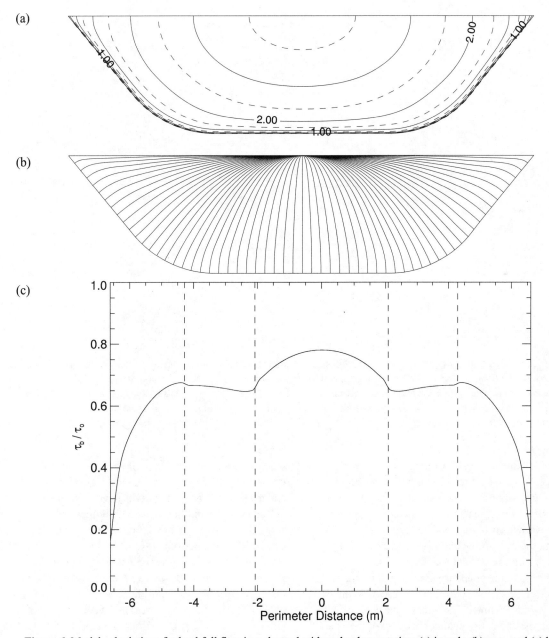

Figure 6. Model calculations for bankfull flow in a channel with no bank vegetation: (a) isovels, (b) rays, and (c) boundary shear stress. The channel is drawn to scale. The solid contours in (a) are at 1 m/s and 0.5 m/s increments and the dashed contours are at 0.25 m/s and 0.75 m/s increments. The boundary shear stress is normalized by the depth-slope product, $\tau_o = \rho g S H$. The dotted vertical lines in (c) denote the extents of the two rounded corners along the boundary.

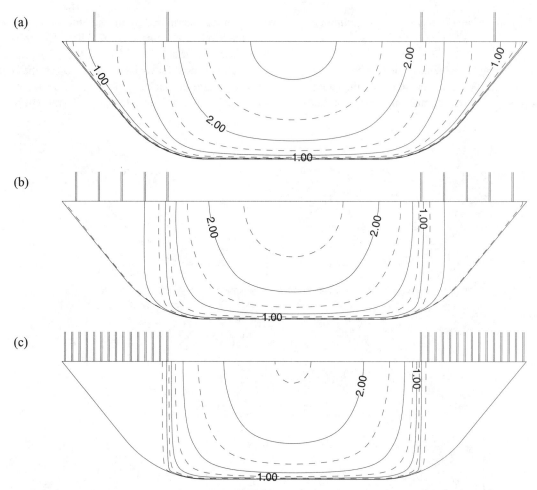

Figure 7. Model calculations of velocity for bankfull flow in three channels with different densities of bank vegetation: (a) sparse ($\alpha = 0.01$ m^{-1}), (b) medium ($\alpha = 0.1$ m^{-1}), and (c) dense ($\alpha = 1.0$ m^{-1}). The vegetation density is depicted using vertical bars above the surface. The distance between the bars corresponds to the mean spacing, λ, of the stems. The innermost stem on either side of the channel is drawn at the edge of the vegetation. The model assumes the vegetation is randomly spaced throughout the vegetated zone.

averaged boundary shear stress to the stress given by the hydraulic radius, $\bar{\tau}_b/(\rho g S R)$, Froude number in the center of the channel, Fr, and Manning's coefficient, n.

Results of model calculations for the case with no vegetation ($\alpha = 0$ m^{-1}) are shown in Figure 6. The results include the isovels, rays, and the ratio of the boundary shear stress (τ_b) to the depth-slope product at the center of the channel ($\tau_o = 26.5$ N/m^2). By normalizing the boundary shear stress in this way, the result provides a direct comparison to the case of an unvegetated infinitel y-wide channel. The banks have a significant effect on flow across the entire channel because of the channel's low width to depth ratio. The boundary shear stress in the center of the channel is reduced by about 20% of what it would be if the channel were infinitely wide. This reduction is counteracted by τ_b on the banks. The dip in τ_b between the

dashed lines on either side of the channel is due to the rounded corners. In the immediate region of the corner both the bed and the banks resist the flow, which reduces velocity and boundary shear stress in that area. This reduction is also apparent from close inspection of the rays emanating from the rounded corner. The area between rays is directly proportional to τ_b as given by (7). As the corner rays approach the surface, they are focused together and, as a result, have less area between them than rays originating from the bed or middle portion of the bank. The boundary shear stress in the corner is particularly sensitive to the shape of the corner. Channels with more distinct corners, such as cut banks, have relatively lower values of τ_b in the immediate vicinity of the corner.

Figure 7 shows model calculations of the velocity for the three cases with vegetation, and Figure 8 shows the boundary

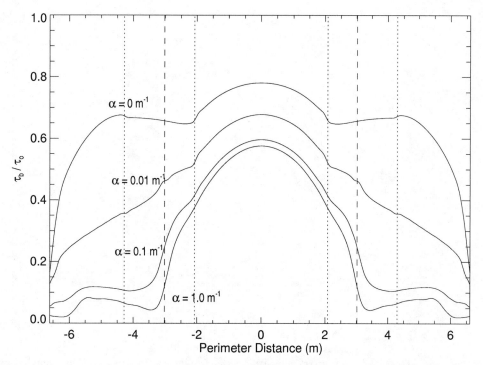

Figure 8. Boundary shear stress for the four bankfull cases. The dotted vertical lines denote the extent of the rounded corners, and the vertical dashed lines denote the interface between the open channel and the vegetation.

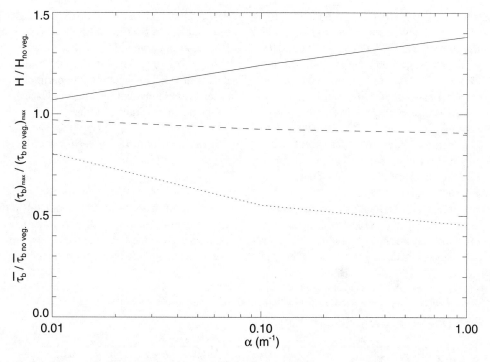

Figure 9. Stage, H (solid line), maximum boundary shear stress, $(\tau_b)_{max}$ (dashed line), and average boundary shear stress, $\bar{\tau}_b$ (dotted line) as a function of vegetation density parameter, α, for the same discharge. The results are normal-ized by the value of each variable in the channel with no vegetation: $H_{no\ veg.} = 1.96$ m, $(\tau_{bno\ veg.})_{max} = 16.8$ N/m^2, and $\bar{\tau}_{b\ no\ veg.} = 13.4$ N/m^2.

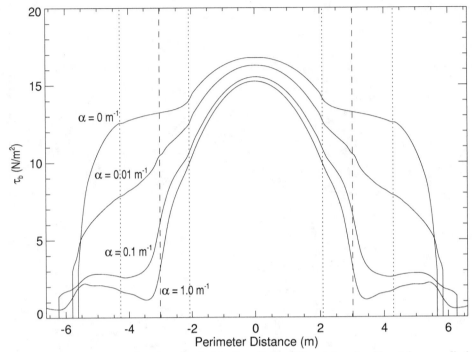

Figure 10. Boundary shear stress as a function of perimeter distance for different stem densities at the same discharge (29.2 m³/s). The total perimeter distance for each case is different due to differences in stage. The vertical lines are the same as in Figure 8.

shear stress results for these cases together with the unvegetated case. These figures show the significant effect rigid bank vegetation has on reducing the velocity and boundary shear stress both within the vegetated zone as well as in the center of the channel. Even for sparse vegetation ($\alpha = 0.01$ m⁻¹), there is a substantial reduction in velocity, corresponding to a 16% decrease in Q relative to the unvegetated case. There is a 40% decrease in Q for the dense case.

The ratio of the perimeter-averaged boundary shear stress to the stress given by the hydraulic radius, $\bar{\tau}_b /(\rho g S R)$, summarizes the total effect of the vegetation on boundary shear stress. The ratio for each case is given in Table 2. For the unvegetated case, the stress on the boundary is the only source of resistance and the ratio is equal to 1. The additional flow resistance provided by the drag on the vegetation reduces the contribution of the stress on the boundary substantially. For the

dense vegetation case, drag on the vegetation provides 64% of the total flow resistance. As is shown in Figure 8, most of the flow resistance provided by the boundary for the vegetated cases is bed stress. The values of this ratio show that for channels with woody vegetation, the perimeter-average boundary shear stress should not be calculated based solely on the channel slope and hydraulic radius.

4.3. Variation of Flow Properties with Vegetation Density for the Same Discharge

The relationship between vegetation density, stage, and boundary shear stress for the modeled Rio Puerco channel is determined by routing a steady flow of the same discharge through each of the four channels in the previous set. The discharge used in these calculations is the bankfull discharge in

Table 3. Vegetation and flow parameters for the constant discharge cases ($Q = 29.2$ m³/s).

Vegetation density	D_s/λ^2 (m⁻¹)	H (m)	R (m)	$(\tau_b)_{max}$ (N/m²)	$\bar{\tau}_b / \bar{\tau}_{b\ no\ veg.}$	Fr	n (s/m^{1/3})
None	0	1.96	1.37	16.8	1	0.49	0.021
Sparse	0.01	2.09	1.45	16.3	0.81	0.47	0.023
Medium	0.1	2.43	1.62	15.5	0.55	0.43	0.030
Dense	1.0	2.70	1.76	15.3	0.45	0.40	0.037

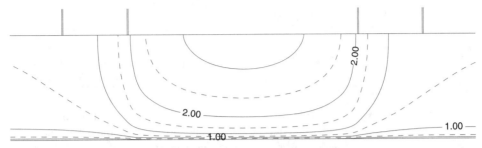

Figure 11. Model calculations of velocity in a sparsely vegetated ($\alpha = 0.01$ m^{-1}) channel without lateral boundaries. The depth and width of the unvegetated center of the channel are also the same.

the previous dense vegetation case ($Q = 29.2$ m^3/s). The stage, maximum boundary shear stress, and average boundary shear stress are plotted as functions of vegetation density parameter, α, in Figure 9. These variables are normalized by the value of that variable in the channel without vegetation. The boundary shear stress distribution for all of the cases is shown in Figure 10, and the vegetation and flow parameters for each case are given in Table 3.

The additional flow resistance of the vegetation causes the stage to increase and decreases the boundary shear stress throughout the channels. The basic shapes of the boundary shear stress distributions shown in Figure 10 are similar to those of the previous set of cases (Figure 8). A comparison of

Manning's coefficient between this set of cases and the previous set shows little change in the value for the three cases where the stage is different (the flow conditions are identical for the dense cases of each set). It should be noted that Manning's equation can not provide the average value of the actual stress on the boundary for the vegetated cases, or the distribution of stress in the channel for any of the cases.

Figure 9 shows that the increased stage in the vegetated channels does not lead to an increase in the maximum boundary shear stress in the middle of the channel. In the dense vegetation case there is a 0.74 m increase in stage over the unvegetated case, but a 9% reduction in boundary shear stress. For unaccelerated flow in an unvegetated infinitely wide chan-

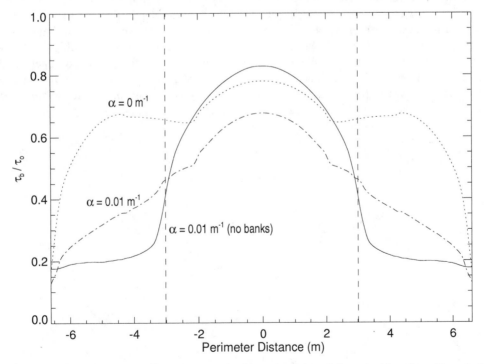

Figure 12. Model calculations of boundary shear stress in a sparsely vegetated channel without lateral boundaries (solid line). The boundary shear stress for the Rio Puerco channel cases with sparse (dot-dashed line) and no vegetation (dotted line) are shown for comparison. The value of $\tau b/\tau o$ a distance ±7m away from the center channel without lateral boundaries is close to its asymptotic value.

Table 4. Vegetation and flow parameters for the overbank channel cases.

Vegetation density	D_s/λ^2 (m^{-1})	Q_{main} (m^3/s)	$\bar{\tau}_{main}$ (N/m^2)	Fr	n_{main} (s/m$^{1/3}$)	$\bar{u}_{flood}$ (m/s)	$\bar{\tau}_{b\,flood}$ (N/m^2)	n_{flood} (s/m$^{1/3}$)
None	0	74.1	20.8	0.52	0.021	0.93	4.90	0.022
Medium	0.1	68.1	20.3	0.48	0.023	0.35	1.04	0.057

nel, an increase in stage would lead to a proportional increase in the boundary shear stress. In this channel, the lateral effects of the vegetation overwhelm the tendency of stage to increase the boundary shear stress. A comparison of the ray fields for the unvegetated and dense vegetation cases, which are not given, would show that although the rays emanating from the bed of the channel in the vegetated cases are longer, the area between them is less than in the unvegetated case because they are focused together by the lateral effects of the vegetation.

4.4. Vegetated Channel Without Lateral Boundaries

The next case considered is a swamp, a channel with woody vegetation and no lateral boundaries. The channel is infinitely wide and populated with sparse vegetation ($\alpha = 0.01$ m^{-1}), except in one unvegetated region. It has the same slope, depth, and width of the unvegetated region as the modeled Rio Puerco channel. Figure 11 and Figure 12 show model calculations of the velocity and boundary shear stress fields for this wetland channel. Both fields approach the conditions of pure floodplain flow with distance away from the center of the channel. If the vegetation were denser, the velocity and boundary shear stress would reach floodplain flow conditions a shorter distance away from the center of the channel.

Figure 12 also compares boundary shear stress in the wetland channel to the boundary shear stress in the unvegetated

and sparsely vegetated channels of the first set. This comparison allows us to isolate the lateral effects of sparse vegetation from the lateral effects of friction on the bank. The figure shows that drag on sparse vegetation in the wetland channel has about the same effect on reducing the mid-channel boundary shear stress as does friction on the banks alone. The figure also shows that the combined lateral effects of drag on sparse vegetation and friction on the channel banks provide an additional 25% reduction in the mid-channel boundary shear stress.

4.5. Overbank Flow With Floodplain Vegetation

The last set of cases considered is for overbank flow in compound channels with and without woody vegetation on the floodplain. The main channel has the same cross-section as the modeled Rio Puerco channel, except that the top of the bank is rounded to smoothly meet the floodplain surface. The depth of the flow is 3.2 m, which is 0.5 m above the bankfull depth. Figure 13 shows model calculations of velocity for the two cases, and Figure 14 shows the results for boundary shear stress. Table 4 lists the vegetation and flow parameters for the main channel and floodplain. Using this table, the discharge for an overbank flow of this depth and with these densities of floodplain vegetation can be calculated for a floodplain of any width.

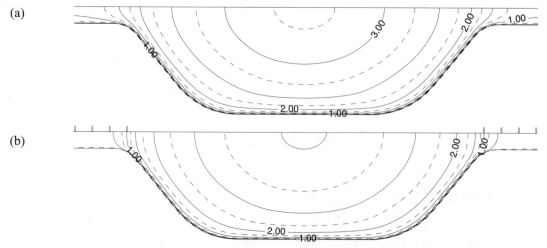

(a)

(b)

Figure 13. Model calculations of velocity for 0.5 m of overbank flow in a channel with (a) no vegetation on the floodplain and (b) medium density vegetation on the floodplain ($\alpha = 0.1$ m^{-1}). The cross-section of the main channel is the same as in Figure 6. With the overbank flow, the cross-sectional area of the main channel is 28.7 m^2.

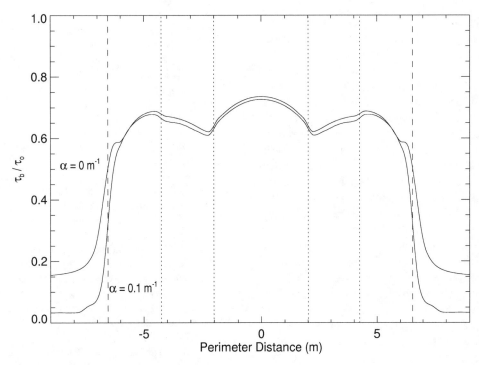

Figure 14. Comparison of boundary shear stress for the two overbank flow cases. The vertical dashed lines denote the interface between the vegetated floodplain and the main channel.

Although the boundary shear stress on the floodplain is substantially less in the vegetated case, there is not a significant difference in boundary shear stress over most of the banks and bed. The shape of the distribution of boundary shear stress for these cases, however, is substantially different from the shape in the bankfull case with no vegetation (Figure 6c). The relative boundary shear stress on the banks in the compound cases is higher, and the effect of the corner on reducing τ_b is more pronounced.

The calculations presented in the preceding four sets of cases can be used together with information on the critical shear stress of the bed and bank material to determine the portions of the boundary capable of transporting sediment. In addition, these calculations can also be used to help explain the patterns of erosion and deposition that occur in channels with high suspended sediment concentrations such as the Rio Puerco. Recent field observations made at three trench cross-sections in the Rio Puerco show that erosion typically occurs on the beds of these channels while deposition occurs on the banks and floodplains [*Kirk Vincent*, personal communication]. The distribution of velocity and boundary shear stress shown in the previous cases indicate that this pattern of erosion and deposition is likely produced by the transport of suspended sediment from the center of the channel outward to the banks and floodplains. Drag on the bank and floodplain vegetation concentrates the flow towards the center of the chan-

nel. The high velocity and boundary shear stress in the center of the channel can erode sediment from the bed and put it into suspension in higher concentrations than in the vegetated zone. This results in a concentration gradient between the center of the channel and the vegetated banks and floodplain. The concentration gradient forces the suspended sediment in the center of the channel to move into the vegetated zone either by diffusion or advection from secondary circulations associated with channel curvature. The lateral transport of suspended sediment may be further enhanced by coherent flow structures associated with the vegetation, such as those observed by *Bennett* [this volume]. The additional sediment supplied to the vegetated regions cannot be supported by the reduced flow through the vegetation and would be deposited on the boundary beneath the vegetation producing channel boundaries at the bulk angle of repose.

5. SUMMARY AND CONCLUSIONS

The calculations presented in this paper demonstrate that including the effects of (1) drag on vegetation and (2) friction on the lateral boundaries is essential for determining the near-bank velocity and boundary shear stress fields in rivers with woody vegetation on the banks and floodplains. This can be done by explicitly calculating the drag on the stems and branches and using a turbulence closure that accommodates lateral boundaries. By comparison to the experimental data set of *Pasche* [1984], a model that includes these effects was shown to reproduce the essential structure of the depth-averaged velocity profile in a channel that has simulated dense vegetation and steep banks. The model was then used to examine the effects of vegetation density and channel geometry on the distribution of velocity and boundary shear stress in a channel modeled after that of the Rio Puerco near Belen, New Mexico.

Drag on woody bank vegetation substantially reduces the perimeter averaged boundary shear stress from that calculated from the product of the slope and the hydraulic radius. The slow velocity within the vegetation reduces the boundary shear stress beneath the vegetation. This increases the lateral velocity gradient near the bank over what it would be with the banks alone, and further reduces the nearby bed stress. The model shows that in channels with sparse vegetation on the banks, both drag on the vegetation and stress on the banks contribute to reducing the near-bank flow and boundary shear stress, while in channels with dense bank vegetation, drag on the vegetation is the dominant effect. The model also shows that sparse vegetation by itself can have effects comparable to a sloping bank alone.

For sediment transport and erosion/deposition calculations it is necessary to know the field of boundary stress, devoid of the effects of form drag. This is called skin friction by *Smith and McLean* [1977]. To determine this field, a model such as the one presented in this paper is both essential and suitable.

Acknowledgements. Sean Bennett, Jonathan Friedman, John Pitlick, and Ellen Wohl reviewed this paper and made many helpful suggestions.

REFERENCES

Barnes, H. H., Roughness characteristics of natural channels, *U.S. Geological Survey Water Supply Paper 1849*, 1967.

Chow, V. T., *Open Channel Hydraulics*, 680 pp., McGraw-Hill Book Company, New York, 1959.

Einstein, H. A., and N. Chien, *Effects of heavy sediment concentrations near the bed on velocity and sediment distribution*, University of California, Berkeley, Institute of Engineering Research, no. 8, 1955.

Elliot, J. G., A. C. Gellis, and S. B. Aby, Evolution of arroyos: Incised channels of the southwestern United States, in *Incised River Channels: Processes, Forms, Engineering and Management*, edited by S. E. Darby, and A. Simon, pp. 153–185, John Wiley, New York, 1999.

Houjou, K, Y. Shimizu, and C. Ishii, Calculation of boundary shear stress in open channel flow, *Journal of Hydroscience and Hydraulic Engineering*, 8, 21–37, 1990.

Kean, J. W., Computation of flow and boundary shear stress near the banks of streams and rivers, unpublished, Ph.D. dissertation, University of Colorado, Boulder, 2003.

Knight, D. W., K. W. H. Yuen, and A. A. I. Al-Hamid, Boundary shear stress distributions in open channel flow, in *Mixing and Transport in the Environment*, edited by K. J. Beven, P. C. Chatwin, and J. H. Millbank, pp 51–87, John Wiley, New York, 1994.

Long, C. E., P. L. Wiberg, and A. R. M. Nowell, Evaluation of von Karman's constant from integral flow parameters, *Journal of Hydraulic Engineering*, 119, 1182–1190, 1993.

López, F., and M. García, *Open-channel flow through simulated vegetation: Turbulence modeling and sediment transport*, Wetlands Research Program Technical Report WRP-CP-10, Waterways Experimental Station, Vicksburg, Mississippi, Aug. 1997.

Nepf, H. M., Drag, turbulence, and diffusion in flow through emergent vegetation, *Water Resources Research*, 35(2), 479–489, 1999.

Pasche, E., Turbulenzmechanismen in naturnahen Fließgewässern und die Möglichkeiten ihrer mathematischen Erassung. Mitteilungen, Institut für Wasserbau und Wasserwirtshaft, TH Aachen, Heft 52, 1984.

Pasche, E., and G. Rouvé, Overbank flow with vegetatively roughened flood plains, *Journal of Hydraulic Engineering*, 11(9), 1262–1278, 1985.

Rattray, M. Jr., and E. Mitsuda, Theoretical analysis of conditions in a salt wedge, *Estuarine and Coastal Marine Science*, 2, 373–394, 1974.

Shimizu, Y., Effects of lateral shear stress in open channel flow, report for the River Hydraulics and Hydrology Laboratory, Civil Engineering Research Institute, Hokaido, Japan, 22 pp., 1989.

Shimizu, Y., and T. Tsujimoto, Numerical analysis of turbulent open-channel flow over a vegetation layer using a k-ε turbulence model, *Journal of Hydroscience and Hydraulic Engineering*, 11(2), 57–67, 1994.

Simões, F. J. M., Three-dimensional modeling of flow through sparse vegetation, in *Proceedings of the Seventh Federal Interagency Sedimentation Conference*, 1(I), pp. 85–92, 2001.

Smith, J. D., and S. R. McLean, Spatially averaged flow over a wavy surface, *Journal of Geophysical Research*, 82, 1735–1746, 1977.

Smith, J. D., On quantifying the effects of riparian vegetation in stabilizing single threaded streams, in *Proceedings of the Seventh Federal Interagency Sedimentation Conference*, 1(IV), pp. 22–29, 2001.

Smith, J. D., and E. R. Griffin, Relation between geomorphic stability and the density of large shrubs on the flood plain of the Clark Fork of the Columbia River in the Deer Lodge Valley, Montana, *U.S. Geological Survey Water Resources Investigations Report 02-4070*, 25 pp., 2002a.

Smith, J. D., and E. R. Griffin, Quantitative analysis of catastrophic transformation from a narrow, sinuous to a broad, straight creek, *U.S. Geological Survey Water Resources Investigations Report 02-4065*, 2002b.

Jason W. Kean and J. Dungan Smith, U.S. Geological Survey, 3215 Marine Street, Suite E-127, Boulder, CO, 80303

A Depth-Averaged Two-Dimensional Numerical Model of Flow and Sediment Transport in Open Channels with Vegetation

Weiming Wu and Sam S. Y. Wang

National Center for Computational Hydroscience and Engineering The University of Mississippi, Mississippi

A depth-averaged two-dimensional (2-D) numerical model for the simulation of flow, sediment transport and bed morphological changes in vegetated open channels is established. The flow model solves the depth-averaged 2-D shallow water equations, with the eddy viscosity being determined by the standard k-ε turbulence model. The vegetation effect is considered by including the drag force exerted by the flow on the vegetation in the momentum equations, and the generation and dissipation of turbulent energy due to the presence of vegetation in the k and ε equations. Because the vegetation density appears in the governing equations, the model is applicable to situations with high vegetation density. The sediment transport model simulates the non-equilibrium transport of nonuniform total load. The governing equations are solved using the finite volume method applied to a curvilinear, non-staggered grid. The model has been tested against measured data of three laboratory experiments. The cross-stream profiles of flow velocity and Reynolds shear stress in a straight flume partially covered with vegetation, the meandering flow pattern in a flume with alternate vegetation zones, and the growth of a vegetated island are well reproduced by the proposed model.

1. INTRODUCTION

Vegetation growing on channel bed, banks and floodplains plays an important role not only in biological processes, but also in morphodynamic processes in streams [*Brookes and Shields*, 1996]. The flow within and close to the vegetation is retarded, whereas the flow away from it may be strengthened, thereby changing the channel morphology. The pollutant transport and aquatic habitat may also be affected by vegetation. With increasing attention to the ecological quality of surface water systems and the restoration of streams to their natural states, research on the vegetation effect on streams has been broadened and accelerated. A lot of field and laboratory studies, and numerical analyses have been carried out by many investigators. *Tsujimoto et al.* [1992]; *Lopez and Garcia* [1996]; *Hodge et al.* [1997]; *Fairbanks and Diplas* [1998]; *Carollo et al.* [2002]; *Stone and Shen* [2002] performed experiments to study the effect of rigid, flexible, submerged or emergent vegetation on mean flow velocity, turbulent characteristics and flow resistance. *Li and Shen* [1973]; *Reid and Whitaker* [1976]; *Dunn et al.* [1996]; *Fathi-Maghadam and Kouwen* [1997]; *Wu et al.* [1999] studied the drag coefficient of vegetation. *Barfield et al.* [1979]; *Okabe et al.* [1997]; *Tsujimoto* [1998]; *Lopez and Garcia* [1998] studied sediment transport and filtration in vegetated channels. *Bennett et al.* [2002] experimentally studied the flow in a channel with alternate vegetation bars, and *Tsujimoto* [1998] performed experiments and a field study about the geomorphological development of a vegetated island in streams. *Shimizu and Tsujimoto* [1994]; *Lopez and Garcia* [2001] established ver-

Riparian Vegetation and Fluvial Geomorphology
Water Science and Application 8

tical two-dimensional models for turbulent flow with vegeta-tion effect, and *Tsujimoto et al.* [1996] reported a depth-aver-aged two-dimensional model for flow and sediment transport in vegetated open channels. These studies have given physi-cal insights and numerical tools for vegetation-related river engineering problems.

However, interaction amongst vegetation, flow, sediment and channel form is very complicated. A robust tool to numer-ically study the whole system is needed. This paper presents the basis of the mathematical model of the flow and sediment transport in vegetated channels. A depth-averaged 2-D numer-ical model is established to supplement the previous studies conducted by *Shimizu and Tsujimoto* [1994]; *Tsujimoto et al.* [1996]; *Lopez and Garcia* [2001]. In particular, the proposed model considers the density of vegetation in the governing equations, and simulates the turbulent flow, nonuniform total-load transport and bed morphological changes in vegetated open channels.

2. CONCEPTS AND DEFINITIONS

2.1. Representation of Vegetation

Natural vegetation is highly irregular in shape, perhaps rigid or flexible, and also probably submerged or unsub-merged. It is very difficult to represent with simple geometry. However, as an approximation, the individual vegetation ele-ment is often conceptualized as a cylinder with a height of h_v and a representative diameter of D. The height is defined as the actual height of the vegetation element without any action of flow. The representative diameter D can be related to the actual volume V of the vegetation element by $D = \sqrt{4V/\pi h_v}$. The wetted projection area of the submerged or unsubmerged vegetation element perpendicular to the flow direction is

$$A_v = \alpha_v D \min(h_v, h) \qquad (1)$$

where h is the flow depth; and α_v is the shape factor of veg-etation. $\alpha_v = 1$ for a rigid cylinder. If the vegetation is irregu-lar or flexible, α_v may have different values. The flexible vegetation will bend under the action of flow, and thus the value of α_v for flexible vegetation may change with flow con-ditions. The shape factor α_v is therefore able to account for the irregularity and somehow the flexibility of vegetation.

2.2. Vegetation Density

Considering a volume of the mixture consisting of a group of vegetation elements and the water, the vegetation density (or volumetric concentration), c, is defined as the ratio of the vegetation volume over the total volume of the mixture of water and vegetation.

The wetted volume of a vegetation element is $\pi D^2 \min(h_v, h)/4$. Therefore, the number of vegetation elements in a unit volume of the mixture is

$$n_v = \frac{4c}{\pi D^2 \min(h_v, h)} \qquad (2)$$

and the projection area of the vegetation elements in a unit vol-ume of the mixture is

$$\lambda_a = n_v A_v = \frac{4\alpha_v c}{\pi D} \qquad (3)$$

2.3. Forces on Vegetation

The vegetation immersed in water may experience a buoyancy force, drag force, virtual mass force, Basset force [*Basset*, 1961], Saffman force [*Saffman*, 1965], and a lateral ("lift") force due to its asymmetric shape. Of these forces, the drag force is the most important force acting on the vegetation. Com-pared to the drag force, the virtual mass force, Basset force and Saffman force are usually negligible [see *Wu and Wang*, 2000], and thus not considered in the present model. The lateral force due to asymmetric shape on a group of vegetation elements is usually ignored because the direction of vegetation shape

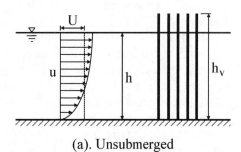 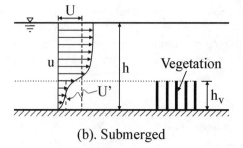

(a). Unsubmerged   (b). Submerged

Figure 1. Sketch of Flow in Open Channels with Vegetation.

distributes randomly. The drag force on the vegetation elements in a unit volume of the mixture is expressed as

$$\vec{f}_d = \frac{1}{2}C_d\rho\lambda_a|\vec{U}|\vec{U} = C_d\rho\alpha_v\frac{2c}{\pi D}|\vec{U}|\vec{U} \qquad (4)$$

where C_d is the drag coefficient, which is related to the flow conditions, vegetation size and shape, among others; ρ is the water density; $\vec{U}$ is the flow velocity vector; $|\vec{U}|$ is the magnitude of flow velocity.

In a depth-averaged model, if the vegetation is unsubmerged, the flow velocity $\vec{U}$ is the velocity averaged over the whole flow depth, as shown in Figure 1(a). If the vegetation is submerged, the flow velocity acting on the vegetation is different from the depth-averaged velocity, but Eq. (4) is still applicable by defining the drag coefficient C_d as

$$C_d = C'_d\,\overline{U}'^2/\overline{U}^2 \qquad (5)$$

where $\overline{U}'$ is the flow velocity averaged only over the vegetation height shown in Figure 1(b), and C'_d is the drag coefficient corresponding to $\overline{U}'$. The velocity $\overline{U}'$ can be determined using *Stone and Shen's* [2002] method.

3. GOVERNING EQUATIONS AND BOUNDARY CONDITIONS

3.1. Governing Equations of Flow

The flow around vegetation is unsteady and three-dimensional, due to the disturbance of vegetation. However, the most considered flow properties in practical engineering are the time-averaged and space-averaged behaviors, rather than the detailed flow features around each single vegetation element. Time-averaging and space-averaging the Navier-Stokes equations, one can derive three-dimensional governing equations for the flow with vegetation effect. Integrating these three-dimensional equations over the flow depth, one obtains the depth-integrated continuity and momentum equations for the flow in vegetated open channels:

$$\frac{\partial[\rho(1-c)h]}{\partial t} + \frac{\partial[\rho(1-c)Uh]}{\partial x} + \frac{\partial[\rho(1-c)Vh]}{\partial y} = 0 \qquad (6)$$

$$\frac{\partial[\rho(1-c)Uh]}{\partial t} + \frac{\partial[\rho(1-c)UUh]}{\partial x} + \frac{\partial[\rho(1-c)UVh]}{\partial y}$$
$$= -\rho g(1-c)h\frac{\partial z_s}{\partial x} + \frac{\partial[(1-c)hT_{xx}]}{\partial x} + \frac{\partial[(1-c)hT_{xy}]}{\partial y} \qquad (7)$$
$$- \tau_{bx} - f_{dx}h$$

$$\frac{\partial[\rho(1-c)Vh]}{\partial t} + \frac{\partial[\rho(1-c)UVh]}{\partial x} + \frac{\partial[\rho(1-c)VVh]}{\partial y}$$
$$= -\rho g(1-c)h\frac{\partial z_s}{\partial y} + \frac{\partial[(1-c)hT_{yx}]}{\partial x} + \frac{\partial[(1-c)hT_{yy}]}{\partial y} \qquad (8)$$
$$- \tau_{by} - f_{dy}h$$

where t is the time; x and y are the horizontal Cartesian coordinates; U and V are the depth-averaged flow velocities in x- and y-directions; z_s is the water surface elevation; g is the gravitational acceleration; T_{xx}, T_{xy}, T_{yx} and T_{yy} are the depth-averaged turbulent stresses; τ_{bx} and τ_{by} are the bed shear stresses determined by $\vec{\tau}_b = \rho(1-c)c_f\vec{U}|\vec{U}|$, with $c_f = gn^2/h^{1/3}$ and n being the Manning's roughness coefficient; and f_{dx} and f_{dy} are the x- and y-components of the drag force on the vegetation exerted by the flow, which is expressed in Eq. (4).

Compared with the governing equations derived by *Tsujimoto et al.* [1996] and *Lopez and Garcia* [2001], Eqs. (6)-(8) include the influence of the vegetation density c, which should be important when the vegetation density is high.

The turbulent stress terms, which are usually important in the situation of complex geometry, are calculated with the aid of the depth-averaged k-ε model employing the Boussinesq's assumption

$$T_{xx} = 2\rho(\nu + \nu_t)\frac{\partial U}{\partial x} - \frac{2}{3}k \qquad (9a)$$

$$T_{xy} = T_{yx} = \rho(\nu + \nu_t)\left(\frac{\partial U}{\partial y} + \frac{\partial V}{\partial x}\right) \qquad (9b)$$

$$T_{yy} = 2\rho(\nu + \nu_t)\frac{\partial V}{\partial y} - \frac{2}{3}k \qquad (9c)$$

where ν is the kinematic viscosity; and ν_t is the eddy viscosity determined by

$$\nu_t = c_\mu\frac{k^2}{\varepsilon} \qquad (10)$$

where c_μ is an empirical constant. The turbulence energy k and its dissipation rate ε are determined with the following transport equations:

$$\frac{\partial k}{\partial t} + U\frac{\partial k}{\partial x} + V\frac{\partial k}{\partial y} = \frac{\partial}{\partial x}\left(\frac{\nu_t}{\sigma_k}\frac{\partial k}{\partial x}\right) + \frac{\partial}{\partial y}\left(\frac{\nu_t}{\sigma_k}\frac{\partial k}{\partial y}\right) \qquad (11)$$
$$+ P_h + P_{kb} + P_v - D_v - \varepsilon$$

$$\frac{\partial\varepsilon}{\partial t} + U\frac{\partial\varepsilon}{\partial x} + V\frac{\partial\varepsilon}{\partial y} = \frac{\partial}{\partial x}\left(\frac{\nu_t}{\sigma_\varepsilon}\frac{\partial\varepsilon}{\partial x}\right) + \frac{\partial}{\partial y}\left(\frac{\nu_t}{\sigma_\varepsilon}\frac{\partial\varepsilon}{\partial y}\right) \qquad (12)$$
$$+ c_{\varepsilon1}\frac{\varepsilon}{k}[P_h + c_{\varepsilon3}(P_v - D_v)] + P_{\varepsilon b} - c_{\varepsilon2}\frac{\varepsilon^2}{k}$$

where $P_{kb} = c_f^{-1/2} U_*^3 / h$; $\quad\quad$ $P_{\varepsilon b} = c_{\varepsilon\Gamma} c_{\varepsilon 2} c_\mu^{1/2} c_f^{-3/4} U_*^4 / h^2$; $P_h = v_t \left[2(\partial U/\partial x)^2 + 2(\partial V/\partial y)^2 + (\partial U/\partial y + \partial V/\partial x)^2 \right]$; U_* is the bed shear velocity; P_v is the generation of turbulence energy due to the presence of vegetation, determined by $P_v = c_{vk} \left(f_{dx} U + f_{dy} V \right) / \left[\rho(1-c) \right]$, with the coefficient being given a value of 1.0 [*Tsujimoto et al.*, 1996 and *Lopez and Garcia*, 2001]; D_v accounts for the dissipation of turbulence energy due to the fluctuation of drag force, and is derived by us from the time-averaged kinetic energy equation as $D_v = \overline{f'_{dx} U'} + \overline{f'_{dy} V'} = 4\beta_v c C_d \alpha_v |\overline{U}| k / [\pi D (1-c)]$, in which the prime " ' " denotes the fluctuating quantity, and the coefficient β_v is given a value of about 2.0; $c_{\varepsilon 1}$, $c_{\varepsilon 2}$, $c_{\varepsilon 3}$, $c_{\varepsilon\Gamma}$, σ_κ, and σ_ε are empirical coefficients in the depth-averaged standard k-ε model, and their values are $c_\mu = 0.09$, $c_{\varepsilon 1} = 1.44$, $c_{\varepsilon 2} = 1.92$, $c_{\varepsilon 3} = 1.33$, $\sigma_\kappa = 1.8$, $\sigma_\varepsilon = 1.3$, and $c_{\varepsilon\Gamma} = 3.6$ for laboratory experiments, or for natural rivers [*Rodi*, 1993; *Tsujimoto et al.*, 1996; *Lopez and Garcia*, 2001].

3.2. Governing Equations of Sediment Transport

The total load is separated as bed load and suspended load. The advection-diffusion equation of suspended load is

$$\frac{\partial \left[(1-c) h S_k \right]}{\partial t} + \frac{\partial \left[(1-c) U h S_k \right]}{\partial x} + \frac{\partial \left[(1-c) V h S_k \right]}{\partial y}$$
$$= \frac{\partial}{\partial x} \left[\varepsilon_s (1-c) h \frac{\partial S_k}{\partial x} \right] + \frac{\partial}{\partial y} \left[\varepsilon_s (1-c) h \frac{\partial S_k}{\partial y} \right] \quad (13)$$
$$+ \alpha \omega_{sk} (1-c) \left(S_{*k} - S_k \right)$$

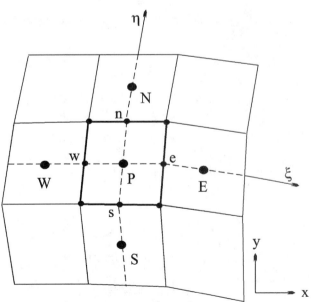

Figure 2. Two-dimensional Control Volume.

where S_k is the depth-averaged concentration of kth size class of suspended load; S_{*k} is the depth-averaged suspended-load concentration under equilibrium conditions or the suspended-load transport capacity; ε_s is the diffusivity coefficient of sediment; α is the non-equilibrium adaptation coefficient of suspended load; ω_{sk} is the settling velocity of sediment particles.

The non-equilibrium bed-load transport equation is

$$\frac{\partial \left[(1-c) \delta \overline{s}_{bk} \right]}{\partial t} + \frac{\partial \left[\alpha_{bx} (1-c) q_{bk} \right]}{\partial x} + \frac{\partial \left[\alpha_{by} (1-c) q_{bk} \right]}{\partial y}$$
$$+ \frac{1}{L} (1-c) \left(q_{bk} - q_{b*k} \right) = 0 \quad\quad (14)$$

where δ is the thickness of bed-load zone; $\overline{s}_{bk}$ is the average concentration of bed load at the bed-load zone; α_{bx} and α_{by} are direction cosines of bed-load movement, which are usually assumed to be along the direction of bed shear stress, but are adjusted when taking into account the influences of the bed slope and secondary flow in curved channels; q_{bk} is the actual transport rate of kth size class of bed load; L is the non-equilibrium adaptation length of total load, whose specification can be referred to *Wu* [2003].

The change in bed elevation by size class is calculated as:

$$\left(1 - p' \right) \left(\frac{\partial z_b}{\partial t} \right)_k = \alpha \omega_{sk} \left(S_k - S_{*k} \right) + \frac{1}{L} \left(q_{bk} - q_{b*k} \right) \quad (15)$$

where p' is the porosity of bed material.

Sediment transport capacity may be different in cases with and without vegetation. However, it is very difficult to resolve this difference, at the present time. As an approximation, the transport capacities of bed load and suspended load in the vegetation zone are determined by the four empirical functions that were calibrated based on the data without including the vegetation. These four formulas are the modified Ackers and White's formula [*Proffit and Sutherland*, 1983], the SEDTRA module [*Garbrecht et al*, 1995], *Wu et al.'s* [2000] formula, and the modified *Engelund and Hansen's* [1967] formula with *Wu et al.'s* correction factor [*Wu and Vieira*, 2002]. The four formulas can be written as general forms:

$$S_{*k} = p_{bk} S_k^{(*)}; \quad q_{b*k} = p_{bk} q_{bk}^{(*)} \quad\quad (16)$$

where p_{bk} is the bed-material gradation in the mixing layer; $q_{bk}^{(*)}$ and $S_k^{(*)}$ are the potential transport capacities of kth size class of bed load and suspended load.

The bed-material gradation usually varies along the vertical. Therefore, the bed material above the non-erodible layer is divided into several layers. The top layer is the mixing layer, in which the temporal variation of bed-material gradation is determined by

$$\frac{\partial(\delta_m p_{bk})}{\partial t} = \left(\frac{\partial z_b}{\partial t}\right)_k + p_{bk}^* \left(\frac{\partial \delta_m}{\partial t} - \frac{\partial z_b}{\partial t}\right) \qquad (17)$$

where δ_m is the thickness of the mixing layer, which is related to the flow and sediment conditions as well as the bed deformation; $\partial z_b / \partial t$ is the total bed deformation rate, $\partial z_b / \partial t = \sum_{k=1}^{N} (\partial z_b / \partial t)_k$; N is the total number of size classes; p_{bk}^* is p_{bk} when $\partial \delta_m / \partial t - \partial z_b / \partial t \le 0$, and p_{bk}^* is the bed-material gradation in the subsurface layer when $\partial \delta_m / \partial t - \partial z_b / \partial t > 0$.

3.3. Boundary Conditions

Near rigid wall boundaries, such as banks and islands, the wall-function approach is employed. The resultant wall shear stress $\vec{\tau}_w$ is related to the flow velocity $\vec{V}_P$ at the center P of the control volume close to the wall by the following relation:

$$\vec{\tau}_w = -\lambda \vec{V}_P \qquad (18)$$

where the coefficient λ is determined as $\lambda = \rho c_\mu^{1/4} k_P^{1/2} \kappa / \ln\left(E y_P^+\right)$ for $11.6 < y_P^+ < 300$. Here $y_P^+ = c_\mu^{1/4} k_P^{1/2} y_P / \nu$. κ is the von Karman constant, E is about 9, and y_p is the distance from the cell center P to the wall.

The turbulence generation P_h and the dissipation rate near the wall are determined by $P_{h,P} = \tau_w^2 / \kappa \rho \nu y_P^+$ and $\varepsilon_P = c_\mu^{3/4} k_P^{3/2} / \kappa y_P$.

At banks and islands, the bed-load transport rate and the gradient of suspended-load concentration normal to the boundary are set to zero, i.e.,

$$q_{bk} = 0 \; ; \; \frac{\partial S_k}{\partial n} = 0 \qquad (19)$$

where n is the direction normal to the boundary.

3.4. Wetting and Drying Technique

During the simulation of unsteady flow in a open channel with sloped banks, sand bars and islands, the computational domain may change from dry to wet or vice versa due to variations in water surface elevation over time. Even for steady flow, the water edges are not known until the computation is finished. In the present model, a threshold flow depth (a small value such as 0.02 m in natural rivers) is used to judge drying and wetting. If the flow depth in a node is larger than the threshold value, this node is considered to be wet, and if the flow depth is lower than the threshold value, this node is dry.

4. NUMERICAL METHODS

4.1. Solution Procedure of Flow

The governing equations are discretized using the finite volume method on a curvilinear, non-staggered grid. In a curvilinear coordinate system Eqs. (6)–(8) and (11)–(13) can be written in the common tensor notation form:

$$\frac{\partial}{\partial t}\left[J\rho(1-c)h\phi\right] + \frac{\partial}{\partial \xi_m}\left[J\rho(1-c)h\left(\hat{u}_m\phi - \Gamma_\phi \alpha_j^m \alpha_j^n \frac{\partial \phi}{\partial \xi_n}\right)\right]$$
$$= J\rho h(1-c)S_\phi \qquad (20)$$

where ϕ stands for 1, U, V, k, ε and S_k, respectively, depending on the equation considered; $\Gamma_\phi = \nu + \nu_t / \sigma_\phi$ is the diffusivity of the quantity ϕ; S_ϕ is the source term in the equation of ϕ; J is the Jacobian of the transformation between the Cartesian coordinate system x_i ($x_1 = x$ and $x_2 = y$) and the computational curvilinear coordinate system ξ_m ($m = 1,2$); $\hat{u}_m = \alpha_i^m U_i$; $\alpha_i^m = \partial \xi_m / \partial x_i$.

Eq. (20) is integrated over the control volume shown in Figure 2. The convection terms in Eq. (20) are discretized by the hybrid upwind/central scheme [Spalding, 1972], the exponential scheme [Spalding, 1972], the QUICK scheme [Leonard, 1979] or the HLPA scheme [Zhu, 1992]. The diffusion terms are discretized by a central scheme. The time derivative terms are discretized by the first-order backward scheme or a third-order implicit scheme. The resulting discretized equations can be written as

$$a_P^{(\phi)} \phi_P^{n+1} = \sum_{k=W,E,S,N} a_k^{(\phi)} \phi_k^{n+1} + S_{\phi P} \qquad (21)$$

where $a_P^{(\phi)}$, $a_W^{(\phi)}$, $a_E^{(\phi)}$, $a_S^{(\phi)}$ and $a_N^{(\phi)}$ are the coefficients in the discretized equation of ϕ; the superscript n is the time step number; $S_{\phi P}$ is the source term in the discretized equation, which includes the source term S_ϕ, the terms evaluated at the previous time level n, and the cross-derivative diffusion terms. The discretized equations are solved by the Strongly Implicit Procedure (SIP) [Stone, 1968].

Spurious numerical oscillation may exist due to the checkerboard splitting in the collocated arrangement, as explained by Patankar [1980] and others. In order to avoid this problem, Rhie and Chow [1983] proposed a momentum interpolation technique to evaluate the cell face variables from the centered quantities. Zhu [1992] built this technique in a general code for two-dimensional flows, which was extended to the depth-averaged simulation of open-channel flow by Wenka [1992] and Minh Duc [1998]. Wu [2003] modified further the formulations of Wenka and Minh

Duc, and established a depth-averaged 2-D model for unsteady open-channel flow. *Wu's* [2003] method is extended to solve the equations (6)–(8) for the open-channel flow with vegetation effect in this study.

From the discretized momentum equations, one can derive the following equation for velocities $U_{i,P}^{n+1}$ (i=1,2):

$$U_{i,P}^{n+1} = \frac{\sum_{k=W,E,S,N} a_k^{(ui)} U_{i,k}^{n+1} + S_{ui}}{a_P^{(ui)}} \tag{22}$$
$$+ D_i^1 \left(p_w^n - p_e^n\right) + D_i^2 \left(p_s^n - p_n^n\right)$$

where $D_i^m = \left[Jh(1-c)\alpha_i^m \right]_P / a_P^{(ui)}$; $p = \rho g z_s$.

The discretized continuity equation reads

$$p_P^{n+1} = p_P^n - g\frac{\Delta t}{\Delta A(1-c_P)}\left(C_e - C_w + C_n - C_s\right) \tag{23}$$

where ΔA is the area of the control volume; C_e, C_w, C_n and C_s are the fluxes at cell faces e, w, n and s. C_e and C_w are defined as $C_f = \left[J\rho(1-c)h\alpha_i^1 U_i\right]_f$ ($f = e,w$), while C_n and C_s are $C_f = \left[J\rho(1-c)h\alpha_i^2 U_i\right]_f$ ($f=n,s$).

Using the momentum interpolation procedure of *Rhie and Chow* [1983], one obtains

$$C_w = C_w^* + a_W^{(p)}\left(p_W' - p_P'\right) \tag{24a}$$

$$C_s = C_s^* + a_S^{(p)}\left(p_S' - p_P'\right) \tag{24b}$$

where p' is the pressure correction, defined as $p' = p^{n+1} - p^*$; $a_W^{(p)} = [Jh(1-c)]_w \left(Q_{1,w}^1\alpha_{1,w}^1 + Q_{2,w}^1\alpha_{2,w}^1\right)$; $a_S^{(p)} = [Jh(1-c)]_s \left(Q_{1,s}^2\alpha_{1,s}^2 + Q_{2,s}^2\alpha_{2,s}^2\right)$; $Q_{i,f}^m$ (i=1, 2; m=1, 2) is defined in *Zhu* [1992]; and the asterisk * stands for the guessed quantities.

Inserting Eqs. (24a) and (24b) into Eq. (23) leads to the equation for pressure correction:

$$a_P^{(p)} p_P' = \sum_{k=W,E,S,N} a_k^{(p)} p_k' + S_p \tag{25}$$

where $a_P^{(p)} = a_W^{(p)} + a_E^{(p)} + a_S^{(p)} + a_N^{(p)} + \Delta A(1-c_P)/(g\Delta t)$; $S_p = -\left(C_e^* - C_w^* + C_n^* - C_s^*\right) - \Delta A(1-c_P)\left(p_P^* - p_P^n\right)/(g\Delta t)$.

Eq. (25) is used to determine the pressure correction and then p. This is the SIMPLE procedure or SIMPLEC procedure, depending on how the parameter $Q_{i,f}^m$ is defined [*Zhu*, 1992].

4.2. Solution Procedure of Sediment Transport

The bed-load and suspended-load transport equations (13) and (14) are discretized by the same finite volume method as used in the flow model. Eqs. (15) and (17) are discretized by finite difference schemes in time at cell centers. The bed-material gradation in Eq. (16) is treated implicitly, following the approach proposed by *Wu and Vieira* [2002] and *Wu* [2003].

The equation set consisting of the discretized suspended-load transport equation, bed-load transport equation, bed change equation and bed-material sorting equation are solved simultaneously in an iteration fashion, which results in a coupled procedure for nonuniform sediment calculation. The advantages of this coupled procedure were discussed in details by *Wu and Vieira* [2002]. It is very stable and has sound physical merits. However, the coupled sediment calculation is decoupled with the flow calculation, in order to simplify the global solution procedure.

5. MODEL TESTS

5.1. Case 1: Flow in a Flume with Vegetation Along One Side

Quasi-uniform flow in a flume covered with unsubmerged vegetation along one side (Figure 3) was studied experimentally by *Tsujimoto and Kitamura* [1995]. The flume was 12 m long and 0.4 m wide, with a slope of 0.0017. The vegetation zone was 0.12 m wide, consisting of bamboos with a diameter of 0.15 cm distributed in a parallel pattern. In experiment runs A1 and B1, the average flow velocity was 0.32 m/s and 0.276 m/s, the flow depth was 0.0457 m and 0.0428 m, and the spacing of vegetation elements was 0.028 m and 0.020 m respectively, as shown in Table 1.

The computational mesh consists of 143 × 45 nodes. The mesh is nonuniformly distributed along the cross section, refined around the interface between the vegetation and non-

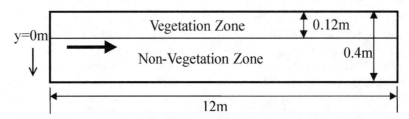

Figure 3. Plan View of *Tsujimoto and Kitamura's* [1995] Experiments.

Table 1. Flow and Vegetation Conditions in *Tsujimoto and Kitamura's* [1995] Experiments.

Experiment Run	Flow Velocity (m/s)	Flow Depth (m)	Flume Slope	Vegetation Material	Vegetation Diameter (cm)	Vegetation Spacing (cm)
A1	0.32	0.0457	0.0017	Bamboo	0.15	2.8
B1	0.276	0.0428	0.0017	Bamboo	0.15	2.0

vegetation zones. The shape factor α_v is given a value of 1.0, and the drag coefficient C_d is set as 1.5 for both experimental runs. Figure 4 shows the comparison of the measured and calculated lateral distributions of flow velocity and Reynolds shear stress $-\rho\overline{U'V'}$. The agreement between the measured and calculated flow velocities is very good. Due to the presence of vegetation, the flow in the vegetation zone is retarded, while the velocity of the main flow is increased. The deviation between the measured and calculated Reynolds shear stresses is larger, but the general trend is well reproduced by the numerical model. A peak value of Reynolds shear stress is found at the computational node close to the interface of the vegetation and non-vegetation zones in the simulation results, and was also observed at the measurement

point close to the interface in the measurements. Because the transverse spacing of computational grid is much smaller than the interval between two measurement points in the experiments, there seems to a be difference between the predicted and measured locations of the peak Reynolds shear stress. However, this difference is actually less than the interval between two measurement points.

5.2. Case 2: Flow in a Flume with Alternate Vegetation Zones

Laboratory experiments on the flow around alternate vegetation zones were performed by *Bennett et al.* [2002] using a tilting recirculating flume that was 16.5 m long and 1.2 m wide. Six semi-circular vegetation zones with equal spacing

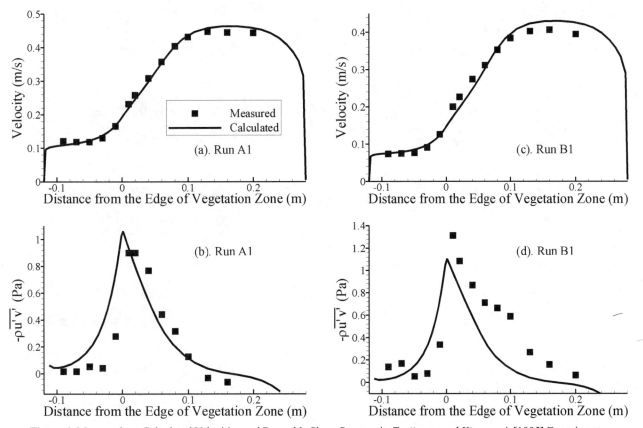

Figure 4. Measured vs. Calculated Velocities and Reynolds Shear Stresses in *Tsujimoto and Kitamura's* [1995] Experiments.

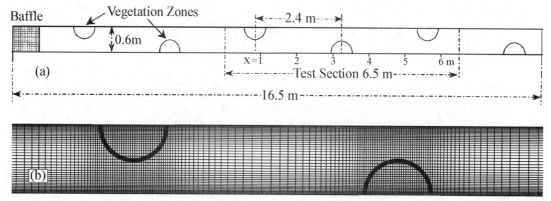

Figure 5. (a) Plan View of *Bennett et al.'s* [2002] Experiments; (b) Computational Mesh around Two Vegetation Zones.

of 2.4 m were distributed alternately in the flume to achieve a meandering pattern, as shown in Figure 5(a). The diameter of a vegetation zone was 0.6 m. The model vegetation was wooden dowel with a diameter of 3.2 mm, laid out in a staggered pattern in each vegetation zone. Five vegetation densities of 0.04%, 0.2%, 0.6%, 2.5% and 10% were used. The flow discharge was 0.0043 m³/s, and the pre-vegetation flow depth was 0.027 m. The slope of the flume was 0.0004. The surface flow velocity was measured by using the Particle Image Velocimetry (PIV) technique.

All the experimental runs with five vegetation densities are calculated by using the proposed model. The computational mesh consists of 461×41 nodes, part of which is shown in Figure 5(b). Finer mesh is used around each vegetation zone.

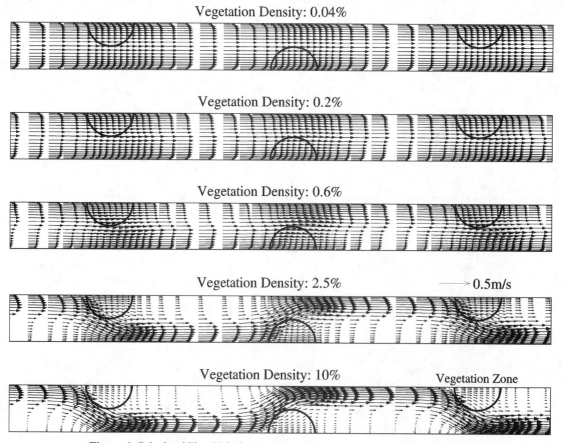

Figure 6. Calculated Flow Velocity Vectors in *Bennett et al.'s* [2002] Experiments.

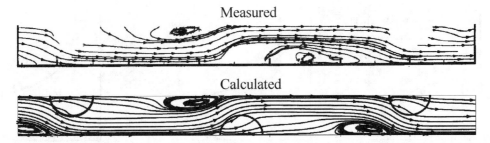

Figure 7. Measured vs. Calculated Streamlines in *Bennett et al.'s* [2002] Experiments (Vegetation Density: 10%).

In the simulation, the shape factor α_v is given a value of 1.0. The drag coefficient C_d is set as 0.8, 1.0, 1.2, 1.8 and 3.0 for the runs with vegetation densities of 0.04%, 0.2%, 0.6%, 2.5% and 10%, respectively, to obtain acceptable agreement between the simulation and measurement. When the vege-

tation density is higher, the flow velocity and Reynolds number Re $=UD/v$ in the vegetation zone are smaller, and thus, according to the relation of C_d and Re for single cylinder, a larger drag coefficient should be used. The used values of drag coefficient are qualitatively reasonable.

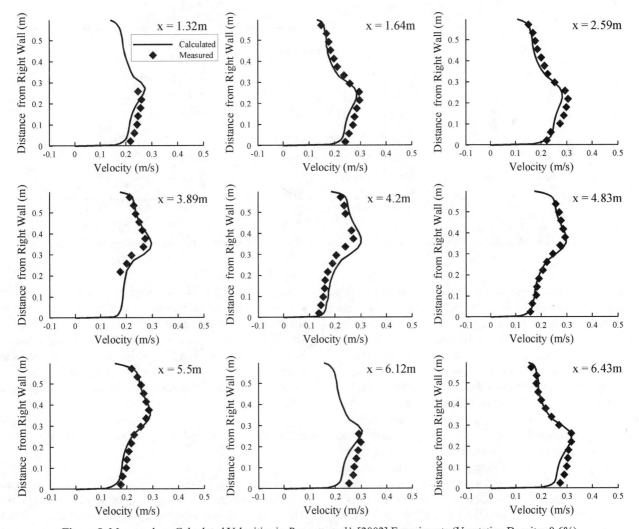

Figure 8. Measured vs. Calculated Velocities in *Bennett et al.'s* [2002] Experiments (Vegetation Density: 0.6%).

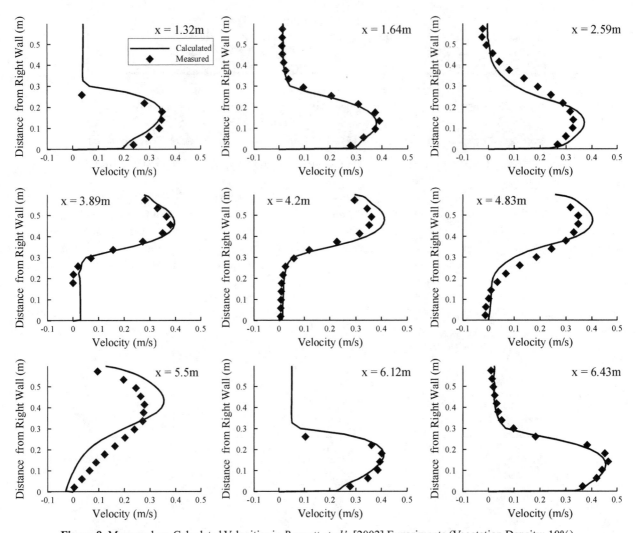

Figure 9. Measured vs. Calculated Velocities in *Bennett et al.'s* [2002] Experiments (Vegetation Density: 10%).

Figure 6 shows the simulated flow vectors. It can be seen that the vegetation curves the flow. The meandering flow pattern becomes more obvious with increasing vegetation density. When the vegetation density is 10%, a recirculation flow occurs downstream of each vegetation zone. The streamlines pass through each vegetation zone, detach from the wall at certain distance downstream of this vegetation zone and then reattach to the wall at the location opposite to the next vegetation zone. Figure 7 shows the comparison of the measured and simulated streamline patterns with vegetation density of 10%. A recirculation flow downstream of each vegetation zone was also observed in the experiment, but the locations and sizes of the recirculation eddies were not stable due to vortex shedding. The sizes of all the simulated recirculation eddies are almost the same, matching the smaller one observed in the experiment.

Figures 8 and 9 show the comparison of the measured and calculated flow velocities along several cross sections for the cases with vegetation densities of 0.6% and 10%. It should be noticed that the measured data are the velocities at water surface taken by PIV, while the calculated results are the depth-averaged flow velocities. The qualitative agreement between the simulation and the measurement is generally reasonable.

5.3. Case 3: Bed Morphological Changes Around a Vegetation Island

The experiment on the expansion of an island with vegetation due to sedimentation was conducted by *Tsujimoto* [1998]. The island was a non-submerged porous body (0.05 m wide and 0.25 m long), located in the center of a 0.4 m-wide straight

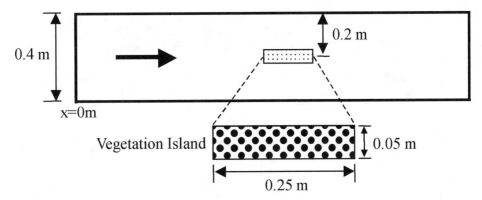

Figure 10. Plan View of *Tsujimoto's* [1998] Experiment.

flume with a sand bed (*d*=1.6 mm), as shown in Figure 10. The properties of the model vegetation were not reported, but the projected area λ_a was determined as 10 m^{-1} in the experiment. The slope of the flume was 1/100, and the flow discharge was 0.003 m^3/s.

The experimental run in which the inflow sediment discharge was at the equilibrium state is simulated here using the proposed model. The computational mesh consists of 122 × 44 nodes, refined around the vegetation island. The computational time step used is 15 seconds. α_v is given a value of 1.0, and C_d is set as 3.0. The simulated flow vectors and velocity magnitude at initial stage (*t*=0) are shown in Figure 11. Due to the effect of vegetation, the flow velocity is reduced inside, upstream and downstream of the vegetation zone, while the flow velocity is increased along two sides of the island. Figure 12 shows the measured and calculated bed elevation changes after 30 minutes. The sediment deposits in front of the vegetation island, and inside

and behind it, and erosion occurs on the two sides of the island. The pattern and magnitude of deposition and erosion are well predicted. Both the experiment and simulation suggest the growth of the vegetation island.

6. DISCUSSION AND CONCLUSIONS

A computational model for the prediction of the flow, sediment transport and channel morphological changes in vegetated open channels has been successfully developed. The flow field is modeled by the depth-averaged, 2-D shallow water equations, in which the drag force terms are added to consider the vegetation effect. The eddy viscosity is determined by the standard *k-ε* turbulence model that includes the generation and dissipation of turbulence energy due to the existence of vegetation. Compared with *Tsujimoto et al.'s* [1996] and *Lopez and Garcia's* [2001] models, the present model considers the density of vegetation in the governing

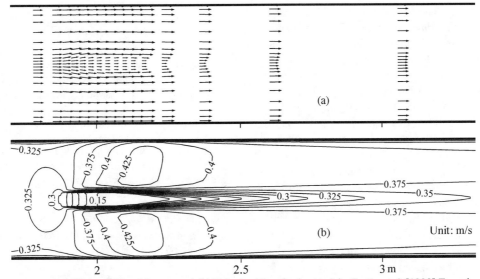

Figure 11. (a) Calculated Flow Vectors and (b) Velocity Magnitude at t=0 in *Tsujimoto's* [1998] Experiment.

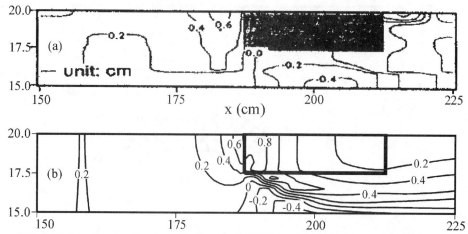

Figure 12. Bed Changes after 30 min in *Tsujimoto's* [1998] Experiment: (a) Measured and (b) Calculated.

equations, and thus should be more applicable in the case of high vegetation density.

A non-equilibrium transport model of nonuniform sediment is adopted, considering bed-load and suspended-load transports, bed elevation changes and bed-material sorting. Because the difference in sediment transport capacities in the situations with and without vegetation effect is not well understood for the time being, the sediment transport capacity in the vegetation zone is determined by using empirical formulas that were calibrated with the data without vegetation effect. Therefore, the effect of vegetation on sediment transport is only indirectly taken into account by using the bed shear stress or the flow velocity adjusted by the vegetation. This treatment can be improved in the future when the formula of sediment transport capacity with vegetation effect is available.

The governing equations of flow and sediment transport are discretized using the finite volume method on a curvilinear, non-staggered grid. The SIMPLE and SIMPLEC algorithms in conjunction with *Rhie and Chow's* [1983] momentum interpolation technique are adopted to achieve the velocity and pressure coupling in the flow model. This method eliminates the possible numerical oscillation when using a non-staggered grid. A coupled calculation procedure is adopted in the sediment module to solve nonuniform sediment transport, bed elevation changes and bed-material sorting. However, the sediment module is still decoupled with the flow module to simplify the whole simulation procedure.

The proposed model has been tested against three sets of experimental data. The calculated transverse distributions of flow velocity and Reynolds shear stress agree reasonably well with the results observed in *Tsujimoto and Kitamura's* [1995] experiments. The meandering flow pattern in a channel with alternate vegetation zones in *Bennett et al.'s* [2002] experiments and the growth of vegetated island in *Tsujimoto's* [1998] experiment are well reproduced by the numerical model.

It should be noticed that to simplify the mathematical modeling of flow in vegetated open channels, the proposed model approximates each vegetation element as a cylinder with a representative diameter D and a shape factor α_v of irregularity and flexibility, and relates the drag force to the flow velocity, the projected area of vegetation and the drag coefficient C_d. However, these parameters vary for different vegetation species. They need to be further studied in cooperation with field investigations and laboratory experiments, to broaden the applicability of the proposed numerical model.

Acknowledgments. The proposed model is a part of the CCHE2D*fvm* model for flow and sediment transport in rivers, which is a result of the research project sponsored by the USDA Agricultural Research Service Specific Research Agreement No.58-6408-2-127 and the University of Mississippi. Mr. Nobuyuki Chiba, Research Assistant of NCCHE, and Dr. Brian D. Barkdoll are thanked for their help and suggestion.

REFERENCES

Barfield, B. J., E. W. Tollner, and Hayes, J. C., Filtration of sediment by simulated vegetation, I. Steady-state flow with homogeneous sediment, *Transactions, Soil and Water Division*, ASCE, 4451, 540–556, 1979.

Basset, A. B., Hydrodynamics, Vol. 2, Dover, New York, pp. 285–292, 1961.

Bennett, S. J., T. Pirim, and B. D. Barkdoll, Using simulated emergent vegetation to alter stream flow direction within a straight experimental channel, *Geomorphology*, 44, 115–126, 2002.

Brookes, A., and Shields Jr., F. D., *River Channel Restoration*, 433 PP., Wiley, Chichester, 1996.

Carollo, F. G., V. Ferro, and D. Termini, Flow velocity measurements in vegetated channels, *J. of Hydraulic Engineering*, ASCE, 128, 664–673, 2002.

Dunn, C., F. Lopez, and M. Garcia, Vegetation-induced drag: An experiment study, in *Proceedings of the North American Water and Environment Congress*, ASCE, New York, 1996.

Engelund, F., and E. Hansen, *A Monograph on Sediment Transport in Alluvial Streams*, Teknisk Vorlag, Copenhagen, Denmark, 1967.

Fairbanks, D. J., and P. Diplas, Turbulence characteristics of flows through partially and fully submerged vegetation, in *Proceedings of the Wetland Engineering and River Restoration* Conference, Denver, CO, 1998.

Fathi-Maghadam, M., and N. Kouwen, Nonrigid, nonsubmerged, vegetative roughness on floodplains, *J. of Hydraulic Engineering*, ASCE, 123, 51–57, 1997.

Garbrecht, J., R. Kuhnle, and C. Alonso, A sediment transport capacity formulation for application to large channel networks, *J. of Soil and Water Conservation*, 50(5), 527–529, 1995.

Hodge, C., P. Diplas, and T. Younos, Resistance to flow through riparian wetlands, *Environmental and Coastal Hydraulics: Protecting the Aquatic Habitat, The 27th Congress of the International Association for Hydraulic and Research*, vol. 2, pp. 913–919, 1997.

Leonard, B. P., A stable and accurate convective modelling procedure based on quadratic interpolation, *Comput. Meths. Appl. Mech. Eng.*, 19, 1979.

Li, R. M., and H. W. Shen, Effect of tall vegetation on flow and sediment, *J. of the Hydraulics Division*, ASCE, 99, 5, 793–814, 1973.

Lopez, F., and M. Garcia, Synchronized measurement of bed shear stress and flow velocity in open channels with simulated vegetation, in *Proceedings of the North American Water and Environment Congress*, ASCE, New York, 1996.

Lopez, F., and M. Garcia, Open-channel flow through simulated vegetation: Suspended load transport modeling, *Water Resources Research*, 34, 2341–2352, 1998.

Lopez, F., and M. Garcia, Mean flow and turbulence structure of open-channel flow through non-emergent vegetation, *J. of Hydraulic Engineering*, 127, 392–402, 2001.

Minh Duc, B., Berechnung der stroemung und des sedimenttransports in fluessen mit einem tiefengemittelten numerischen verfahren, *Doctoral Dissertation*, The University of Karlsruhe, 1998.

Okabe, T., T. Yuuki, and M. Kojima, Bed-load rate on movable beds covered by vegetation, *Environmental and Coastal Hydraulics: Protecting the Aquatic Habitat, The 27th Congress of the International Association for Hydraulic and Research*, Vol. 2, pp. 809–814, 1997.

Patankar, S. V., *Numerical Heat Transfer and Fluid Flow*, Hemisphere, New York, 1980.

Proffit, G. T., and A. J. Sutherland, Transport of nonuniform sediment, *J. of Hydraulic Research*, 21, 33–43, 1983.

Reid. O. R., and E. R. Whitaker, Wind-driven flow of water influenced by a canopy, *J. of the Waterways Harbors and Coastal Engineering Division*, ASCE, 102, 61–77, 1976.

Rhie, C. M., and W. L. Chow, Numerical study of the turbulent flow past an airfoil with trailing edge separation, *AIAA J.*, 21, 1525–1532, 1983.

Rodi, W., *Turbulence Models and Their Application in Hydraulics*, 3rd Ed., IAHR Monograph, Balkema, Rotterdam, 1993

Saffman, P. G., The lift force on a sphere in a slow shear flow, *J. Fluid Mechanics*, 22, 385–400, 1965.

Shimizu, Y., and T. Tsujimoto, Numerical analysis of turbulent open-channel flow over a vegetation layer using a k-e turbulence model, *J. of Hydroscience and Hydraulic Engineering*, JSCE, 11, 57–67, 1994.

Spalding, D. B., A noval finite-difference formulation for differential expressions involving both first and second derivatives, *Int. J. Num. Meth. Eng.*, 4, 1972.

Stone, B. M., and H. T. Shen, Hydraulic resistance of flow in channels with cylindrical roughness, *J. of Hydraulic Engineering*, ASCE, 128, 500–506, 2002.

Stone, H. L., Iterative solution of implicit approximation of multidimensional partial differential equations. *SIAM J. on Numerical Analysis*, 5, 530–558, 1968.

Tsujimoto, T., Development of sand island with vegetation in fluvial fan river under degradation, *Water Resource Engineering `98*, Edited by S. R. Abt, J. Young-Pezeshk and C. C. Watson, pp. 574–579, ASCE, 1, 1998.

Tsujimoto, T., and T. Kitamura, Lateral bed-load transport and sand-ridge formation near vegetation zone in an open channel, *J. of Hydroscience and Hydraulic Engineering*, JSCE, 13, 35–45, 1995.

Tsujimoto, T., T. Kitamura, and S. Murakami, Basic morphological process due to deposition of suspended sediment affected be vegetation, *The 2nd International Symposium on Habitat Hydraulics*, 395–405, 1996.

Tsujimoto, T., Y. Shimuzu, T. Kitamura, and K. Okada, Turbulent open channel flow over a bed covered by rigid vegetation, *J. of Hydroscience and Hydraulic Engineering*, JSCE, 10, 13–25, 1992.

Wenka, T., Numerische Berechnung von Stroemungsvorgaengen in naturnahen Flusslaeufen mit einen tiefengemitelten Modell, *Doctoral Dissertation*, The University of Karlsruhe, 1992.

Wu, F. C., H. W. Shen, and Y. J. Chou, Variation of roughness coefficient for unsubmerged and submerged vegetation, *J. of Hydraulic Engineering*, ASCE, 125, 934–942, 1999.

Wu, W., Depth-averaged 2-D numerical modeling of unsteady flow and nonuniform sediment transport in open channels, *Under review by J. of Hydraulic Engineering*, 2003.

Wu, W., and D. A. Vieira, One-dimensional channel network model CCHE1D 3.0—Technical manual, *Technical Report No. NCCHE-TR-2000-1*, National Center for Computational Hydroscience and Engineering, The University of Mississippi, 2002.

Wu, W., and S. S. Y. Wang, Mathematical model for liquid-solid two-phase flow, *International J. of Sediment Research*, 15, 288–298, 2000.

Wu, W., S. S. Y. Wang, and Y. Jia, Nonuniform Sediment Transport in Alluvial Rivers, *J. of Hydraulic Research*, IAHR, 38, 427–434, 2000.

Zhu, J., FAST2D: A computer program for numerical simulation of two-dimensional incompressible flows with complex boundaries, *Report No. 690*, Institute of Hydromechanics, Karlsruhe University, 1992.

Weiming Wu and Sam S. Y. Wang, National Center for Computational Hydroscience and Engineering, The University of Mississippi, Carrier Hall 102, University, MS 38677.

Numerical Modeling of Bed Topography and Bank Erosion Along Tree-Lined Meandering Rivers

Marco J. Van De Wiel

Institute of Geography and Earth Sciences, University of Wales, Aberystwyth, UK

Stephen E. Darby

School of Geography, University of Southampton, Southampton, UK

It is generally acknowledged that riparian vegetation influences the geomorphological dynamics of rivers. However, the precise nature of the impact depends on a wide range of ecological and geomorphological processes, making it difficult to isolate the effects of specific individual factors in a field setting. As a result, there are many uncertainties concerning the effects of riparian vegetation on channel morphology, riverbank erosion and meander migration. In this study an alternative approach, using controlled numerical experiments, is developed to address these issues. The numerical model used herein is based on a coupled fluvial-geotechnical model, mRIPA, which combines a two-dimensional flow and sediment transport model with a geotechnical bank-stability analysis. In this study mRIPA is developed further by introducing additional submodels designed to account for the hydraulic and geotechnical effects of riparian vegetation. The new version of mRIPA is used to simulate a range of vegetation scenarios to explore the possible influence of vegetation density and biophysical characteristics of vegetation on geomorphological change. Results show that riparian trees can have a considerable impact on channel planform evolution and, to a lesser extent, bed topography. In particular, it is found that vegetation is usually stabilizing on the reach scale, although it can be locally destabilizing. Vegetation density and root structure are identified as the parameters that exert most influence over channel morphology.

1. INTRODUCTION

In recent years an increasing body of research has focused on the role of riparian vegetation within fluvial geomorphic systems. This attention is not surprising given the significance of riparian vegetation to the ecological functioning of river systems [*Nilsson*, 1992; *Naiman et al.*, 1993], and its potential for use in river stabilization and restoration [*Brookes and Shields*,

Riparian Vegetation and Fluvial Geomorphology
Water Science and Application 8
10.1029/008WSA19

1996]. These research efforts are providing results that seem to converge on a number of key points. Specifically, empirical studies have documented that channels with dense bank vegetation are narrower (and deeper) than equivalent channels with less dense vegetation [*Charlton et al.*, 1978; *Hickin*, 1984; *Hey and Thorne*, 1986; *Huang and Nanson*, 1997, 1998]. Furthermore, analytical and experimental models have highlighted that riparian vegetation can be a significant factor in the establishment of planform [e.g. *Millar*, 2000; *Gran and Paola*, 2001]. These results are entirely consistent with established views of the way in which riparian vegetation actually interacts with individual fluvial process mechanisms. Thus, the potential beneficial impacts on river stability are believed to be derived from: (1) the positive influence of root reinforcement and improved bank drainage [*Thorne*, 1990; *Coppin and Richards*, 1990] on bank stability, and (2) the influence of enhanced vegetative roughness in reducing near-bank flow velocity [*Li and Shen*, 1973; *Kouwen and Li*, 1980; *Kouwen*, 1988; *Freeman et al.*, 2000].

In fact, much of the existing knowledge of the impacts of vegetation on bank stabilization is derived from studies on hill slopes which have focused on the shear strength of tree roots [*Gray*, 1974; *Waldron*, 1977; *Wu et al.*, 1979, 1988]. It is not clear whether this research is transferable to the specific slope profiles, sedimentary settings, and vegetation assemblages associated with natural river banks [*Abernethy and Rutherfurd*, 2001]. Fortunately, more recent research is now starting to provide much needed additional data on physical root properties and their distribution within the soil, within the context of natural river banks [*Abernethy and Rutherfurd*, 2001; *Collison et al.*, 2001; *Simon and Collison*, 2002]. This work has highlighted the importance of the hydrological effects of vegetation, which can be either stabilizing or destabilizing, depending on canopy and rooting characteristics, as well as local site conditions [*Simon and Collison*, 2002]. It also appears that the destabilizing impact of tree surcharge is relatively low [*Abernethy and Rutherfurd*, 2000; *Simon and Collison*, 2002], at least in the limited range of river bank environments examined so far. In terms of the effects of vegetation on flow hydraulics, sustained research has now led to a relatively well-developed understanding of the physical properties underlying the roughness for stiff emergent vegetation [*Li and Shen*, 1973; *Petryk and Bosmajian*, 1975; *Pasche and Rouve*, 1985; *Fathi-Moghadam and Kouwen*, 1997], flexible submerged vegetation [*Kouwen and Li*, 1980; *Kouwen*, 1988], and flexible emergent vegetation [*Kouwen and Fathi-Moghadam*, 2000; *Freeman et al.*, 2000]. Although the roughness of riparian vegetation is highly relevant to bank erosion processes in terms of reducing flow velocity, the related capacity to redirect flow towards the center of the channel may be equally important [*Thorne and Furbish*, 1995; *Bennett et al.*, 2002] and might have unforeseen impacts. For example, redirection of the flow towards the center of the channel may well promote bank stability locally, but the resultant redistribution of channel bed topography could feasibly lead to enhanced bank erosion rates downstream.

The adoption of riparian vegetation as an environmentally acceptable alternative to conventional stream stabilization techniques has not yet become widespread. In part this is because the use of riparian vegetation presents certain practical difficulties to river managers, including negative impacts on flow conveyance, and difficulties in inspecting and maintaining the structural integrity of flood embankments [*Hemphill and Bramley*, 1989; *Darby and Thorne*, 1994]. Nonetheless, part of the reluctance to develop the application of riparian vegetation

treatments is also related to the fact that key scientific questions still remain to be resolved. First, although the work undertaken on natural riverbanks is highly valuable, as yet a limited range of bank environments have been investigated. The empirical nature of this research therefore makes it difficult to determine if the conclusions drawn so far are applicable in a wider range of environments. Second, the complexity of vegetated river systems, where there are many interacting factors that vary in significance both spatially and temporally, implies that it is difficult to isolate the influence of singular factors using existing empirically-based analyses. Third, most existing studies have focused on the effects of vegetation at essentially the site scale. However, the local effects of vegetation can be expected to have an influence on flow and sediment processes downstream, due to spatial and temporal feedbacks in the fluvial sediment transfer system. It is therefore not yet clear whether the potential benefits of vegetation can be scaled up from the local to reach-scale.

To address these issues, we herein develop and apply a numerical model to investigate the effects of a limited range of riparian trees on lateral channel migration and bed topography in a meandering river. An important advantage of the simulation modeling approach is that it affords relatively easy control of individual parameters so that it becomes possible to isolate the effects of singular change across a range of scenarios. Furthermore, unanticipated outcomes arising from non-linear interactions and feedbacks between variables can also be analyzed. Finally, numerical models are readily applied at the scale of the river reach. Naturally, simulation modeling also has disadvantages, and not least amongst these is that users must be confident that a model is reliable enough to predict the response of vegetated rivers accurately. The model used in this study is described in more detail in the next section. It is based on a two-dimensional depth-averaged model (mRIPA) of bed topography and bank erosion for single-thread meandering rivers that has recently been developed and tested [*Darby et al.*, 2002]. In this study mRIPA is extended by incorporating new insights derived from empirical investigations into the effects of riparian vegetation on flow resistance and bank erosion. The extended model is then used to undertake a simple set of numerical experiments to demonstrate the impact on reach-scale channel morphology of variations in biophysical parameter values associated with different species of simulated riparian vegetation.

2. METHODS

2.1. The mRIPA Model

The simulations presented in this study are based on mRIPA, a two-dimensional depth-averaged model of bed topography and bank erosion for meandering rivers with erodible cohesive banks [*Darby et al.*, 2002]. The original version of mRIPA has three key features that are relevant here. First, it couples a two-dimensional depth-averaged model of flow and bed topography with a mechanistic bank erosion model. Second, it simulates the deposition of failed bank material debris at, and its subsequent removal from, the toe of the bank. Finally, the governing conservation equations are implemented in a moving, boundary-fitted, curvilinear coordinate system that can be both orthogonal or non-orthogonal. This simplifies grid generation in curved channels that experience bank deformation, allowing the complex shapes associated with natural meanders to be simulated. mRIPA's numerical logic decouples the computations of flow, bed topography and planform change, based on the assumption that these processes operate on quite different time scales. Nevertheless, the submodels are allowed to interact through their effects on channel morphology and sediment transport. This is important in addressing the combined effects of vegetative flow resistance and biomechanical bank reinforcement on channel morphology. For the current study, mRIPA is extended to incorporate the effects of riparian trees (*i.e.* non-flexible emergent vegetation) on water flow, sediment transport and bank stability. Full details of the flow, sediment transport and bank erosion submodels that comprise mRIPA are available in the original paper [*Darby et al.*, 2002], but a short overview is provided here for completeness.

2.1.1. Flow model. mRIPA predicts the two-dimensional flow velocity field by solving the equations for the continuity of mass and momentum of water, which in depth-averaged, curvilinear form are [*Mosselman*, 1992]:

$$\frac{\partial \left(hU_s \right)}{\partial s} + \frac{\partial \left(hU_n \right)}{\partial n} + \frac{hU_s}{R_s} + \frac{hU_n}{R_n} = 0 \quad (1)$$

$$U_s \frac{\partial U_s}{\partial s} + U_n \frac{\partial U_s}{\partial n} + \frac{U_s U_n}{R_s} - \frac{U_n^2}{R_n}$$
$$+ g \frac{\partial \left(z_b + h \right)}{\partial s} + \frac{gU_s^2}{hC^2} = 0 \quad (2a)$$

$$U_s \frac{\partial U_n}{\partial s} + U_n \frac{\partial U_n}{\partial n} + \frac{U_s U_n}{R_n} - \frac{U_s^2}{R_s}$$
$$+ g \frac{\partial \left(z_b + h \right)}{\partial n} + \frac{gU_s U_n}{hC^2} = 0 \quad (2b)$$

where s and n represent the streamwise and transverse axes of the curvilinear coordinate system; U_s and U_n denote the asso-

ciated velocity components; h is the flow depth; z_b is the elevation of the channel bed; R_s and R_n respectively denote the divergence of the transverse and streamwise coordinate lines; g is the gravitational acceleration and C is the Chezy roughness coefficient. The convective influence of the secondary flow deforms the horizontal distribution of the primary flow. This effect can be represented in depth-averaged models via an additional acceleration or friction term [*Kalkwijk and De Vriend*, 1980]. The streamwise friction term in equation (2a) is, therefore, replaced by [*Mosselman*, 1992; *Darby et al.*, 2002]:

$$\frac{gU_s^2}{hC^2} \mapsto \frac{gU_s^2}{hC^2} + \frac{k_{sn}U_s^2}{h} \frac{\partial}{\partial n} \left(\frac{i_s h}{U_s} \right) \quad (3)$$

where k_{sn} denotes the secondary flow convection factor, which is a function of Chezy C [*Olesen*, 1987] and, hence, vegetative properties (see below), and i_s represents the secondary flow intensity, which is a function of the local channel curvature and flow field [*Mosselman*, 1992; *Darby et al.*, 2002]. The additional term accounts for the effects of secondary flow and the transverse exchange of momentum that follows from the redistribution of the primary flow. In mRIPA, equations (1), (2) and (3) are expanded for non-orthogonal, curvilinear coordinate systems and solved for U_s and U_n. A full derivation of the solution and its numerical implementation is given elsewhere [*Mosselman*, 1991; *Darby et al.*, 2002].

In the extended version of mRIPA, the effects of vegetation on flow velocity are represented by the Chezy roughness coefficient in equations (2) and (3). The vegetative roughness term can be determined from basic physical principles. Many researchers [e.g. *Petryk and Bosmajian*, 1975; *Freeman et al.*, 2000; *Kouwen and Fathi-Moghadam*, 2000] have expressed vegetative roughness in a form similar to:

$$C_{veg} = \sqrt{\frac{2g}{C_d A_v}} \quad (4)$$

where C_d is a species-dependent drag coefficient, and A_v is the projected area of the vegetation normal to the flow. The problem with the application of equation (4) is that few data are available to parameterize the values of C_d and A_v. However, for the relatively simple case of mature trees, the stems can be approximated as rigid cylinders, so that A_v is given by:

$$A_v = \sigma_{v,n} d_{vb} H_v^* \quad (5)$$

where $F_{v,n}$ denotes the lateral spacing of vegetation elements, d_{vb} is the stem diameter and H is *the projected submerged height of the plant (i.e.* the flow depth in the case of emergent mature trees). The drag coefficient, C_d, depends on the biomechanical properties of the trees (Section 2.2.3).

2.1.2. Sediment transport model. Changes in bed topography are estimated using the continuity equation for sediment movement, which for orthogonal, curvilinear grids reads [*Mosselman*, 1992]:

$$\frac{\partial z_b}{\partial t} + \frac{\partial S_s}{\partial s} + \frac{\partial S_n}{\partial n} + \frac{S_s}{R_n} + \frac{S_n}{R_s}$$
$$= S_{RB}\delta_{RB}(s,n) + S_{LB}\delta_{LB}(s,n) \qquad (6)$$

where t denotes time; S_s and S_n respectively are the streamwise and transverse components of the sediment transport vector, S; and the right hand side contains source terms reflecting the inputs of erosion products from the banks: S_{RB} and S_{LB} denote the volumetric inflow rate of eroded sediment from the right and left banks, respectively, while the function δ describes how the eroded material is distributed over the cross-section [*Mosselman*, 1992]. The magnitude of the sediment transport vector, $|S|$, can be derived from any suitable transport equation. In this study, a generic power law is applied, which relates the sediment transport flux to the local flow velocity, using:

$$|S| = k_s U^{p_s} \qquad (7)$$

in which k_s and p_s are calibration parameters. In the mRIPA model, this relation is adjusted by a correction factor to account for the effects of the longitudinal slope [*Olesen*, 1987]:

$$|S| = \left(1 - k_{ls}\frac{\partial z_b}{\partial s}\right)k_s U^{p_s} \qquad (8)$$

where k_{ls} is a weighting coefficient for the influence of the longitudinal slope. For orthogonal coordinate systems, the transverse components, S_n and S_s, are calculated using:

$$S_s = |S|\cos\psi_s \qquad (9a)$$
$$S_n = |S|\sin\psi_s \qquad (9b)$$

where ψ_s denotes the angle between the sediment transport vector and the s-axis, and is derived from the local flow field and bed topography [*Olesen*, 1987; *Mosselman*, 1992]:

$$\psi_s = \frac{U_n}{U_s} - k_{si}\frac{i_s}{U_s} - k_{ts}\frac{\partial z_b}{\partial n} \qquad (10)$$

where k_{si} and k_{ts} are weighting coefficients for the influence of secondary flow and transverse slope, respectively. The first two terms on the right hand side of equation (10) represent the effect of the flow on bed load transport direction, and the last term represents the effect of the local bed topography.

2.1.3. Bank erosion model. There are two components to the bank erosion model, corresponding to lateral fluvial erosion and geotechnical mass-wasting. Lateral erosion of the bank by hydraulic forces may be expressed using a simple excess shear formula [*Ariathurai and Arulanandan*, 1978]:

$$\left|\frac{\partial n_b}{\partial t}\right| = k_b\left(\frac{\tau_b - \tau_{bc}}{\tau_b}\right)^{p_b} \quad \text{if } \tau_b > \tau_{bc} \qquad (11a)$$

$$\left|\frac{\partial n_b}{\partial t}\right| = 0 \quad \text{if } \tau_b \le \tau_{bc} \qquad (11b)$$

where $\partial n_b/\partial t$ represents the rate of lateral retreat, τ_b is the shear stress applied on the bank, τ_{bc} is the critical shear stress for bank sediment entrainment, k_b denotes the erodibility of the bank material, and p_b is an exponent, which is usually taken to equal unity [*Ariathurai and Arulanandan*, 1978; *Mosselman*, 1992]. Formally, hydraulic entrainment on the bank face is governed by the three-dimensional flow field. However, for mildly curved channels it may be represented relatively well by the longitudinal shear stress, since the secondary flow components are driven by the streamwise flow and are small with respect to the latter [*Mosselman*, 1992]. The longitudinal shear stress, τ_s, is expressed as:

$$\tau_s = \frac{\rho g U_s^2}{C^2} \qquad (12)$$

in which ρ is the fluid density. The shear stress exerted on the banks, τ_B, is then obtained as a simple function of the longitudinal shear stress applied on the bed, τ_s:

$$\tau_b = k_\tau \tau_s \qquad (13)$$

where k_τ is a proportionality coefficient, which is normally taken to equal 0.75 [*Lane*, 1955]. The sediment which has been entrained from the banks is distributed over the channel, which is expressed mathematically by the source terms, S_{RB} and S_{LB}, on the right hand side of equation (6) [*Darby et al.*, 2002].

To account for the erosion of bank materials by mass-wasting a conventional geotechnical stability analysis for planar failure is expanded to account for changes in soil cohesion and bank load induced by riparian vegetation. According to *Darby and Thorne* [1996], the factor of safety, N_{FS}, for planar failures is given by:

$$N_{FS} = \frac{cL_f + \left(W_f \cos\beta_f - F_{pp} + F_{cp}\cos\alpha_{cp}\right)\tan\phi}{W_f \sin\beta_f - F_{cp}\sin\alpha_{cp}} \quad (14)$$

where L_f and β_f respectively denote the length and angle of the incipient failure plane; W_f represents the weight of the incipient failure block; c is the soil cohesion; F_{pp} and F_{cp} are the soil pore pressure and the hydrostatic confining pressure, respectively; α_{cp} is the angle at which the hydrostatic confining pressure is transmitted over the failure plane; and ϕ is the internal friction angle of the soil. The bank collapses when $N_{FS} < 1$. Subsequent to failure, coarse bank material (≥ 10 mm) is deposited on the bed, very fine bank sediment (< 0.062 mm) is washed out of the system, and the remainder is added to the sediment load, via the source terms, S_{RB} and S_{LB}, on the right hand side of equation (6) [*Darby et al.*, 2002]. To account for the impact of riparian vegetation, two new terms are added to equation (14): the additional cohesion, c_v, induced by the presence of the tree roots, and the surcharge of the vegetation on the failure block, W_v:

$$N_{FS} = \frac{(c + c_v)L_f + \left[\left(W_f + W_v\right)\cos\beta_f - F_{pp} + F_{cp}\cos\alpha_{cp}\right]\tan\phi}{\left(W_f + W_v\right)\sin\beta_f - F_{cp}\sin\alpha_{cp}} \quad (15)$$

Although this expansion is in itself fairly straightforward, the derivation of the terms c_v and W_v is less obvious. Tree roots which intersect the failure plane add shear strength to the soil. The increase in soil shear strength, Δs_r, can be represented by [*Wu et al.*, 1979]:

$$\Delta s_r = A_r^* T_r \left(\sin\theta_s + \cos\theta_s \tan\phi\right) \quad (16)$$

where A_r^* denotes the average root-area-ratio, a measure of root density (see below); T_r is the mean tensile strength of the roots; and θ_s is the angle of distortion of the shear zone at the moment of failure. The bracketed term is not very sensitive to variations of θ_s and ϕ, it is usually taken to be a constant, equal to 1.2. In practice, the increase in shear strength can be treated as an increase in cohesion [*Wu et al.*, 1979], so that:

$$c_v = \Delta s_r = 1.2\, A_r^*\, T_r \quad (17)$$

Approximate values for the mean tensile strength of the roots, T_r, can be found in the literature [e.g. *Greenway*, 1987], or can be obtained experimentally [*Abernethy and Rutherfurd* 2001; *Collison et al.*, 2001]. The parameterization of T_r in our simulations is discussed in more detail in Section 2.2.3.

At any point in the soil, the local root-area-ratio, A_r, is defined as the proportion of the soil area that is occupied by roots, for an arbitrary section of soil. The root-area-ratio varies along the bank in all spatial dimensions, and depends on the vegetation density and the specific rooting properties of an individual tree or assemblage of trees. It is usually maximal directly underneath the plant's stem, and diminishes in both the lateral and vertical dimensions [*Abernethy and Rutherfurd*, 2001]. Natural vegetation may be subject to preferential root growth in a particular direction, which may lead to a skewed root density distribution. For modeling purposes, however, a generalized symmetrical conical distribution is assumed, defined by a rate of lateral decline and a rate of vertical decline. Further assuming that the decline is exponential [*Abernethy and Rutherfurd*, 2001], this can be represented as:

$$A_r = e^{a_0 + a_1 L_{xH} + a_2 L_{xV}} \quad (18)$$

where L_{xH} and L_{xV} are the horizontal and vertical distance to the plant stem, respectively; and a_0, a_1 and a_2 are non-positive coefficients describing the rate of decline. Physically, this declining function is incorrect, since it never reaches a value of zero, implying an infinite extent of the tree roots. It is, therefore, necessary to assume a critical value, A_{rmin}, below which the root-area-ratio can be neglected. Arbitrarily, a value of $A_{rmin} = 0.0001$ has been chosen for all species, indicating the presence of 1 mm^2 of roots in a 10 cm x 10 cm section of soil as the minimal density of roots to influence soil strength. The coefficients a_i can then be calculated as:

$$a_0 = \ln A_{r\max} \quad (19)$$

$$a_1 = \frac{\ln A_{r\min} - a_0}{L_{rH}} \quad (20)$$

$$a_2 = \frac{\ln A_{r\min} - a_0}{L_{rV}} \quad (21)$$

where L_{rV}, L_{rH}, A_{rmax} are three case-specific root parameters, respectively denoting maximal root depth, maximal horizontal root extent, and maximal root density. Parameterization of these variables is discussed in Section 2.2.3. The average

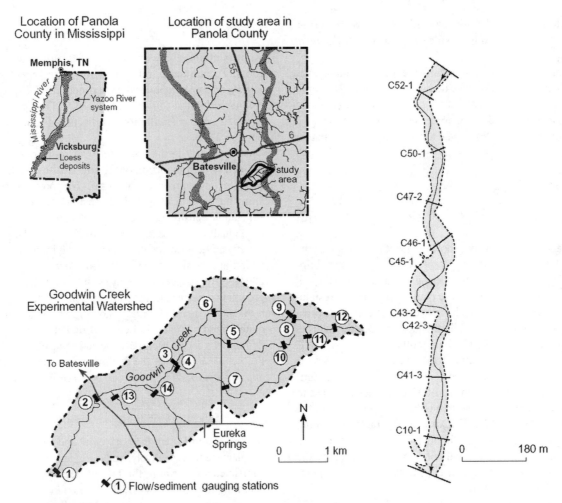

Figure 1. Location of the control scenario reach within the Goodwin Creek Experimental Watershed (Reprinted from *Darby et al.* [2002], with permission from AGU). The modeled reach stretches from cross-sections C50-1 to C41-3.

root-area-ratio, A^*_r, is then obtained by integrating equation (18) over the width of the failure plane, L_f:

$$A^*_r = \frac{1}{L_f} \int_{L_f} A_r \, dx \qquad (22)$$

The second new term in equation (15) is the surcharge term, W_v, which is given by the weight of the tree, distributed across the near-surface root area:

$$W_v = \frac{\rho_v g}{\pi L_{rh}^2} V_v \qquad (23)$$

where ρ_v denotes the mass density of the tree and V_v is the volume of the tree, which can be approximated using [*De Vries*, 1974]:

$$V_v = \frac{\pi}{8} H_v \left(d_{vb}^2 + d_{vt}^2 \right) \qquad (24)$$

where H_v is the height of the tree; and d_{vb} and d_{vt} are the stem diameters, at the base and the top of the tree, respectively. To obtain the surcharge for multiple trees, the surcharges of all individual trees needs to be aggregated.

2.2. Numerical Experiments

The aim of this study is to apply the model in a series of numerical experiments to investigate the effects of riparian trees on channel bed morphology and planform adjustment in a meandering river. The design of these numerical simulation experiments involves several elements. First, selecting a natural reach of channel that serves as a control to quantify the

Table 1. Input data used in simulations.

Symbol	Description	Value
Q	Flow discharge (m^3/s)	7.5 [b]
C	Chezy flow resistance	10 [c]
k_{sn}	Secondary flow convection factor [a]	0.45 [d]
i_s	Secondary flow intensity [a]	2.93
k_{ls}	Weighting coefficient for the influence of the longitudinal slope on sediment transport [a]	12
k_{ts}	Weighting coefficient for the influence of the transverse slope on sediment transport [a]	0.67
k_s	Coefficient in sediment transport equation [a]	8.2×10^{-3}
p_s	Exponent in sediment transport equation [a]	5
k_b	Coefficient in fluvial bank erosion equation [a]	1.0×10^{-6}
p_b	Exponent in fluvial bank erosion equation [a]	1
τ_{bc}	Critical shear stress for fluvial entrainment of bank material (Pa) [a]	17.5
c	Bank material cohesion (Pa)	12000
ϕ	Bank material friction angle (degrees)	20

[a] Calibrated parameter [*Darby et al.*, 2002]
[b] The flow discharge used here is a steady, representative discharge, equivalent to the dominant discharge value estimated by *Bingner* [pers. comm., 1997]. Based on the flow duration data from Station 2 (figure 1), this means that the model simulation period of 360 days used here is equivalent to the real time elapsed during the November 1982 to May 1988 period, which was used in the calibration procedure [*Darby et al.*, 2002].
[c] This is the baseline value for control scenario, and is adjusted at the bank nodes in the vegetation scenarios. The baseline value is calculated using the continuity equation, from measured discharge at Station 2 (figure 1) and the known cross-sectional geometry near Station 2 (C50-1; figure 1).
[d] This is derived from the initial Chezy value, in accordance with *Olesen* [1987].

simulated bed topography and bank erosion in the absence of vegetation. Next, identifying specific attributes of vegetation to investigate in a range of modeling scenarios. And finally, parameterizing these selected attributes.

2.2.1. Control experiment. A control simulation, in which vegetation is assumed to be absent, is necessary to provide a reference set of data against which data from other simulations, in which vegetation is analyzed, can be compared. The control simulation was undertaken on a short (707 m) reach of Goodwin Creek, Mississippi, USA (see Figure 1 for the reach location). The bank materials along this reach are cohesive (Table 1) and relatively resistant to fluvial entrainment, so that bank erosion and planform adjustment occur predominantly through mass-wasting of deeply incised banks [*Murphey and Grissinger*, 1985]. This reach of Goodwin Creek was selected by *Darby et al.* [2002] for use in the validation analysis for two main reasons. First, the channel is very active, allowing for the possibility of modeling channel changes within a logistically simple timescale. Moreover, the study reach is located within the Goodwin Creek Experimental Watershed and has been intensively monitored by United States Department of Agriculture, Agricultural Research Service, personnel based at the National Sedimentation Laboratory in Oxford, Mississippi

[*Murphey and Grissinger*, 1985; *Blackmarr*, 1995; *Kuhnle et al.*, 1996]. In fact, Goodwin Creek is a rare example of a dataset from an actively adjusting channel that has sufficient spatial-temporal resolution to allow accurate model initialization and parameterization, while simultaneously enabling comparison between modeled and measured channel morphology during adjustment.

The topography of the channel was represented on a numerical grid consisting of a total of 451 nodes, with an initial configuration as shown on Figure 2. Model boundary conditions (flow discharge and sediment inflow at upstream boundary) were set in accordance with flow discharge and sediment loads recorded at the measuring flume (Station 2) located at the head of the study reach (see Figure 1 for location). Values of the various remaining adjustable coefficients and parameters (Table 1) were selected in relation to the results of model calibration for two laboratory flume experiments [*Darby et al.*, 2002]. The simulation period spans a five and a half year interval from the onset of hydrological monitoring at Station 2 in November 1982 until May 1988. Extending the simulation period beyond May 1988 is not practical because bank protection works were constructed in the reach at this time, rendering planform adjustment simulations pointless.

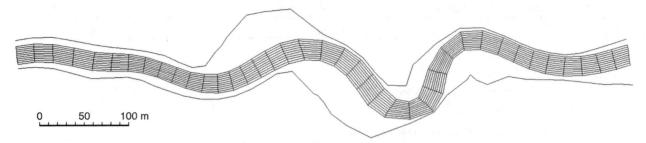

Figure 2. Numerical grid for the Goodwin Creek reach. The grid represents the initial (Nov 1982) channel bed and is constructed within the constraints set by *Mosselman* [1992]. The two lines parallel to it show the initial (Nov 1982) position of the bank tops. Flow is from left to right.

Comparison of the model simulations against observed channel changes shows that mRIPA is capable of representing the overall bed topography, although it does underpredict transverse bed slopes [*Darby et al.*, 2002]. In addition, mRIPA correctly predicts the location of observed bank retreat, but the magnitude of retreat is underestimated. Although we recognize that there are some deficiencies in the model simulation relative to the observed data, we believe that the general pattern of simulated bed topography and bank erosion is broadly consistent with observations. It is therefore appropriate to use the Goodwin Creek simulation as a base case for qualitative comparisons against different vegetation scenarios. The channel morphology simulated in the control case is discussed in Section 3. The next step is to describe the vegetation scenarios and their respective parameterizations.

2.2.2. Vegetation scenarios. From a river management perspective, two aspects of vegetation are readily available for consideration in any bioengineering design: species and density. We have therefore selected these aspects of vegetation as the variable parameters for the numerical simulations. A total of seven scenarios are simulated here (Table 2), consisting of the control scenario for unvegetated conditions, and six scenarios with different vegetation covers, com-

prising three vegetation species at two different densities. It must be emphasized that these vegetation scenarios are hypothetical, each simulated species being characterized through parameterization of certain biophysical properties (see Section 2.2.3). The nature of this parameterization is such that all trees of a particular species are assumed to be mature, identical and static (*i.e.* do not change over time). Real vegetation, however, is diverse, complex and dynamic, even for homogeneous stands, as the properties of individual trees will vary as a function of age, health, season and local conditions such as soil material, pH and microclimate. Representing the simulated species via a set of unvariable characteristics facilitates interpretation of the results in terms of identifying the vegetation properties that exert most influence on channel morphology. Although each scenario therefore represents a collection of biophysical properties rather than a genuine species (a key point emphasized by our use of quotation marks around simulated species names), the vegetation data used for the parameterization in the simulations is broadly based on real riparian species: *Betula pendula*, *Populus nigra* and *Salix fragilis*. It is not relevant that these three species do not actually occur naturally in Goodwin Creek, since the focus is on the differences between the biophysical properties of each virtual species rather than the actual values themselves. A second aspect of vegetation cover investigated in the numerical experiments is the vegetation density. In each of the scenarios the vegetation is positioned in wide strips along the bank, parallel to the river, starting at the bank toe and extending on to the floodplain. The spacing is the same in both the lateral and streamwise direction and is determined through the vegetation density. For the current study we use two densities, termed "low" and "high", which correspond to an average tree spacing of 10 m and 5 m, respectively.

2.2.3. Vegetation parameterization. The hypothetical tree species for the simulations are "*Betula pendula*", "*Populus nigra*" and "*Salix fragilis*". The biophysical properties of

Table 2. Overview of simulations.

Scenario	Species	Vegetation Density
1	none	-
2	"*Betula pendula*"	low
3	"*Betula pendula*"	high
4	"*Populus nigra*"	low
5	"*Populus nigra*"	high
6	"*Salix fragilis*"	low
7	"*Salix fragilis*"	high

Table 3. Vegetation parameters used in mRIPA.

Parameter	Symbol	*"Betula pendula"*	*"Populus nigra"*	*"Salix fragilis"*
Stem height (m)	H_v	20	30	25
Stem diameter at base (m)	d_{vb}	0.8	1.1	0.7
Stem diameter at top (m)	d_{vt}	0.1	0.15	0.1
Wood mass density (kg/m^3)	ρ_v	960	880	705
Vertical root depth (m)	L_{rV}	0.5	2	1
Horizontal root extent (m)	L_{rH}	2	4	2
Maximal root density (-)	A_{rmax}	0.8	0.8	0.8
Root tensile strength (kPa)	T_r	37000	10000	18000
Drag coefficient (-)	C_d	0.9	1	1.2

these trees as used in the numerical experiments are presented in Table 3. Where possible, parameterization of the vegetation properties is based on available empirical data, although many of the published values are only indicative, as they depend on tree age and health, as well as local soil conditions and climate.

The above-ground properties of the trees are the physical dimensions, defined by stem diameter and tree height, and the wood mass density. While values for the stem height and the diameter at the base of the stem (at breast height) have been reported in the literature for mature trees [*Rehder*, 1947; *Kozlowski*, 1971; *Forest Products Research Laboratory*, 1956], there are no data available for the stem diameter near the top. The latter value has been estimated, arbitrarily, as being between 10% and 15% of the stem diameter at the base of the tree. Wood mass density for living vegetation depends on the dry wood mass density and the amount of water within the tree. As the latter of these is variable and can contribute between 10% and 80% of the tree weight [*Husch*, 1963; *Forest Products Research Laboratory*, 1956], the values used here are approximated based on reported wood mass densities for live vegetation [*Forest Products Research Laboratory*, 1956]. Empirical data on the rooting properties of vegetation is generally more scarce, due to the practical difficulties associated with below-ground measurements and the dependency on soil properties. The parameterization of vertical root depth and horizontal root extent is based on a qualitative description for root structures in clay soils [*Greenway*, 1987], where we have parameterized the terms "shallow", "moderately deep" and "deep" as 0.5, 1.0 and 2.0 m respectively. Horizontal root extent is taken to be 2.0 m, unless it was specified as being "widespread", in which case it is parameterized as 4.0 m. The maximal root-area-ratio is arbitrarily set equal to 0.8 for all species. Although this is much higher than reported root-area-ratio values [e.g. *Abernethy and Rutherfurd*, 2001; *Simon and Collison*, 2002], a

value of 0.8 is chosen to represent the contact of the tree trunk and the soil, where most of the soil will be occupied by roots. The exponentially declining root-area-ratio model (Section 2.1) ensures that the average root-area-ratio is considerably lower. The values for root tensile strengths are set according to published values for mean tensile strengths [*Greenway*, 1987]. In parameterizing the drag coefficient, the trees are assumed to be rigid cylinders. In this case the drag coefficient is of the order of 1.0 [*Petryk and Bosmajian*, 1975], although published values for different species range from 0.2 to 3.0, depending on the spatial pattern of the vegetation [*Li and Shen*, 1973; *Klaassen and Van der Zwaard*, 1974; *Petryk and Bosmajian*, 1975]. The values adopted here are taken to be close to 1.0, with minor deviations to account for inter-species variations in roughness.

We recognize that there are many assumptions and approximations in this crude parameterization of the biophysical properties of simulated vegetation. This emphasizes the point that the modeled species represent a collection of biophysical properties, rather than an actual species. Indeed, it might be very difficult to find a single real *Betula pendula* which conforms to all the properties as listed in Table 3. Nonetheless, the parameterization of the three hypothetical species covers a range of biophysical parameter values and is, therefore, adequate for the purposes of the current investigation.

3. RESULTS

Table 4 summarizes the morphological characteristics of the channel at the end of the simulation, for all scenarios. The following paragraphs will describe the results in more detail, in terms of both the evolution of bed topography and channel planform change. The connections between these two morphological characteristics, and the extent to which they are affected by specific vegetation parameters, are discussed in Section 4.

Table 4. Summary of results.

Scenario	Vegetation Species	Vegetation Density (-)	Floodplain Area Loss (m²)	Maximal Bank Retreat (m)	Maximal Bed Scour (m)	Maximal Bed Deposition (m)
1	none	-	585	9.63	0.23	0.66
2	"Betula pendula"	low	584	9.67	0.25	0.63
3	"Betula pendula"	high	588	9.99	0.26	0.62
4	"Populus nigra"	low	601	9.97	0.27	0.62
5	"Populus nigra"	high	441	9.6	0.53	0.75
6	"Salix fragilis"	low	574	9.66	0.25	0.63
7	"Salix fragilis"	high	316	7.79	0.39	0.66

3.1. Bed Topography Change

As mRIPA is a rigid lid flow model, flow depths represent the bed topography. Figure 3 illustrates modeled flow depths at the end of the simulation for the control scenario, *i.e.* for unvegetated conditions. Although there are some small point bars in the upper parts of the reach, the major topographical features are located in the lower part of the reach. Most striking of these are a well-defined point bar (LPB) along the right bank, between cross-sections C43-2 and C42-3, together with the associated thalweg along the left bank, and an elongated bar (EB) near the right bank between cross-sections C42-3 and C41-3 (see Figure 1 for cross-section locations).

From Table 4 it appears that the channel bed topography is only significantly perturbed from the control scenario in the high density scenarios for "*Populus nigra*" and "*Salix fragilis*" (scenarios 5 and 7), where maximal bed scour (relative to the control scenario) is increased by 130% and 70% respectively. Maximal channel deposition seems hardly affected at all, with only a 13% increase (relative to the control scenario) in scenario 5. However, summarizing the results with a single maximal value indicator hides spatial effects. The location of maximal scour, *i.e.* the thalweg between

cross-sections C43-2 and C42-3, does not necessarily coincide with the location of maximal change in scour (relative to the control scenario), which, for most scenarios, is located along EB. Similarly, the location of maximal bed deposition, *i.e.* LPB, does not necessarily correspond with the maximal change in bed deposition (relative to the control scenario), which, for most scenarios, is located opposite EB. This spatial aspect of vegetation-induced morphological change of the channel bed is illustrated, for scenarios 3, 5, 6 and 7, in Figure 4. The results for scenarios 2 and 4 are broadly similar to those of scenario 3 and are not shown here. The graphs show the differences in bed morphology relative to the control scenario, Figures 4B and 4D clearly show a significant change in bed topography (e.g. 0.3 m and more) induced by the high density assemblages of "*Salix fragilis*" and "*Populus nigra*" (scenarios 7 and 5), especially towards the downstream end of the reach. Large parts of the elongated bar (EB), observed in the control scenario, are eroded and more sediment is deposited along the left bank, resulting in a shallower transverse bed slope. The deposition pattern along LPB is also affected, with markedly less deposition beyond the apex of the bend. The differences are smaller in the upper parts of the reach for both scenarios. Although the channel is generally more incised throughout most of

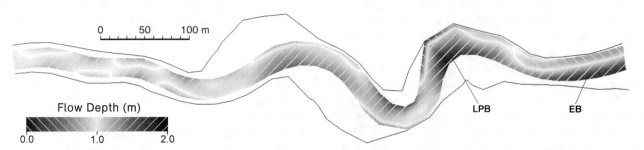

Figure 3. Modeled flow depths for the control scenario at the end of the simulation (May 1988). Shading is chosen to emphasize pools and bars. LPB (large point bar) and EB (elongated bar) show the location of two distinct morphological features which are explained in detail in the text. Flow is from left to right.

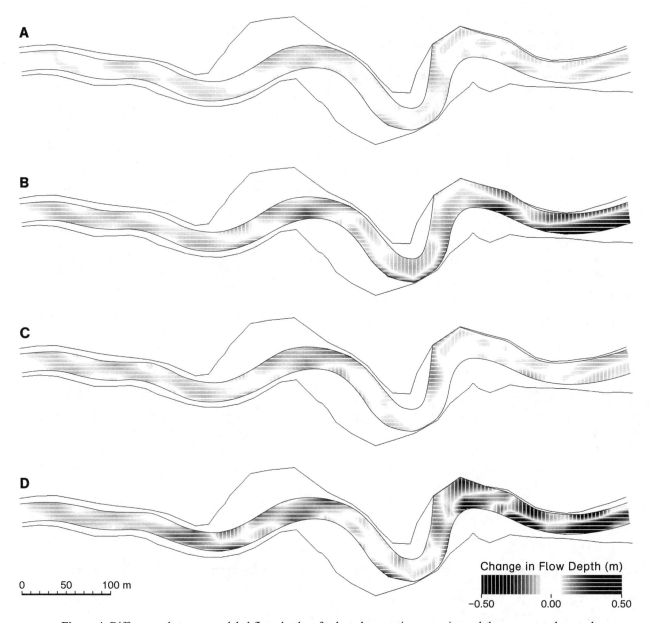

Figure 4. Differences between modeled flow depths of selected vegetation scenarios and the unvegetated control scenario. **A**: Scenario 6: low density "*Salix fragilis*". **B**: Scenario 7: high density "*Salix fragilis*". **C**: Scenario 3: high density "*Betula pendula*". **D**: Scenario 5: high density "*Populus nigra*". Flow is from left to right.

the upper reach, there are some areas of increased deposition, mainly on the existing point bars. The effects of vegetation on bed morphology are clearly less pronounced in scenarios 3 and 6 (Figures 4A and 4C), as well as in scenarios 2 and 4 (not depicted). However, the difference is not just one of scale. The spatial pattern of bed topography change is also altered, in particular for scenarios 3 and 4, where EB is accentuated, rather than eroded. The possible causes for these changes are discussed in detail in Section 4.

3.2. Planform Change

Planform change is quantified in terms of both maximal bank retreat and total floodplain area loss (Table 4). While maximal bank retreat has a strong initial appeal as an indicator, it refers to a single location along the reach and, as with bed topography change, hides important spatial effects. Floodplain area loss is a more helpful statistic, summarizing the total bank retreat along the entire reach. Figure 5A shows the

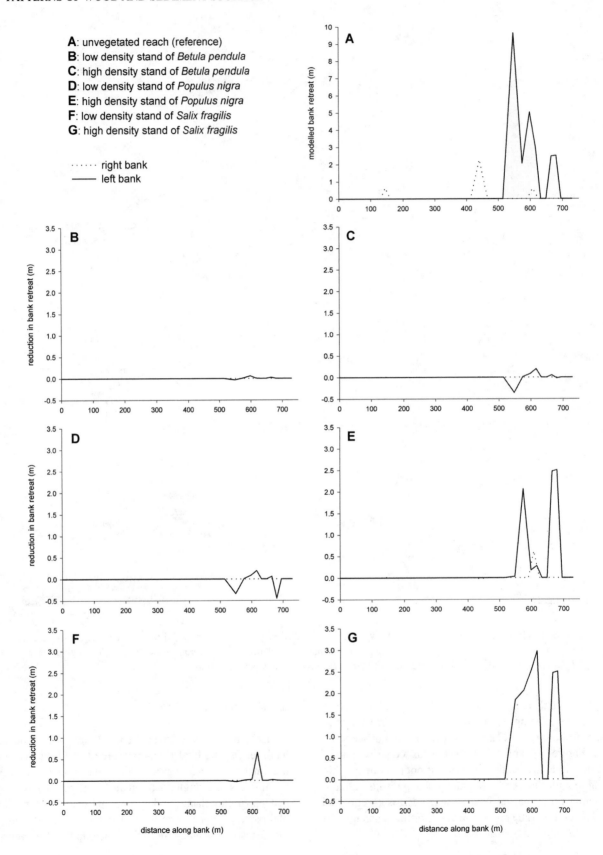

A: unvegetated reach (reference)
B: low density stand of *Betula pendula*
C: high density stand of *Betula pendula*
D: low density stand of *Populus nigra*
E: high density stand of *Populus nigra*
F: low density stand of *Salix fragilis*
G: high density stand of *Salix fragilis*

· · · · · right bank
——— left bank

modeled bank retreat along the reach for the unvegetated control scenario. Most of the upper reach is stable, while the floodplain loss is mainly located along the left bank towards the end of the reach, with two peaks of bank retreat at 550 m and 600 m along the bank, *i.e.* opposite LPB, and a smaller one about 675 m along the bank, *i.e.* opposite the EB.

From Table 4 it can be seen that each vegetation scenario has a different impact on total floodplain area loss. In scenarios 2,3,4 and 6, there is little overall impact on floodplain area loss (less than 3% change relative to the control scenario), while scenarios 5 and 7 show a significant reduction in floodplain area loss relative to the control scenario (25% and 46%, respectively). It is noteworthy that the maximal bank retreat is less affected, suggesting there is a strong spatial aspect to the impact of vegetation on bank stability in these scenarios. This is illustrated in Figures 5B to 5G, which show spatial differences in modeled bank retreat between each of the vegetated scenarios and the unvegetated control scenario. There are notable reductions in bank retreat along the left bank for scenarios 5 and 7 (Figures 5E and 5G), where bank retreat is reduced by 2 to 3 m in two distinct zones, located approximately 520-620 m and 650-700 m along the reach, respectively. The high density cover of "*Salix fragilis*" (scenario 7; Figure 5G) induces the highest impact, effectively eliminating the last two peaks in bank retreat at 600 m and 675 m, and slightly reducing the first one at 550 m. High density "*Populus nigra*" (scenario 5; Figure 5E) also removes the last of the three erosion peaks, but does not affect the peaks in bank retreat at 550 m and 600 m. Nonetheless, it does reduce the bank retreat at about 575 m along the left bank, in between the first two peaks. In scenarios 2, 3, 4 and 6 the effects are less pronounced, although still spatially variable. Most pointedly, there are places where the bank retreat along the left bank has increased. For the high density "*Betula pendula*" and low density "*Populus nigra*" assemblages (scenarios 3 and 4; Figures 5C and 5D), the net effect over the reach is such that total floodplain area loss is actually increased (Table 4). In both these scenarios there is enhanced bank retreat at 550 m along the left bank, *i.e.* at the location of the first peak in bank retreat (Figure 5A), while in scenario 4 the last of the three erosion peaks is also increased. The significance of this is not the magnitude of the enhanced bank retreat, which is rela-

tively small (0.35 and 0.45 m), but the fact that it occurs at all. The possible causes and implications will now be discussed.

4. DISCUSSION

Two-dimensional, depth-averaged flow models are, by definition, incapable of accurately representing the full three-dimensional flow field, which may have repercussions regarding the model's ability to simulate morphological change. *Darby et al.* [2002] have identified mRIPA's limitations in this context. They concluded that the model has a tendency to underpredict transverse slopes and cannot accurately simulate the detailed variation of boundary shear stress in regions of rapidly varying bed topography. Nonetheless, mRIPA appears capable of predicting the macroscale morphological features of river channels [*Darby et al.*, 2002]. The results presented herein focus on the effects of riparian vegetation on such macroscale morphological features.

4.1. Effects of Vegetation on Channel Morphology

The impact of simulated riparian vegetation on bed topography is closely related to the effects vegetation has on bank retreat, and vice versa. A significant change in bed morphology is notable only in those scenarios where bank retreat is also significantly affected, *i.e.* for the high density stands of "*Populus nigra*" and "*Salix fragilis*" (scenarios 5 and 7). Most of the change in bed topography occurs in the lower 150 m of the reach, downstream of the bend with the highest rates of bank retreat, while only minor changes in bed topography are observed upstream of this bend. This leads us to believe that most of the bed topography change is directly linked to the upstream bank retreat, either due to the different sediment supply from the banks, or to the change in channel curvature and its effect on the flow field. Likely, it is a combination of both, although we cannot isolate one effect from the other. This subsequently affects the transverse bed slope, secondary flow, sediment transport and the distribution of scour and deposition further downstream. Additional evidence for the relation between upstream bank retreat and bed topography evolution may be obtained by looking at the pattern of change rather than the magnitude. In the two scenarios where total floodplain area loss is increased (scenarios 3 and 4), *i.e.* where more sediment is supplied to the channel, the elongated bar (EB) and the associated thalweg are accentuated. Similarly, in scenarios 5, 6 and 7, where the floodplain area loss is reduced, there is a general trend towards less scour along the left bank and less sediment deposition along the right bank in the lower parts of the reach, although the details of the change in sedimentation pattern are different for each scenario. However, this might be a circular argument because part of the change

Figure 5. Spatial pattern of differences in modeled bank retreat at the end of the simulations (May 1988) for the different vegetation species and densities. **A**: Modeled bank retreat in unvegetated conditions (control scenario). **B-G**: Differences in bank retreat for vegetation scenarios, expressed as a reduction in bank retreat. Negative values indicate an increase in bank retreat. Note the different scale on the Y-axis for figure **A**.

in floodplain area loss is likely to be the result of the redistribution of bed scour. Nonetheless, the impact of vegetation on bank retreat seems to be a significant factor in influencing downstream bed morphology, although it cannot be the only one. The smaller changes in bed topography in the upper parts of the reach can only be explained in terms of vegetative roughness, as there is no discernible bank retreat in this part of the channel. The roughness of the riparian vegetation reduces the near bank flow, which in turn affects the secondary flow pattern (via the influence of the adjusted Chezy coefficient in equations (1) (2) and (3)). This change in helical flow alters the distribution of scour and deposition across the channel. Again, the impact is spatially distributed, as interaction with the flow causes the effects of vegetation to propagate downstream.

Unfortunately, it is precisely this complex interaction between flow, sediment transport and bank erosion processes, and the extent to which vegetation influences each of those, that prohibits a single, simple answer as to how vegetation will affect channel morphology. What is clear is that vegetation is capable of affecting river morphology significantly. Critically, the impact of vegetation is spatially variable. The local effects of vegetation can propagate downstream due to the interactions between channel planform, flow and bed response. This implies that at-a-site analysis by itself is not sufficient to determine the net beneficial or adverse impact of riparian vegetation on channel morphology, even in terms of bank stability. Instead, the spatial variability of the effects of vegetation highlights the complexity and intricacy of fluvial systems at the reach scale. Empirical research, numerical studies and river management studies all need to consider the interactions between bank mechanisms, flow processes and riparian vegetation, in order to represent the spatial diversity of fluvial morphology and processes.

4.2. Identification of Relevant Vegetation Parameters

The spatial variability of the impact of riparian vegetation complicates the identification of relevant vegetation parameters in terms of their influence on bed topography, bank stability and channel planform change. The influence of vegetation not only depends on the vegetation properties themselves, but also on how the local bed topography has been affected by processes upstream. However, a few general trends can be observed. First, for each of the three simulated species, the overall effect of vegetation on bed topography and floodplain area loss is greater in the high density scenarios. This highlights vegetation density as an obvious key parameter. High density stands of "Salix fragilis" and "Populus nigra", in particular, significantly reduce floodplain area loss. Second, although "Betula pendula" has the highest tensile strength of the species simulated here, its impact on floodplain area loss is markedly less than that of the other two "species". Instead, "Salix fragilis" and "Populus nigra" have lower tensile strengths, but have deeper and more extensive root networks with higher average root-area-ratios. Hence, it seems that root structure is more important than root tensile strength. This is in agreement with the observations of *Simon and Collison* [2002], who observe that most mechanical reinforcement is induced by species with high root-area-ratios. Third, for high density stands of "Betula pendula" and low density stands of "Populus nigra" (scenarios 3 and 4), the balance between stabilizing and destabilizing vegetation effects varies from point to point, thus resulting in reduced bank retreat in some places and accelerated bank retreat at others. The overall effect, however, is an increase in total floodplain area loss. It appears that this occurrence of accelerated bank retreat is very much related to the redistribution of scour in these scenarios. Although the surcharge of the vegetation is likely to contribute to the local instability, it appears to be a less important factor. This has also been suggested by *Abernethy and Rutherfurd* [2000]. Finally, the effect of vegetative roughness appears to be especially complex. While the reduced flow velocity resulting from the increased roughness reduces fluvial entrainment rates, its main impact is via the alteration of the flow pattern and the subsequent redistribution of scour and deposition (see above).

5. CONCLUSION

The extended version of mRIPA has been used to conduct a set of numerical experiments in which the impacts of hypothetical riparian vegetation assemblages on the morphology of a meandering river are evaluated. The results indicate that, depending on species and spatial density, riparian trees can have a considerable impact on vegetative roughness, bank stability, and thus channel planform evolution, floodplain area loss and, to a lesser extent, bed topography. Higher density stands of vegetation are usually, but not always, more effective in stabilizing river banks. Of the few virtual species included in this investigation, the biophysical properties of "Salix fragilis" appear to deliver an effective reduction in floodplain area loss. In general, the bank properties of the vegetation appear to have a higher impact on morphological processes than the flow properties. The root structure of the trees, in particular, seems to be a controlling factor in reducing floodplain area loss, more so than root tensile strength, above-ground physical dimensions or vegetative roughness. Although the flow properties of vegetation have less direct impact, the interaction between flow and bank processes means that the local effects of vegetation can propagate downstream. The resulting spatial variability emphasizes the complex nature of fluvial biomorphological systems. In certain vegetation scenarios the total floodplain area loss actually increases, as the downstream redistribution of scour and deposition locally enhances bank retreat. This has significant consequences for predictive modeling studies, as it implies that at-a-site analyses are insufficient to determine the net impact of riparian vegetation in meandering rivers. Instead, an integrated spatially-distributed modeling approach, with coupled fluvial-geotechnical analyses is needed. This study illustrates the potential of such an integrated numerical modeling approach in river studies.

Acknowledgments. The original development of mRIPA was funded by a grant (GR/M46532/01) from the UK Engineering and Physical Sciences Research Council, and was undertaken in collaboration with Dr. Erik Mosselman (Delft Hydraulics) and Prof. Andrei Alabyan (Moscow State University). The work to extend and apply mRIPA to vegetated rivers was undertaken during MJVDW's PhD thesis, which was sponsored by a High Performance Computing bursary from the University of Southampton.

REFERENCES

Abernethy, B., and I. D. Rutherfurd, Does the weight of trees destabilize riverbanks?, *Regulated Rivers Research and Management, 16*, 565–576, 2000.

Abernethy, B., and I. D. Rutherfurd, The distribution and strength of riparian tree roots in relation to riverbank reinforcement, *Hydrological Processes, 15*, 63–79, 2001.

Ariathurai, R., and K. Arulanandan, Erosion rates of cohesive soils, *Journal of the Hydraulics Division, ASCE, 104*, 279–283, 1978.

Bennett, S. J., T. Pirim, and B. D. Barkdoll, Using simulated emergent vegetation to alter stream flow direction within a straight experimental channel, *Geomorphology, 44*, 115–126, 2002

Blackmarr, W. A., *Documentation of hydrologic, geomorphic and sediment transport measurements on the Goodwin Creek Experimental Watershed, Northern Mississippi, for the period 1982 to 1993*, Research Report 3, National Sedimentation Laboratory, US Department of Agriculture, Oxford, Mississippi, 1995.

Brookes, A., and F. D. Shields, *River Channel Restoration: Guiding Principles for Sustainable Projects,* John Wiley, Chichester, UK, 1996.

Charlton, F. G., P. M. Brown, and R. W. Benson, *The Hydraulic Geometry of Some Gravel Rivers in Britain*, Hydraulics Research Station, Wallingford, UK, 1978.

Collison, A. J. C., N. L. Pollen, and A. Simon, Mechanical reinforcement and enhanced strength of streambanks: contribution of common riparian species, *EOS Transactions, AGU, Fall Meeting Supplement, 82*, H13A-0232, 2001.

Coppin, N. J., and I. G. Richards, *Use of Vegetation in Civil Engineering*, Butterworths, London, UK, 1990.

Darby, S. E., A. Alabyan, and M. J. Van De Wiel, Numerical simulation of bank erosion and channel migration in meandering rivers, *Water Resources Research, 38*, 1163, doi:10.1029/2001 WR000602, 2002.

Darby, S. E., and C. R. Thorne, Fluvial maintenance operations in managed alluvial rivers, *Aquatic Conservation: Marine and Freshwater Ecosystems, 5*, 37–54, 1994.

Darby, S. E., and C. R. Thorne, Numerical simulation of widening and bed deformation of straight sand-bed rivers. I: Model development, *Journal of Hydraulic Engineering, ASCE, 122*, 184–193, 1996.

De Vries, D. G., Multi-stage line intersect sampling, *Forestry Science, 20*, 129–133, 1974.

Fathi-Moghadam, M., and N. Kouwen, Non-rigid, non-submerged vegetative roughness on floodplains, *Journal of Hydraulic Engineering, ASCE, 123*, 51–57, 1997.

Forest Products Research Laboratory, *A Handbook of Hardwoods*, Forest Products Research Laboratory, UK, 1956.

Freeman, G. E., W. J. Rahmeyer, and R. R. Copeland, *Determination of resistance due to shrubs and woody vegetation*, Report TR-00-25, US Army Corps of Engineers, Engineering Research and Development Center, Coastal and Hydraulics Laboratory, Vicksburg, Mississippi, 2000.

Gran, K., and C. Paola, Riparian vegetation controls on braided stream dynamics, *Water Resources Research, 37*, 3275–3283, 2001.

Gray, D. H., Reinforcement and stabilization of soil by vegetation, *Journal of the Geotechnical Division, ASCE, 100*, 695–699, 1974.

Greenway, D. R., Vegetation and slope stability, in *Slope Stability*, edited by M. G. Anderson, and K. S. Richards, pp. 187–230, John Wiley, Chichester, UK, 1987.

Hemphill, R. W., and M. E. Bramley, *Protection of River and Canal Banks*, Butterworths, London, UK, 1989.

Hey, R. D., and C. R. Thorne, Stable channels with mobile gravel beds, *Journal of Hydraulic Engineering, ASCE, 112*, 671–689, 1986.

Hickin, E. J., Vegetation and river channel dynamics, *Canadian Geographer, 28*, 111–126, 1984.

Huang, H. Q., and G. C. Nanson, Vegetation and channel variation: A case study of four small streams in southeastern Australia, *Geomorphology, 18*, 237–249, 1997.

Huang, H. Q., and G. C. Nanson, The influence of bank strength on channel geometry: An integrated analysis of some observations, *Earth Surface Processes and Landforms, 23*, 865-876, 1998.

Husch, B., *Forest Mensuration and Statistics*, Ronald Press, New York, NY, 1963.

Kalkwijk, J. P. T., and H. J. De Vriend, Computation of the flow in shallow river bends, *Journal of Hydraulic Research, 18*, 327–342, 1980.

Klaassen, G. J., and J. J. Van der Zwaard, Roughness coefficients of vegetated flood plains, *Journal of Hydraulic Research, 12*, 43–63, 1974.

Kouwen, N., Field estimation of the biomechanical properties of grass, *Journal of Hydraulic Research, 26*, 559–568, 1988.

Kouwen, N., and M. Fathi-Moghadam, Friction factors for coniferous trees along rivers, *Journal of Hydraulic Engineering, ASCE, 126*, 732–740, 2000.

Kouwen, N., and R.-M. Li, Biomechanics of vegetative channel linings, *Journal of the Hydraulics Division, ASCE, 106*, 1085–1103, 1980.

Kozlowski, T. T., *Growth and Development of Trees*, Academic Press, New York, N.Y., 1971.

Kuhnle, R. A., R. L. Bingner, G.R. Foster, and E. H. Grissinger, Effect of land use change on sediment transport in Goodwin Creek, *Water Resources Research, 32*, 3189–3196, 1996.

Lane, E. W., Design of stable channels, *Transactions of the American Society of Civil Engineers, 120*, 1234–1260, 1955.

Li, R.-M., and H. W. Shen, Effect of tall vegetation on flow and sediment, *Journal of the Hydraulics Division, ASCE, 99*, 793–814, 1973.

Millar, R. G., Influence of bank vegetation on channel patterns, *Water Resources Research, 36*, 1109–1118, 2000.

Mosselman, E., *Modeling of River Morphology with Non-orthogonal Horizontal Curvilinear Coordinates*, Report 91-1, Technische Universiteit Delft, Delft, Netherlands, 1991.

Mosselman, E., *Mathematical Modeling of Morphological Processes in Rivers with Erodible Cohesive Banks*, PhD thesis, Technische Universiteit Delft, Delft, Netherlands, 1992.

Murphey, J. B., and E. H. Grissinger, Channel cross-section changes in Mississippi's Goodwin Creek, *Journal of Soil and Water Conservation, 40*, 148–153, 1985.

Naiman, R. J., H. Decamps, and M. Pollock, The role of riparian corridors in maintaining regional biodiversity, *Ecological Applications, 3*, 209–212, 1993.

Nilsson, C., *Conservation Management of Riparian Communities: Ecological Principles of Nature Conservation*, Elsevier Applied Science, London, UK, 1992.

Olesen, K. W., *Bed Topography in Shallow River Bends*, Report 87-1, Technische Universiteit Delft, Delft, Netherlands, 1987.

Pasche, E., and G. Rouve, Overbank flow with vegetatively roughened flood plains, *Journal of Hydraulic Engineering, ASCE, 111*, 1262–1278, 1985.

Petryk, S., and G. Bosmajian, Analysis of flow through vegetation, *Journal of Hydraulic Engineering, ASCE, 101*, 871–884, 1975.

Rehder, A., *Manual of Cultivated Trees and Shrubs*, MacMillan, New York, NY, 1947.

Simon, A., and A. J. C. Collison, Quantifying the mechanical and hydrological effects of riparian vegetation on streambank stability, *Earth Surface Processes and Landforms, 27*, 527–546, 2002.

Thorne, C. R., Effect of vegetation on riverbank erosion and stability, in *Vegetation and Erosion*, edited by J.B. Thornes, pp. 125–144, John Wiley, Chichester, UK, 1990.

Thorne, S. D., and D. J. Furbish, Influences of course bank roughness on flow within a sharply curved river bend, *Geomorphology, 12*, 241–257, 1995.

Waldron, L. J., Shear resistance of root permeated homogeneous and stratified soil, *Soil Science Society of America Journal, 41*, 843–849, 1977.

Wu, T. H., P. E. Beal, and C. Lan, In-situ shear test of soil-root systems, *Journal of Geotechnical Engineering, ASCE, 114*, 1376-1394, 1988.

Wu, T. H., W. P. McKinnell, and D. N. Swanton, Strength of tree roots and landslides on Price of Wales Island, Alaska, *Canadian Geotechnical Journal, 16*, 19-33, 1979.

Marco J. Van De Wiel, Institute of Geography and Earth Sciences, University of Wales, Aberystwyth, SY23 3DB, UK.

Stephen E. Darby, School of Geography, University of Southampton, Southampton, SO17 1BJ, UK.